JOHN GREGORY AND THE INVENTION OF PROFESSIONAL
MEDICAL ETHICS AND THE PROFESSION OF MEDICINE

Philosophy and Medicine

VOLUME 56

Editors

H. Tristram Engelhardt, Jr., *Center for Medical Ethics and Health Policy, Baylor College of Medicine and Philosophy Department, Rice University, Houston, Texas*

S. F. Spicker, *Massachusetts College of Pharmacy and Allied Health Sciences, Boston, Mass.*

Associate Editor

Kevin Wm. Wildes, S.J., *Department of Philosophy, Georgetown University, Washington, D.C.*

Editorial Board

The titles published in this series are listed at the end of this volume.

JOHN GREGORY AND THE INVENTION OF PROFESSIONAL MEDICAL ETHICS AND THE PROFESSION OF MEDICINE

LAURENCE B. MCCULLOUGH

Center for Medical Ethics and Health Policy,
Baylor College of Medicine,
Houston, Texas

KLUWER ACADEMIC PUBLISHERS
DORDRECHT / BOSTON / LONDON

A C.I.P Catalogue record for this book is available from the Library of Congress.

ISBN 0-7923- 4917-2

Published by Kluwer Academic Publishers
PO Box 17, 3300 AA Dordrecht, The Netherlands

Sold and distributed in North, Central and South America
by Kluwer Academic Publishers,
PO Box 358, Accord Station, Hingham, MA 02018-0358, USA

In all other countries, sold and distributed
by Kluwer Academic Publishers,
PO Box 322, 3300 AH Dordrecht, The Netherlands

Printed on acid-free paper

Printed in Great Britain.

FOR LINDA, ALWAYS

TABLE OF CONTENTS

PREFACE

The best things in my life have come to me by accident and this book results from one such accident: my having the opportunity, out of the blue, to go to work as H. Tristram Engelhardt, Jr.'s, research assistant at the Institute for the Medical Humanities in the University of Texas Medical Branch at Galveston, Texas, in 1974, on the recommendation of our teacher at the University of Texas at Austin, Irwin C. Lieb. During that summer Tris "lent" me to Chester Burns, who has done important scholarly work over the years on the history of medical ethics. I was just finding out what bioethics was and Chester sent me to the rare book room of the Medical Branch Library to do some work on something called "medical deontology." I discovered that this new field of bioethics had a history.

This string of accidents continued, in 1975, when Warren Reich (who in 1979 made the excellent decisions to hire me to the faculty in bioethics at the Georgetown University School of Medicine and to persuade André Hellegers to appoint me to the Kennedy Institute of Ethics) took Tris Engelhardt's word for it that I could write on the history of modern medical ethics for Warren's major new project, the *Encyclopedia of Bioethics*. Warren then asked me to write on eighteenth-century British medical ethics. I had learned already from Chester Burns and Tris Engelhardt about Thomas Percival and antebellum American medical ethics, but that's all that I knew. By then, I was on my Post-Doctoral Fellowship at the Hastings Center and so I went into New York City to the New York Academy of Medicine and looked in their catalogue under the history of medical ethics and, going through the centuries, came in the eighteenth-century cards to this fellow named John Gregory. I didn't know it then but this book started that day. I now present it to the reader as a labor of love. I have come to be in awe of Gregory's intellectual accomplishments and I hope to convey some of my respect – and, indeed, affection – for him in the pages that follow.

I have received magnificent support from my colleagues and academic institutions over the many years of the preparation of this book, starting with John McDermott and James Knight at Texas A&M University, where I had my first teaching position. Warren Reich encouraged and supported my interest in and writing on Gregory and other topics in the history of medical ethics, as did Baruch Brody when I came to the Center for Medical Ethics and Health Policy at the Baylor College of Medicine in Houston, Texas, in 1988. Baruch supported with Center funds a crucial research trip to Scotland and England in 1991, during which I identified and read many of the manuscript sources that appear in this book. This research trip was also supported by a Travel Grant from the National Endowment for the Humanities. In addition, Baruch supported my application and found the funding for a sabbatical leave during the 1995-1996 academic year, during which I completed the research for and writing of this book. This sabbatical leave was also supported by an American Council of Learned Societies Fellowship that added substantially to my time off for full-time research that year. Additional travel funds for research during my sabbatical year were provided by a Travel Grant from the American Philosophical Society in Philadelphia. This combination of institutional and extramural support made it possible for me to concentrate for a year on my work on and writing about Gregory, making much easier the work of the past year of putting the manuscript into its final form.

My work, especially on manuscript materials and rare books, was greatly facilitated by truly splendid colleagues on the professional staffs of libraries and rare book and manuscript collections at the Universities of Aberdeen, Edinburgh, and Glasgow, the Royal College of Physicians and the Royal College of Surgeons in Edinburgh, the Royal College of Physicians and Surgeons of Glasgow, the National Library of Scotland, the Royal College of Physicians and the Royal College of Surgeons in London, the John Rylands Library of the University of Manchester in England, the Wellcome Institute for the History of Medicine in London, McGill University in Montreal, the Huntington Library in California, the College of Physicians in Philadelphia, the National Library of Medicine in Bethesda and the Library of Congress in Washington, DC, the Humanities Research Center and Perry-Casteñeda Library at the University of Texas at Austin, the Blocker History of Medicine Collections in the Moody Medical Library of the University of Texas Medical Branch at Galveston, and Rice University's Fondren Library and the Texas Medical Center Library in Houston. Ms. Hannah Glass provided research support at the Osler Library at McGill University in Montreal, Quebec, Canada.

I especially want to thank Colin McLaren and his colleagues, Iain Beavan, Mary Murray, and Myrtle Anderson-Smith for their superb assistance and good cheer while I worked feverishly at the University of Aberdeen on the manuscript materials both of the Gregory Collection and of the Aberdeen Philosophical Society and James Beaton of the Royal College of Physicians and Surgeons in Glasgow for his bringing to my attention materials that play a major role in this book. Michael Barfoot, in a magnificent display of collegiality, put me onto a Gregory manuscript at the Royal College of Surgeons in Edinburgh that, because it was mis-catalogued under his son, James', name, I might well have missed. I learned from these colleagues that Texas hospitality is topped by Scottish hospitality. These individuals provide moral exemplars in which their countrymen and countrywomen should take considerable, and justified, pride.

Manuscript material that appears in this book, often for the first time anywhere, is included with the permission of the institutions that own or house them. I am grateful to the following institutions for granting this permission: the Universities of Aberdeen and Edinburgh; the Royal College of Physicians of Edinburgh; the Royal College of Surgeons of Edinburgh; the Royal College of Physicians and Surgeons of Glasgow; the National Library of Scotland; the Wellcome Institute for the History of Medicine; McGill University; the College of Physicians of Philadelphia; and the National Library of Medicine. Manuscript and other historical materials are presented in their original style of spelling, punctuation, capitalization and other elements of style.

I have had the opportunity over the past decade or so to present papers and seminars on Gregory and these played a major role in the development of this book. I want particularly to thank colleagues for invitations to present my work at Union College (on three separate occasions), the Wellcome Institute for the History of Medicine, the University of Michigan, the University of Pittsburgh, the Cleveland Clinic, the New York Consortium for the History of Medicine at the New York Academy of Medicine, and the History of Medicine Lecture Series of my medical school. I have also developed some of the ideas that appear in Chapter Four in my "Ethical Challenges of Physician Executives" course that I teach regularly for the American College of Physician Executives. Finally, as my students and residents at the Baylor College of Medicine can tell you, I have tried out many of the ideas in this book on them. They have been excellent teachers.

I have benefitted enormously from the work of other scholars, to whom I make reference in the pages that follow. I stand on the shoulders of giants. The scholarly work of Dorothy Porter and Roy Porter (1989), Guenther Risse (1986), and Sylvia Harcstark Myers (1990), the reader will soon discover, plays a major role in the chapters that follow. I especially want to thank Lewis Ulman, whose book on the Aberdeen Philosophical Society (Ulman, 1990) opened the door for me – and for every other scholar who comes after him – to the Aberdeen Philosophical Society and the Society's manuscript records at the University of Aberdeen.

T. Forcht Dagi read the manuscript in its penultimate version and provided many valuable suggestions. By following them I have strengthened this book. Stuart Spicker, the co-editor of the *Philosophy and Medicine* series, provided invaluable editorial direction and suggestions in the preparation of the final version of the manuscript. Robert Baker also read and provided excellent criticisms of a near-final version. He urged me not to hesitate to let my "voice" come through and so I didn't. I thank him for pushing me in this direction. One should always, my father correctly taught us, act on the advice of friends. Bob is a very good friend and so I did.

H. Tristram Engelhardt, Jr., sought this book for the *Philosophy and Medicine* series. I am grateful to have his confidence and even more grateful for his steadfast friendship and unmatched collegiality of many, wonderful years.

In December of 1989 another series of accidents (Robert Baker is responsible for them) found me on a street-corner near the Wellcome Institute being chided by one of its scholars, Christopher Lawrence, for the abstract and therefore irresponsible way philosophers write the history of ideas. I took this chiding very much to heart. (His words, were, I now recall, perhaps a bit stronger on these points). I hope that Dr. Lawrence will agree with me that this is not the typical book that historians of philosophy write.

I was trained in the history of philosophy at the University of Texas at Austin by Ignacio Angelelli, whom I revere and love as my Doktorvater. Ignacio taught me textual scholarship by precept and example of the first order. Neither of us would, I think, have expected in 1975, when I left The University, that I would write a book on the history of medical ethics. I hope that Ignacio will approve of where his training has taken me.

I was taught by my first professor of philosophy, Daniel O'Connor, that just because figures in the history of philosophy are dead doesn't mean that we're smarter or have better answers to the philosophical questions that they addressed or that we want to address. Quite the opposite; these dead thinkers may be far smarter and more successful philosophically than we are – the enduring lesson of studying the histories of ideas and of philosophy. Gregory, I intend to show in the pages that follow, is a compelling case in point.

My wife of more than two decades, Linda Quintanilla, has supported me unstintingly throughout the many years of this project and copied out more than a few manuscripts during our 1991 research trip to Scotland and England. That I met her is the most beautiful accident of all. Being married to Linda is the best thing that I ever will do. This book is for her.

Houston, Texas
August, 1997

CHAPTER ONE

AN INTRODUCTION TO JOHN GREGORY'S
MEDICAL ETHICS

The design of the professorship which I have the honour to hold in this university, is to explain the *practice of medicine*, by which I understand, the art of preserving health, of prolonging life, and of curing diseases. This is an art of great extent and importance; and for this all your former medical studies were intended to qualify you (Gregory, 1772c, p. 2).

I. GREGORY'S CONTRIBUTIONS TO THE HISTORY OF MEDICAL ETHICS AND THE HISTORY OF MEDICINE

A forty-seven year old, previously healthy man presents to his family physician with a chief complaint of serious fatigue – he feels very tired all the time, even after sleeping – and frequent urination, including a history of urination every ninety minutes to two hours during the day and night for the past four to five days. His physician, who has cared for this patient and his wife for six years, expresses her concern that the patient does not look at all well. After a physical examination the physician takes a urine sample, for evaluation in her office laboratory. The physician returns to the examining room and tells the patient that he has red blood cells in his urine. This finding and his history indicate that something may be wrong with his kidneys and that he should see a nephrologist immediately.

Later that same day, the patient sees the nephrologist, who orders twenty-four hour urine collection and obtains blood for laboratory analysis. Two days later, the nephrologist tells the patient that he is spilling a great deal of calcium in his urine, that his serum calcium is abnormally high, and that his BUN and creatinine levels indicate renal failure, with a loss of about half of normal function.

The nephrologist then explains that the differential diagnosis of these findings includes a very serious form of cancer, multiple myeloma, which attacks bones and, if it is present, would explain the high levels of calcium in the patient. The mortality rate from multiple myeloma is quite significant, with a high cumulative mortality. To rule out a diagnosis of multiple myeloma, the nephrologist sends the patient for a Galium scan and skeletal survey, diagnostic tests that will indicate whether inflammatory processes are occurring in the patient's bones.

This rapid chain of events focused the patient's mind. He could, he thought, be facing a very serious disease, a difficult regimen of treatment, and possibly death – none of which he was expecting at his age. During the next forty-eight hours, of testing and waiting, the patient's mood swung between fairly well disciplined and outright, uncontrolled fear.

Two days later, the patient was greatly relieved to learn that the imaging test results were normal; there was no detected inflammatory process in the patient's skeletal system. The nephrologist explained that this did not mean that there was no chance of multiple myeloma being present, but, as a practical matter, this was no longer in the differential diagnosis. After further work-up to detect the cause(s) of the patient's hypercalcemia and renal failure the nephrologist – who presented the patient's case twice at nephrology section conferences, trying to achieve a definitive diagnosis – reached the diagnosis of idiopathic renal failure secondary to hypercalcemia of unknown origin. The patient was put on an empiric regimen of a reduced-calcium diet and oral steroid medication and the patient's hypercalcemia and renal failure resolved within eight weeks.

This patient had come to depend on physicians – as every other patient does – including a family physician, a nephrologist, the nephrologist's colleagues, a nuclear medicine specialist, and a radiologist. Like all patients, this patient needed to be confident that these physicians were competent. The patient assumed that his physicians possessed an adequate science of medicine and could reliably employ it in clinical judgment and decision making. Like all patients, this patient also needed to be confident that the primary concern of these physicians was the patients's well being. The patient assumed that his physicians would be focused primarily on the medically appropriate management of his condition, and not primarily on their own interests in income, prestige, or power. The patient needed to be confident that he would not be used by these physicians for their own purposes but would be cared for properly. In other words, like all patients, this patient needed to be able to trust this physicians – intellectually and morally. I was this patient and I felt the need for this intellectual and moral trust acutely, very acutely indeed.

Patients assume that they can have such trust in their physicians, just as I did. When we are not overly sanguine, we invest such trust prudentially, aware that some physicians can sometimes abuse our trust. Nonetheless, if we want the benefits of allopathic or osteopathic medicine, we have to trust our physicians. As patients, we assume that our physicians will act with both intellectual and moral integrity in taking care of us, i.e., that our physicians are professionals.

Allen Buchanan (1996) has recently identified five elements of an "ideal" conception of a profession: "special knowledge of a practical sort;" a commitment to preserve and enhance that knowledge; a commitment to "achieving excellence in the practice of the profession;" an "intrinsic and dominant commitment to serving others on whose behalf the special knowledge is applied;" and "effective self-regulation by the professional group" (Buchanan, 1996, p. 107). The first three components of this conception form the basis for our intellectually trusting physicians when we become their patients. The fourth and fifth components are the key to our being able to trust our physicians morally: that they are committed primarily to protecting and promoting our interests when we become their patients rather than protecting and promoting their own interests in such matters as income, job security and advancement, prestige, fame, and power. As Buchanan puts it:

> To say that the commitment is intrinsic is to say that it is not exclusively instrumental, that is, derived from other motives, such as the desire for personal gain. The commitment is dominant in the sense that, at least in many cases, it overrides other desires or commitments with which it may come into conflict (Buchanan, 1996, p. 107).

The assumption that we can trust our physicians must be based on both the concept and actual social practice of the physician as a professional in the intellectual and moral senses of the term, i.e., as the fiduciary of the patient.

As patients, we have become so used to the expectation that our physicians will conduct themselves as fiduciary professionals that we might be tempted to think that the concept and social practice of the physician as moral fiduciary of the patient and in whom the patient could therefore have trust have existed for a very long time. Edmund Pellegrino, for example, claims that the "ineradicability of trust has been a generative force in professional ethics for a long time" (Pellegrino, 1991, p. 69).

Not so, as the reader of this book will soon discover in Chapter Two. There was a time in the English-speaking world, just a little more than two centuries ago, when patients could not trust their physicians intellec-

tually or morally (Risse, 1986; Porter and Porter, 1989). Although there
was much talk of a "profession" in the eighteenth century (Wear, Geyer-
Kordesch, and French, 1993), medicine as a profession in the intellectual
and moral senses of the term did not yet exist. Before the eighteenth
century physicians and surgeons lacked a stable body of knowledge and
they lacked an ethics to guide the appropriate use of then nascent scientific knowledge and power that such knowledge was about to create. In
short, the concept of the physician as a professional in the intellectual and
moral senses of the term and therefore the concept of the profession of
medicine in its intellectual and moral senses did not exist, at least in the
English-speaking world, until the eighteenth century.

In the latter third of that century an altogether remarkable Scotsman,
John Gregory (1724-1773), invented these concepts. Gregory invents the
concept of medicine as a fiduciary profession in response to what he took
to be *unprofessional* attitudes and practices among physicians and surgeons. In the true Baconian spirit, Gregory set out to improve medicine so
that it could contribute more effectively and reliably to the relief of man's
estate. Indeed, doing so, we shall see, was life-long commitment and
endeavor, both shaping Gregory's medical ethics at its core. Trust has
been a "generative force in professional ethics" for only a little over two
centuries, which is *not* a "long time" in the history of ideas.

Gregory addressed topics in medical ethics and philosophy of medicine
in a series of lectures preliminary to his lectures on the theory and practice of medicine, as well as in the latter lectures. He regularly lent his
students his lecture notes, which were very complete, and so students
were able to record them verbatim. Somehow (a topic reserved for Chapter Three) a student version of these medical ethics and philosophy of
medicine lectures found their way into the press, in 1770, as *Observations
on the Duties and Offices of a Physician; and the Method of Prosecuting
Enquiries in Philosophy* (Gregory, 1770), anonymously. Gregory then,
under his own name, published, in 1772, his own version, as *Lectures on
the Duties and Qualifications of a Physician* (Gregory, 1772c), to repair
the "negligent dress" in which they had first appeared (Gregory, 1772c,
"Advertisement," n.p.). He addresses topics in medical ethics and philosophy of medicine in these books, in his lectures on theory and practice
of medicine, and in his other works with a method that blends an abiding
commitment to the value of clinical experience, elements of a Baconian
method and philosophy of medicine, a substantive, well-known philosophical method, and an ethical concept of a *profession*. The *Lectures*

appeared in numerous subsequent editions in Britain (Gregory, 1788, 1805, 1820) and in the United States (Gregory, 1817).[1]

While he does not use the term, Gregory forged the concept of medicine as fiduciary profession. The concept of the physician as fiduciary means that "as fiduciary (1) [the physician] must be in a position to know reliably the patient's interests, (2) should be concerned primarily with protecting and promoting the interests of the patient, and (3) should be concerned only secondarily with protecting and promoting the physician's own interests" (McCullough and Chervenak, 1994, p. 12). In forging this concept, Gregory created an intellectual legacy that continues to develop and be put into social practice, but today faces ethical challenges from the new managed practice of medicine (Chervenak and McCullough, 1995). Gregory's medical ethics provides us with powerful tools to address these challenges, as we shall see in Chapter Four.

I hope to persuade the reader in the pages that follow that all of us patients and our physicians, whose integrity as professionals validates our intellectual and moral trust in them, stand in Gregory's intellectual and moral debt, although – with rare exceptions – we don't know that we do. The purpose of this book is to correct this stunning deficit in our knowledge of John Gregory's place in the history of medical ethics and, therefore, the history of medicine. I shall argue that Gregory was a pivotal figure: before him there was no professional medical ethics worthy of the name in the English language; after him there was. The history of English-language medical ethics and therefore history of medicine, I shall argue, both pivot on his lectures on medical ethics at the University of Edinburgh in the 1760s, the opening sentences of which introduce this chapter.

II. GREGORY'S CONTRIBUTIONS TO BIOETHICS

The field of bioethics developed in the 1950s and 1960s – its naming came in the 1970s (Reich, 1995d) – largely innocent of its roots in the history of medical ethics. New and unprecedented ethical challenges, it was thought at the time, arose for physicians, patients, institutions, and society. Given the cultural and moral pluralism of a society such as the United States, bioethics had to become a secular enterprise, if it hoped to be successful in academia, especially medical schools, in the profession of medicine, and in the policy arena. Philosophical methods came to the fore in the 1970s, thus displacing those of religious studies and theology that had preceded philosophical methods in the 1950s and 1960s. All of this was at the time thought to be new.

It was not. Gregory was, in fact, the first in the English-language literature to employ philosophical methods to address ethical challenges in medicine and to do so in a self-consciously secular fashion. Gregory thus writes the first philosophical, secular medical ethics in the English language.[2] In doing so, Gregory invented philosophical, secular medical ethics as it is now practiced more than two centuries later in the United States and other countries around the world under the rubric of "bioethics." In the course of inventing philosophical, secular medical ethics, Gregory also laid the conceptual, secular foundations for the profession of medicine as an intellectual and moral enterprise, the basic elements of which Buchanan so nicely captures (without appreciating their historical origins and relative youth). In his work on medical ethics Gregory argues within a serious, powerful, well-known, philosophical, secular tradition – Scottish moral sense philosophy in general and David Hume's philosophy in particular – for what that life ought to be. Hume's concept of sympathy is at the very core of Gregory's medical ethics, as we shall see in Chapters Two and Three.[3] Gregory, self-consciously and with considerable effect, began the process of the cultural transformation of medicine in Great Britain – and therefore in British America and soon thereafter the new United States of America – from a commercial enterprise in which self-interest figured prominently, even dominated (the opposite of Buchanan's fourth component of the professional ideal), to a moral life of service to patients and society.

These enormous intellectual accomplishments – inventing philosophical, secular, medical ethics and forging the intellectual and moral concept of medicine as a fiduciary profession and thereby inventing professional medical ethics and, therefore, the profession of medicine as we know it – would be enough to secure for Gregory a permanent and prominent place in the history of medical ethics and therefore the history of bioethics. To these, I will show in this book, Gregory added the accomplishment of writing the first feminine medical ethics in the history of medical ethics, anticipating by two centuries current methods of bioethics, the advocates of which do not know this history.

In the use of the phrase, 'feminine medical ethics', I follow Rosemarie Tong's distinction between feminine and feminist ethics (Tong, 1993). *Feminine ethics* is based on a feminine consciousness that "regards the *gender* traits that have been traditionally associated with women ... as positive *human* traits" (Tong, 1993, p. 5), while *feminist ethics* emphasizes a political and social agenda to identify and redress the subordination of women to men (Tong, 1993, p. 6). As we shall see in Chapter Two, Gregory held a feminist position on matters such as marriage for love rather than convenience or economic security, views that mark a sharp departure from those of his contemporaries. His feminist views about the social roles of women led him to adopt a feminine medical ethics, as we shall see in the next two chapters.

Gregory's feminine philosophical method – utilized throughout his medical ethics – emphasizes the virtues of tenderness and steadiness as the expression of the properly functioning moral sense of sympathy. Women of learning and virtue became epitomized in Gregory's mind by Elizabeth Montagu and her Bluestocking Circle (Myers, 1990). As we shall see in Chapter Two, these extraordinary intellectual women provide the exemplars for Gregory of these feminine virtues that together should define the professional character of the physician and thus control and direct clinical judgment and conduct. Gregory proposes this feminine medical ethics to his students, all of whom were men, marking him as a progressive thinker, even a radical, by the standards of the day. As we shall see in Chapter Three, there is some indication on the texts that Gregory's students strained against his progressive ideas. Gregory's feminine medical ethics, we shall see in Chapter Four, differs crucially from contemporary feminine ethics and bioethics and therefore it avoids a problem that plagues some forms of contemporary feminine ethics and bioethics, namely, that they threaten or even undermine feminist ethics (Jecker and Reich, 1995). At the same time, Gregory's feminine medical ethics anticipates in important ways contemporary feminine approaches to bioethics, particularly the ethic of care that has recently come to prominence in the recent literature (Jecker and Reich, 1995).

Gregory developed the scope and content of his medical ethics in response to problems in the practice of medicine, the management of medical institutions, and in the medical research of his time. Gregory took up issues of concern for practicing physicians, making his medical ethics deliberately clinical. In the chapters that follow the reader will encounter a leading thinker of the Scottish Enlightenment, who wrote a medical ethics that is at once professional, secular, philosophical, feminine, and clinical (McCullough, 1998). He did what we in bioethics now do, two centuries before we thought of doing it.

III. PLAN OF THIS BOOK

In this book I provide an historical and philosophical account of these extraordinary intellectual accomplishments of *the* medical ethicist of the Scottish Enlightenment. As the reader will soon discover, Gregory anticipates bioethics, particularly virtue-based and care-theory-based bioethics, as well as a very great deal of the agenda of bioethics, including the commitment to philosophy as a central intellectual discipline of the – not so new, after all – field of bioethics.

Gregory wrote his medical ethics more than two hundred years ago. We should, therefore, not read Gregory as if he were our contemporary (an unfortunate trend in recent work in the history of philosophy). Instead, we should set Gregory's work in its historical context, so that it can be understood – as much as we can reconstruct it two centuries later – both as Gregory conceived and wrote it and as it was probably understood by his contemporaries: Edinburgh medical students, fellow physicians, and intellectuals. I therefore turn in the second chapter to a detailed examination of Gregory's intellectual development. I will show that Gregory – with a self-consciousness that typifies him as a major, but neglected, figure of the Scottish Enlightenment – drew broadly on and responded to developments and changes that were occurring in the national identity of Scotland, in Scottish society and culture, in the Scottish Enlightenment, in the self-understanding and role of women of intellectual ability and accomplishment, in Baconian science and medicine, medical practice, and in moral sense philosophy.

Having established the historical context of his work in medical ethics, I provide, in Chapter Three, a philosophical account of the method and content of Gregory's medical ethics. In the course of doing so I plan to show – in detail and in its fuller historical context – how Gregory's medical ethics is professional, secular, philosophical, clinical, and feminine. Before Gregory, the relationship between the sick and their physicians was largely a business relationship, a patient-physician relationship initiated by the patient contracting for the physician's services (Porter and Porter, 1989). This relationship lacks all five components of Buchanan's professional ideal. Physicians had addressed their *obligations* to their patients either in theological terms or only in a very cursory manner (French, 1993b, 1993c; Nutton, 1993; Wear, 1993). No thoroughgoing philosophical, secular account of the obligations of the physician as a fiduciary existed that might serve as the basis for a morally authoritative physician-patient relationship, a professional relationship of service by the physician to the patient. Gregory provides such an account, in which he argues – on the basis of Hume's concept of sympathy gendered feminine and on the basis of his own feminist commitments to women of learning and virtue as moral exemplars – for the intellectual and moral virtues requisite in the physician as a true professional. These virtues define the social role of being a physician; this social role, in turn, creates the social role of being a patient. Gregory thus provides a philosophical, secular account of the physician-patient relationship as a professional relationship in its ethical sense, namely, the physician as fiduciary of the patient. The effect of this was to invent both professional medical ethics and the profession of medicine in its intellectual and moral senses. It will become clear in Chapters Three and Four that Gregory's medical ethics should neither be equated with nor reduced to etiquette, as some have mistakenly argued that it should (Leake, 1927; Berlant, 1975; Waddington, 1984). Gregory wrote what his contemporaries counted – and we should count – as philosophically substantive medical ethics.

The scope of Gregory's medical ethics is very broad and is driven by his concerns about existing problems in medicine – its "deficiencies," he calls them – that need to be improved, by identifying and proposing a means to remove these deficiencies. In other words, the scope of Gregory's medical ethics is a function of his Baconian, Scottish Enlightenment commitment to improve medicine. Medicine, like other human activities, has its functions – the right exercise of medicine's three capacities (described in the passage that opens this chapter) in service to the "ease and conveniency" of life – and these functions can be made to work correctly when they mal-function and made to work better when they function well. Gregory nowhere that I can find thinks in terms of the perfection of medicine – the full and complete realization of some *telos* or end of medicine – but rather of the constant and steady improvement of its capacities, attentive always to their limits. Gregory appears to be convinced that this improvement had to start at the very beginning: correcting judgments and behaviors of then contemporary practitioners that were – from the rigorous perspective afforded by his method – deficient and therefore in need of remedy, because they originated in self-interest, not a life of intellectual and moral service to science and patients.

Gregory addresses topics that were – and, for the most part, still are – of considerable clinical ethical importance. These include conflicts of interest, the governance of the patient by the physician, the care of patients with "nervous ailments," changes in practice style as the physician ages, confidentiality – especially concerning female patients, sexual abuse of female patients (only in *Observations*), temperance and sobriety, laying medicine open (reflecting his commitment to Baconian diffidence and its cardinal virtue of openness to conviction), truth-telling (particularly in the case of grave illness), abandonment of dying patients, cooperation with clergy, consultation (which does *not* involve etiquette or the mutual pursuit of self-interest), relationships between younger and older physicians (reflecting the problem of intense market-place competition), regard for older writers and medical writings, the boundaries between medicine and surgery and between medicine and pharmacy (which were hotly contested, indeed, as they are yet again in our day), formality of dress (again, as the reader will discover, not entirely a matter of etiquette), singular manners (addressed to the problem of the man of put-on, purchased, false manners – a major problem at that time, as we shall see in Chapter Two, and re-appearing nowadays in the guise of "customer service training" for physicians), avoiding a reaction of disgust to unpleasant clinical situations, time management, servility to one's social superiors who are patients, secrets and nostrums, disclosure of the composition of secret remedies and nostrums to patients (in a treatment far different from our understanding of informed consent), the physician's responsibility when patients die, medicine and religion, experiments on patients, animal experimentation, and the obligations of professors of medicine.[4]

In his lectures on the institutions of medicine Gregory touches briefly on the definition and clinical determination of death. Because, medicine has no clinically or scientifically reliable definition, clinical criteria should be the most conservative, he argues. Gregory wrote at a time when fear of premature burial – not premature transplant of unmatched organs – concerned many people and his account of the definition and determination of death addresses directly and effectively this clinical and social concern.

These topics display an emphasis on primary care – no surprise given the state of medical science and clinical practice in his time – and also hospital-based issues, such as the abuse of patients. Gregory's topic list may strike the reader as quaint, even without ethical significance. This would make the mistake of making the past a "prisoner of the present" (MacIntyre, 1984), because Gregory wrote his medical ethics at a time when much that we today take for granted quite simply did not exist –

e.g., that physicians keep confidences, especially about their female patients; that clinical investigators not abuse human subjects of research; that physicians on call should diligently limit alcohol consumption; that physicians should follow institutional policies and procedures for consultation; or that physicians and medical students should adopt acceptable modes of dress in the office and hospital setting. Gregory could take nothing for granted in these matters. His problem list, as we shall see in Chapters Two and Three, presented real, substantive ethical challenges, just as we believe that our problem list in bioethics does. In the course of Chapter Three the reader should study Gregory's clinical ethical topics in the spirit in which Gregory addressed them: the improvement of medicine by identifying and correcting its deficiencies. Gregory's problem list was impressive for his time; so too, was his philosophical and clinical response to it.

My goal in Chapters Two and Three will be that of any historian: to get the past right, as much as possible, "to get the facts right and to make sense" (Vann, 1995, p. 1). In attempting to make sense of Gregory's medical ethics, I will read Gregory as an eighteenth-century thinker whose categories of thought and philosophical methods may not be wholly familiar or even congenial to those of contemporary philosophy. My reading will not make Gregory a prisoner of the present (MacIntyre, 1984, p. 33). At the same time, I will, in these two chapters, resist turning Gregory into one more item in a "set of museum pieces" (MacIntyre, 1984, p. 31).

In Chapter Four I provide a philosophical assessment of Gregory's medical ethics. There I will take seriously the possibility that present medical ethics does not defeat past medical ethics, at least in Gregory's case. Indeed, I shall argue that present medical ethics and bioethics do not enjoy "rational superiority" over the past (MacIntyre, 1984, p. 47) and so I intend to put bioethics into critical dialogue with its past.

I begin this philosophical assessment with an account of how his contemporaries understood and received his medical ethics. Some appear to have insufficiently appreciated its philosophical character and failed to have anticipated its substantial influence on medical ethics, the second aspect of my assessment of Gregory's medical ethics. This influence includes translations into French (Gregory, 1787), Italian (Gregory, 1789), and German (Gregory, 1778) and Thomas Percival's *Medical Ethics* (1803). Percival writes the first English-language work on *institutional medical ethics*, in particular, the ethics of a new medical institution, the Royal Infirmary. Percival conjoins Gregory's virtues of tenderness and steadiness to theological virtues of condescension and authority and to the moral realism of Richard Price (1948). The authors of the "medical police" or codes of medical ethics in the state medical societies of the new United States appeal directly to Gregory as one of their sources. This influence culminates in the American Medical Association Code of Ethics of 1847 (Bell, 1995; Bell, et al., 1995; Hays, 1995).

The third aspect of my assessment of Gregory's medical ethics includes five ways in which it remains important for contemporary bioethics. First, I will show that Gregory's "Enlightenment project" (which I will describe) succeeds, contra Alasdair MacIntyre's (1981, 1988) argument that such a project dooms itself to failure, and H. Tristram Engelhardt's related claim that "content-full" bioethics must also fail (1986, 1996).

Second, I will explore the persistence in Gregory's work of pre-modern ideas in a medical ethics that employs what is thought now to be a distinctively modern philosophical method – Hume's moral philosophy. Hume, we shall see, was not through-and-through modern in his method, either. Gregory's account of the physician-patient relationship as a professional relationship, in which the physician assumes fiduciary obligations to the patient, rests upon a medieval, Scottish Highland, and moral-aristocratic concept of paternalism, i.e., an asymmetrical social relationship founded on obligations of service rooted in hierarchical social roles. This moral-aristocracy of paternalism was designed precisely to protect those in the lower social roles in hierarchies of knowledge and power, as we shall see. This pre-modern, anti-egalitarian aspect of medical paternalism was not – and is still not – appreciated by its critics; it also has important implications for the notion of the physician and patient as "moral strangers" to each other (Rothman, 1991; Engelhardt, 1986, 1996). Moreover, this medieval, pre-modern idea can, I will argue, help us to see what is at stake and help us to respond to in the new managed practice of medicine – a much larger, increasingly global phenomenon than simply managed care in the United States.

Third, I provide a "Gregorian critique" of contemporary work on virtue-based bioethics, as well as a bioethical theory that appeals to concepts of care. Reading contemporary virtue-based bioethics through the perspective of Gregory's texts exposes the thin moral psychology on which such bioethics rests. Moreover, contemporary virtue-based bioethics omits mention of moral exemplars, a serious omission for any virtue-based ethics. Too, care theories tend to regard feminine qualities, e.g., nurturance, as natural and always well functioning. Gregory helps us to see that this assumption is not true. Only trained capacities – not capacities in their untutored state – provide the normative basis required by care theories. In other words, care theory requires a well developed, virtue-based account of properly functioning capacities *for* care, especially if care theory hopes to escape the feminist claim that care theory appeals to the very "virtues" that patriarchy uses to suppress and oppress women.

Fourth, in the literature on the affective dimensions of the physician-patient relationship and of the medical humanities, the concept of *empathy* has displaced that of sympathy. The latter, it is thought, involves too much risk of undisciplined affective response to patients, a problem that, it turns out, Gregory identifies and effectively addresses. Empathy, as it is currently understood, also harkens back to Adam Smith's (1976) account of sympathy, an unappreciated historical source that differs in important ways from Hume and Gregory's account of sympathy. Finally, accounts of the normative dimensions of empathy based on relational or affiliative moral psychology link the concept of empathy to feminine theories of care. Both are then open to the charge of abetting the social and political forces that feminist ethics has been developed to combat. Gregory's feminine ethics, I will argue, escapes this debilitating criticism.

Fifth, Gregory's medical ethics was very much the product of the Scottish Enlightenment. Similarly, Percival's medical ethics was the product of a different national Enlightenment, to the south, in England. These national differences already count for a great deal in the work of the two major figures of eighteenth-century British medical ethics. In other words, methodological diversity was introduced at least two centuries ago as an apparently fixed feature of medical ethics, just as it seems to be of bioethics today. Yet Gregory would claim a transnational intellectual authority for his philosophical method. Would he be justified in doing so? This interesting question can be extended into contemporary bioethics. Should we take seriously the possibility that bioethics will always be national (in the cultural sense) bioethics?

IV. CONCLUSION

This book joins and is intended to complement the growing literature on the history of medical ethics (Baker, Porter, and Porter, 1993; Wear,

Geyer-Kordesch, and French, 1993; Baker, 1995a; Haakonssen, 1997) and the entries on the history of medical ethics in the *Encyclopedia of Bioethics* (Reich 1978, 1995c) that initiated – very much to Warren Reich's credit – the process of providing bioethics with its history. The present book contributes to this literature the first full-length scholarly examination of the work on medical ethics (and philosophy of medicine) of a single, major figure, John Gregory. Roy Porter has recently identified a need that the present book is intended to address:

> We remain surprisingly ignorant ... of the backgrounds against which Gregory and Percival were writing, in terms of both ethical theory and the informal rules of medical practice (Porter, 1993, p. 252).

I expect that more such studies will be forthcoming as scholars in both the history of medicine and bioethics turn their attention to the creation of a history for a field, bioethics. The need for such work is urgent: bioethics still lacks an understanding of itself as having a history and therefore lacks status as a fully mature field of the humanities.

The reader is about to encounter a remarkable eighteenth-century intellect, Baconian scientist and physician, moral philosopher, husband and parent, life-long feminist, and, to his core, Scotsman. Gregory, we shall see, is at once a thorough-going Enlightenment and pre-Enlightenment, even medieval thinker. There is no contradiction in his combining these, as we also shall see. Quite the opposite; his combining these is essential to professional medical ethics and, in Gregory's case, make him a very appealing intellectual forbear and exemplar for bioethicists. I have endeavored in the chapters that follow to make Gregory a three-dimensional figure by placing him in the political, cultural, social, intellectual, academic, and practice milieus of his times, thus avoiding (I hope) the abstract approach that typifies so much work in the history of philosophy.

I also place Gregory in critical dialogue with bioethics. Gregory lived, taught, wrote, and died more than two centuries ago. Some of the things that he believed to be the case we now struggle to believe; some we even reject. As the reader will discover in this book, Gregory addresses the ethics of a medicine that is – unhappily – remarkably like our own. And he does so, I will argue, in ways that may be more successful than our own.

CHAPTER TWO

JOHN GREGORY'S LIFE AND TIMES:
AN INTELLECTUAL HISTORY

[Gregory] possessed a large share of the social and benevolent affections, which, in the exercise of his profession, manifested themselves
in the many nameless, but important attentions to those under his care;
attentions which, proceeding in him from an extended principle of humanity, were not squared to the circumstances or rank of the patient,
but ever bestowed most liberally where they were most requisite. In the
care of his pupils he was not satisfied with a faithful discharge of his
public duties. To many of these, strangers in the country, and far removed from all who had a natural interest in their concerns, it was a
matter of no small importance to enjoy the acquaintance and counte-
nance of one so universally respected and esteemed. Through him they
found an easy introduction to an enlarged and elegant society; and,
what to them was still more valuable, they experienced in him a friend
who was ever easy of access, and ready to assist them to the utmost
with his counsel and patronage (Tytler, 1788, pp. 81-82).

I. SETTING GREGORY IN CONTEXT

We should, as cautioned in the previous chapter, not read Gregory with
twentieth-century sensibilities, even though he does at times address
concerns that persist into our time. Instead, we should set Gregory in
context, as best we can across a gap of more than two centuries. Thus, I
do not claim for this chapter the provision of an account of Gregory's life
and intellectual development just as he lived and experienced them; to do
so would constitute an impossible undertaking. I do provide an account of
Gregory's life and intellectual development as best as I can reconstruct
them from the extant documentary evidence and secondary sources.
    I set out to accomplish this goal by using key events in Gregory's life
as a template for the introduction of relevant historical considerations. I
begin with the larger political and social context into which Gregory was
born in 1724 (he lived until 1773), since central concepts of Scottish
moral sense philosophy – which come to play a key role in Gregory's

medical ethics – developed, in part, as a response to that context. I then
turn to an account that, for the most part, follows a chronological ordering
of major events or turning points in Gregory's life.

I begin with Gregory's birth into one of the most distinguished academic families – perhaps the most distinguished – in the history of Scotland. Women enjoyed some prominence in the history of this family, a
fact that may have influenced Gregory's views toward women and his
"feminine medical ethics" and feminism.

I then turn to contextual considerations of eighteenth-century developments in the philosophies of science and of medicine and in the newly
emerging science of physiology – nervous system physiology in particular. Gregory encountered these as a medical student in Edinburgh and
Leiden in the early 1740s, the earliest period for which manuscript
sources for his medical ethics and philosophy of medicine exist. I then
turn to his first, brief appointment at the University of Aberdeen, as
Professor of Philosophy in the late 1740s and to his first medical practice
in that city.

From Aberdeen, Gregory moved to London in the mid-1750s to practice medicine, after his marriage to Elisabeth Forbes. In discussing this
period of his life I draw on recent work on the social history of medicine
to discuss the relations between doctors and their patients, especially the
medical care of women. In London, Gregory made the acquaintance of
Mrs. Elizabeth Montagu, the organizer or "Queen" of the Bluestocking
Circle. The women of learning and virtue with whom Gregory made
acquaintance during this crucial period of his life, Mrs. Montagu especially, had a lasting intellectual impact on him. They became an important
inspiration and source for his feminine medical ethics and feminist views
on the role of women, especially regarding marriage.

Gregory returned to the University of Aberdeen, as Professor of
Medicine, in 1756. There, with his cousin Thomas Reid and others,
Gregory founded the Aberdeen Philosophical Society in 1758, whose
members embraced and contributed to the heady philosophical developments of Scottish moral sense philosophy and to the Scottish Enlightenment. The members of the Society, Gregory included, wrestled mightily
with David Hume's *Treatise on Human Nature* (1978). *Contra* Hume's
later concern, the *Treatise* did not "fall stillborn from the press" in Aberdeen. I trace their response – Gregory's in particular – to Hume's concept
of 'sympathy', which comes to play a major role in Gregory's medical
ethics, as described in Chapter Three. In the deliberations of the "Wise

Club," as the Society came familiarly to be known, Gregory developed key aspects of his philosophy of science, medicine, and religion, as well as moral philosophy.

While in Aberdeen, Gregory's wife of nine years died. He loved her very deeply with complete devotion. His lament for his loss upon her death, in a letter to Mrs. Montagu, provides a window onto Gregory's personal and intellectual life – and how he joined them in an integrated way – that merits separate consideration. Early and frequent loss of loved ones – parents, siblings, spouses, and children – defined daily life in eighteenth-century Scotland and Gregory's experience of such loss shaped his medical ethics in important ways.

Gregory capped his professional career with his appointment as His Majesty's Physician in Scotland, and his academic career with an appointment as Professor of Medicine at the University of Edinburgh in the mid-1760s. He delivered his medical ethics lectures in the University's medical school, beginning in the closing years of the 1760s and continuing until his death in 1773. He also taught in the Royal Infirmary of Edinburgh, an institution I shall later describe in some detail.

Gregory wrote five books, all started after his wife's death (Lawrence, 1971, Vol. I, p. 157). I close this chapter with a careful examination of three of these works. He first contributes a work on philosophy, with particular reference to philosophy of science and of religion, *A Comparative View of the State and Faculties of Man with Those of the Animal World*, first published in 1765 (Gregory, 1765). This book gained no small acclaim for Gregory and probably helped him obtain his academic position at the University of Edinburgh. Then he wrote a moral guide for his daughters – in the correct, as its turned out, anticipation of his premature death – *A Father's Legacy to His Daughters* (Gregory, 1774). This book, published posthumously in 1774, was reprinted well into the nineteenth century; it became a popular *vade mecum* for young women aspiring to refinement. The third, his clinical lectures, *Elements of the Practice of Physic for the Use of Students*, appeared in 1772 (Gregory, 1772b). These three works – along with his life's work and activities – set the stage for his work in medical ethics, which I examine in detail in the Chapter Three.

II. SCOTTISH NATIONAL IDENTITY AND THE SOCIAL PRINCIPLE

Scotland possessed neither sovereignty nor an independent government when Gregory was born in 1724. Not all Scots, however, accepted the political reality of the Act of Union of 1707 as final or normative. In part this resistance had its roots in events of the previous century. David Daiches summarizes the effects of those events succinctly:

> The political and religious turmoil of the seventeenth century raised all sorts of questions about Scotland's position, identity, culture and (most of all) religion, as well as about Scotland's relation to England. Under Cromwell, Scotland was forced into a Commonwealth which included the whole of the British Isles, but she recovered her status as an independent kingdom at the restoration of 1660 (Daiches, 1986, p. 11).

King William initiated the project of the union of Scotland with England, which Queen Anne and Parliament in London moved forward formally as early as 1702. The Crown's motivation came in good measure from its concern about impending war with France, which came in the same year (Mackie, 1987, pp. 257-258). A commission from both kingdoms negotiated the terms of union over several years and the Union took effect in 1707. The terms of the Act of Union included recognition of the Hanoverian succession, freedom of trade, and uniformity in currency, and other matters pertaining to trade. Scottish law retained considerable autonomy. The Act itself did not speak to "the religious establishment of either kingdom" (Ferguson, 1990, p. 48).

Scotland was thus politically and economically united with but not wholly absorbed culturally by England. Political sovereignty ended, but not national identity and some social institutions through which it could be fostered, the law courts and the Church. J.D. Mackie summarizes the Union in this way:

> The Act of Union was a remarkable achievement. It made the two countries one and yet, by deliberately preserving the Church, the Law, and the Judicial System, and some of the characteristics of the smaller kingdom, it ensured that Scotland should preserve the definite national identity which she had won for herself and preserved so long. It realized some of the desires of both countries. To England it gave security, in the face of French hostility, for the Hanoverian succession and for the constitutional settlement of the Revolution [of 1688]; to Scotland it gave a guarantee of her Revolutionary Settlement in Church and State,

and an opportunity for economic development which was sorely needed (Mackie, 1987, p. 263).

The Union did not enjoy wide support and, indeed, was unpopular in many quarters in Scotland. Many believed that Scotland had been sold (Mackie, 1987, p. 264), with the future of the country "mortgaged" to England (Ferguson, 1990, p. 49).

In the course of affirming the Act of Union, the Scottish Parliament added provisions "guaranteeing the security of the presbyterian establishment" which was "declared to be an integral part of the treaty" (Ferguson, 1990, p. 50). This did not sit well with those of other faiths, especially the Jacobites. Episcopalians and Roman Catholics joined in the Jacobite series of uprisings against the Union, beginning in 1715. These sporadic movements coalesced in 1745 with the return of Prince Charles from France and the raising of an army of rebellion mainly from the Highlands and the North East (Ferguson, 1990, p. 150). At first successful, the rebellion came to disaster on April 16, 1746, at Culloden: "Culloden was a slaughter rather than a battle" (Ferguson, 1990, p. 153).

William Ferguson points out that, because so much of the English army included Scotsmen, the Jacobite rebellion might more properly be thought of as a civil war than as a war of independence (Ferguson, 1990, p. 153). However we should describe it, the Jacobite uprising and its subsequent failure, Mackie claims, "had little effect on the development of Scotland" (Mackie, 1987, p. 282), because of the many changes already occurring in Scottish society. The Union came to be accepted by the latter half of the eighteenth century.

The eighteenth-century hope for a separate nation had been extinguished at Culloden, but not the hope for a perceived reality of a distinct and separate national identity. That identity invoked the central, *premodern, medieval concept of paternalism*, the obligation of clanchief to clansmen and clanswomen, carried into the obligations of the patriarch landowner to care for those working and living on his land. This feudal, aristocratic notion involved an obligation to house and feed one's serfs. Serfs, in turn, claimed as a matter of right the fulfillment of this obligation. The concept of paternalism in contemporary bioethics bears little resemblance to the eighteenth-century Scottish social institution and concept of paternalism. This will become important in Chapter Four, when I examine Gregory's importance for contemporary bioethics.

This very old way of life, with roots deep in medieval times, underwent decline after Culloden. Ferguson summarizes these changes:

Much more than Jacobitism died at Culloden. Thereafter the disintegration of the old Highland society, already advanced in some quarters,
was accelerated. The patriarchal authority of the chiefs and great terri-
torial magnates was gradually transformed into landlordism. The de-
militarisation of Highland life broke the ties of mutual interest and
idealised kinship which had bound chiefs and clansmen and paved the
way for a new social relationship in which landlords came to regard
their people solely as tenants and cottars. But economic needs and
some lingering remnants of the old paternalist regime postponed for
half a century, and sometimes longer, the harshest consequences of this
social readjustment (Ferguson, 1990, p. 154).

The old social ties that had bound people together and that had, to a great
extent, fostered a sense of national identity in which the whole country
shared, began to unravel. These old ties were natural in a direct sense:
they were based on kinship. "[C]lans were stark realities, held together by
ideas of kinship fortified by elaborate genealogies in which truth vied
with fiction" (Ferguson, 1990, p. 91). Such natural ties – fictional ties
were sometimes just as natural and powerful as biological ties – were
discoverable in experience and readily idealized to sustain the idea of
national identity. As these natural ties of kinship began to dissolve, the
concept of national identity after the loss of the nation became problematic. Many questions were in the air, ranging from the meaning of the
Jacobite rebellion to which language should be used, Scots or English
(Daiches, 1986, p. 12). What resources were available to meet this challenge to, even crisis of, national identity?

The land itself was diverse, but the kinship traditions of the Highlands
had been mirrored in the Lowlands as well (Ferguson, 1990, p. 91). With
the rise of the great urban centers of Glasgow and Edinburgh and, to
some extent, Aberdeen, land and land-based clans no longer served to
bind people together. In particular, the Highlands, once so physically and
culturally dominant (Mackie, 1987, p. 13), became less so. This distinction between Highlands and Lowlands grew more pronounced and very
little of the natural kinship ties formed city life or shaped the newly-
emerging commercial world of manufacture and trade. Ferguson nicely
describes the Scotland that emerged by the end of the seventeenth century.

Two different conceptions of society confronted each other. That of the
Lowlands was struggling out of the old feudal mould; it came to centre

more and more upon the individual, and this process was speeded up by the religious crises of the seventeenth century .... The Highlands reflected few of these features (Ferguson, 1990, pp. 91-92).

The Lowland trend accelerated as the eighteenth century progressed and began to affect the Highlands.

Religion could not serve to bind people together or produce a single national identity. While the Presbyterian Church gained ascendancy with the provisions added to the Act of Union, this did not settle – but exacerbated – the religions fault lines that ran through the country, which often paralleled geographic lines. Religious pluralism and the universal tolerance of religious diversity that it implies were far off in the future – they may still be for our world. There was a trend toward moderation, but this "did not happen quickly or easily" (Daiches, 1986, p. 14). More than theological matters were at stake; political power was distributed by the Union along religious lines. As the century progressed, fissures opened in the Presbyterian faith community, which splintered, sometimes acrimoniously.

Politics provided an even more fragile reed on which to support national identity. Politics before and after the Union were riven with corruption. Ferguson's summary is pointed:

> A diseased electoral system (*pace* all the special pleading about opening careers to talents, as in the rather exceptional case of Edmund Burke) had a rotting effect on the body politic, and this effect was most evident in administration (Ferguson, 1990, pp. 158-159).

The official legislative and administrative organs of government became more and more detached from the people. National identity could not build on such corrupt and indifferent foundations.

The military effort to restore the nation had failed. Land and the kinship and clan-based ties fostered by land began to lose their hold. Religion might as often serve as a centrifugal force as a uniting force and, while religious moderation began to increase, religious pluralism in the sense of stable, mutual toleration to a great extent did not exist (Mackie, 1987, pp. 300ff). Appeals to religion, therefore, would divide rather than unite. Politics and the legislative and executive organs of government could not fill the gap.

Other social changes only added to the force of the challenge to national identity. As the eighteenth century progressed Scotland experienced significant changes in its previously largely agricultural economy.

Agriculture improved as a result of the introduction of new farming methods, reflecting the spirit of Enlightenment improvement that affected so many social institutions and practices (Mackie, 1987, pp. 289-290). As a consequence, agriculture became less labor-intensive; unemployment in agriculture rose; and people began to migrate to the cities. Economic benefits accrued from these improvements in agriculture, including increased rents, the raising of grand mansions, and the planting of trees in great numbers (Ferguson, 1990, p. 173). Nonetheless, the changes in agrarian society affected the Highlands, reinforcing the deterioration of clan and kinship ties as people moved off the land.

Life on the land was hard, make no mistake. However, the *tradition of paternalism* acted as a buffer, even a sort of social safety net against the worst harshness, especially against recurrent poverty (Ferguson, 1990, p. 78). To some extent paternalism shielded the wage-earner – a growing social class – in agriculture and the coal mines, which were located in the countryside. This tradition began to break down in the cities, especially in cotton cloth manufacturing. Paternalism had bound higher and lower classes together; now gaps began to open as the interests of the owners of capital and their workers began to diverge.

> Such concentration [of capital in the cotton mills, the emerging factory model] contributed to the growth of an industrial proletariat, opening up new and deeper social gulfs than had hitherto existed. Not that the wage earner was new; but in agriculture and in the coalfields paternalism frequently offset exploitation. In the fiercely competitive world of cotton manufacture paternalism was at a discount (Ferguson, 1990, p. 185).

Roads and canals were constructed to move raw material and manufactured goods. While these served to connect the country, they also promoted the growth of commerce and city life, two contexts in which paternalism – based on natural, kinship ties – grew strained and then nonexistent, because commerce and city life began to foster competition and individuality.

Manners, the routine expectations of social behavior that define social classes, were "rough and ready" (Ferguson, 1990, p. 85). Moreover, with the rise of multiple social strata – replacing the simpler, agrarian, feudal-hierarchical society – manners (already diverse and uncertain) destabilized (Lochhead, 1948). How was a gentleman to act toward someone who worked in a cotton mill, and vice versa, when they met? No settled answer to this question existed as the age of manners ended. This becomes a non-trivial question when the gentleman is a university-educated physician and the mill worker a patient at the Royal Infirmary. Genuine manners and false pretense also became hard to distinguish, as we shall see.

The rise of new social classes also meant that land no longer provided the single or even main source of wealth; capital increasingly became the basis of wealth and capital became more and more independent of land. Land ownership and tenant farming on a feudal model, based on paternalism, created relatively stable social relationships of superior and inferior in the countryside. This gave way in the cities to social, hierarchical relationships structured, not always by a mutuality of obligation and interest, but more often by a divergence of interest in the form of competition and therefore sundered obligation of the superior (the factory or colliery owner) and the inferior (the wage earner). The risk of exploitation became a fixed feature of these new social relations, including the patient-physician relationship.

The political and social developments outlined above combined to create a crisis of national identity. Society was becoming increasingly economically and socially stratified. Little seemed to hold these strata together and so disintegration and fragmentation threatened to rend the fabric of society. Ferguson argues that the "only bridges were the law and the aristocracy" (Ferguson, 1990, p. 92).

Aristocracy provides the bridge of steady, well-founded voluntary obligations of superior to inferior, to protect and promote the interests of social inferiors and to minimize the effect of "interest," i.e., mere self-interest. Paternalism, Ferguson suggests, was strained but not altogether absent form society, a legacy of the feudal, aristocratic past that persisted into the eighteenth century.

Scottish law was based on principles: general, discoverable rules that authoritatively govern the conduct of human beings.

> [Scottish] law was philosophical in its bent; it put its faith in principles rather than in collections of dry precedents [as did English common law]; and it too ranged far and wide over the condition of man (Ferguson, 1990, p. 209).

Scottish philosophers, including Hume, wrote on topics in law, such as marriage, and in this way, Ferguson suggests, law deeply influenced the formation of Scottish philosophy in the eighteenth century (Ferguson, 1990, p. 209).

Hume and the other Scottish philosophers of the eighteenth century based their philosophy on a moral sense, a natural, "built-in" regard for the interests of others. The moral sense "involves this social-intellectual communication with 'other minds' (called by Reid 'social acts of the mind', and by Smith 'sympathy')" (Davie, 1991, p. 66). This social principle, as it was also commonly called by the Scottish philosophers, grounded law and civil order because it was the most general principle. 'Principle' here means a discoverable law of nature, in this case human nature, that both describes the natural and proper function of a thing and shapes the very nature of that thing. 'Principle' functions as a term of art in the robust scientific realism of the Scottish Enlightenment's "science of man" (Wright, 1991, pp. 309-310). The social principle describes a natural regard for the interests of others and shapes human nature at its core. The social principle forms the essence of human nature, an essence discoverable by experience.

Thomas Reid's account poignantly illustrates this view. In his *Essays on the Active Powers of the Mind*, Reid describes *the social principle* in the following terms:

> And, when we consider ourselves as social creatures, whose happiness or misery is very much connected with that of our fellowmen ... from these considerations, this principle leads us also, though more indirectly, to the practice of justice, humanity, and all the social virtues (Reid, 1990, p. 49).

In *Practical Ethics* Reid describes justice and humanity as embracing the "whole of social duty. The first implying abstaining from all Injury, the second that we do all the good in our power" (Reid, 1990, p. 138), echoing in a striking fashion the aristocratic, feudal concept of paternalism. Reid elaborates on the principle of humanity:

> It is I think to the honour of the British Nation that in our Language all the Amiable and Benevolent Virtues which prompt us to do good to our fellow Creatures are summed up in the Word Humanity, which implys their being the proper Characteristicks of a Man. Homo sum & nihil humanum a me alienum puto. This noble sentiment is interwoven into our Language. And indeed as man is by his very Constitution a Social Animal, & is not born for himself alone but for his friends his family his Country; he who has no social and benevolent Dispositions is surely Defective in one of the Noblest and best Parts of human Nature, as really Deficient in what belongs to the Nature of Man as if he

were without hands and feet, or without the sense & Understanding of a Man (Reid, 1990, p. 139).

This social principle falls as a species under the larger genus of principles that are discoverable upon careful observation and reflection – what the Scottish philosophers in their shorthand call 'experience'. They make no appeal to religion or any other contended source of knowledge, because they do not require such grounds. Instead, they appeal to a way of knowing that is open to everyone willing to submit their inquiry to the intellectual discipline that yields highly reliable results, namely the Baconian, scientific way of knowing. Thus, the social principle that binds us naturally together can be established on grounds independent of clan and kinship, city or country, or the claims of political officials, including the English king, as well as secular grounds, independent of religious insight and experience.

As noted above in reference to Reid, this social principle retains a key element of aristocracy in Scotland, namely, paternalism as obligatory from the superior – in knowledge, power, social standing, or any other respect – to the inferior. The social principle blunts interest (i.e., self-interest) and causes one to act for the benefit of others. When someone properly develops and exercises the social principle, other-regarding behavior becomes habitual in that individual's life. Such habitual other-regarding behavior marks the virtuous man or woman. Moral sense theory thus recasts a key element of the old Scottish national identity, aristocratic other-regarding paternalism, writing it large for everyone. In a socially increasingly stratified society – stratified by wealth, knowledge, political power, land, religion, and other factors – each person will be at times socially superior and at other times socially inferior. In the first role, he or she acts on the basis of the social principle as a paternalist in the best of its aristocratic sense. In the second role, he or she expects to be protected by the paternalism of the social principle.

Kinship and clan ties have indeed been idealized, but not as clansmen once did, to kith and kin. Rather, the social principle operates to bind together *all* those in proximity to each other. This includes face-to-face relationships as well as, apparently, those of cultural contiguity – including city life and marked by a national border. For those farther away the social principle also operates, but more weakly. The social principle thus more closely binds Scots to each other than to the English. John Millar (1735-1801), the Scottish political philosopher and law professor, wrote in just this vein, arguing that the state evolved from "rude tribes" and their chiefs, through the union of the tribes into a kingdom (Millar, 1990). This emphatically is *not* a social contract theory in which individuals, animated by self-interest, largely sufficient unto themselves in a state of nature, and strangers to each other – i.e., without existing ties of any kind – *consent* to form the state.

Hume also addresses the topic of national identity, for example, in his essay, "Of National Character" (Hume, 1987a). There he explains this phenomenon in terms of the "moral causes" that bind people together into a national identity. This concept of "moral cause" involves "a sympathy or contagion of manners" (Hume, 1987a, p. 204). Scotland shares "physical causes" with England – geography and climate, for example – but the two countries have separate national identities (Hume, 1987a, p. 207). Hume explains how this occurs:

> The human mind is of a very imitative nature; nor is it possible for any set of men to converse often together, without acquiring a similitude of manners, and communicating to each other vices as well as virtues. The propensity to company and society is strong in all rational creatures; and the same disposition, which gives us this propensity, makes us enter deeply into each other's sentiments, and causes like passions and inclinations to run, as it were, by contagion, through the whole club or knot of companions. Where a number of men are united into one political body, the occasions of their intercourse must be so frequent, for defence, commerce, and government, that, together with the same speech or language, they must acquire a resemblance in their manners, and have a common or national character, as well as a personal one, peculiar to each individual (Hume, 1987a, pp. 202-203).

The social principle thus addresses in a powerful way – and in a way thought at the time to be convincing because true on plainly observable grounds – the crisis of Scottish national identity that emerged at the end of the seventeenth century and gathered force in the eighteenth century.

Family ties epitomize the social principle, but the social principle is not limited to such ties. The social principle, as it were, writes clan and family ties and obligations large. This way of thinking appears, for example, in Hume's essay, "Of the Origin of Government":

> Man, born in a family, is compelled to maintain society, from necessity, from natural inclination, and from habit. The same creature, in his farther progress, is engaged to establish political society, in order to administer justice; without which there can be no peace among them,

nor safety, nor mutual intercourse (Hume, 1987b, p. 37).

The echoes of Hobbes are plain but Hume is no Hobbes, for whom interest, or self-interest, alone creates the rationale for social and political order. Self-interest justifies creating the Leviathan to promote mutual self-interest by preventing the war of all against all (Hobbes, 1968).

As a matter of observable fact, Hobbes was wrong, as was Locke, for that matter, the Scots political philosophers such as Millar would say. Instead, human beings start and have their very identity only in families or "rude tribes"; they are naturally other-related and other-regarding. They then act on their natural inclination, other-regarding behavior writ to the social level, and cultivate this inclination into habit as essential to their moral improvement. Hume, along with the other moral sense theorists, was a scientific Scotsman and so rejects social contract theory.

> The traditions, conventions and customs of their native land also influenced their thought. One and all they rejected the social contract theory. To men steeped in the ideas of kinship the social contract must always have appeared as a fanciful notion. Adam Ferguson, himself a Highlander, knew that the so-called primitive society was in fact complex and bound by ties of blood; in insisting that 'mankind are to be taken in groupes, as they have always subsisted' he reasoned well within his own experience. And since experience was the touchstone of their system this is significant. The very first point that David Hume, such a supposedly detached thinker, made in his fragment of autobiography was that he came of a good family ... (Ferguson, 1990, pp. 209-210).

Social contract theory assumes human beings come atomistically, as self-contained, independent individuals with no natural, built-in ties of any kind to each other. All social ties are therefore constructed and, since they threaten self-interest and independence, must be justified, just as Hobbes taught. Moreover, Locke's primitive society, which exists before the state is created, was English, not Scottish – Locke's and England's problem, not Scotland's. The Hobbesian and Lockean assumptions of isolated human beings are deeply discordant with Scottish history and traditions of family and clan, and therefore to Scottish moral sense theorists.

> For Smith, as for Hume and many other moral philosophers of the period, men and women were not in any sense made, much less did they develop, in isolation from one another: their very constitution rendered

them social entities through and through (Tomaselli, 1991, p. 233).

Moral sense philosophy therefore understands human nature to be consti-
tuted of discoverable principles, chief among them the social principle
that makes us, indeed, social animals by a built-in, plainly observable,
natural instinct.

Gregory's concern in his medical ethics will be with the face-to-face
relationship of doctor and patient. For Gregory, also steeped, as we shall
see, in this same tradition, humans are inherently *social* creatures. They
are morally formed, moved, and to be judged by the social principle. For
face-to-face relationships, Gregory needs an account of how at the micro
level the social principle forms character and moves the individual to act
in the interests of another. "Humanity" or sympathy, as developed by
Hume, does this work in Gregory's medical ethics (Tytler, 1788, p. 81).
Sympathy, as it were, writes Scottish national identity and its residual
aristocratic concept of paternalism small. Each person of "humanity"
therefore has a "natural interest" (Tytler, 1788, p. 82) in others, both in
the national clan of Scots at the macro level and in the professor-student
and physician-patient relationships at the micro level (Tytler, 1788, pp.
81-82).

III. GREAT EXPECTATIONS, 1724-1742:
THE "ACADEMIC GREGORIES"

Agnes Grainger Stewart, a nineteenth-century biographer of John Gregory, calls the male members of the family the "Academic Gregories" and
justly so (Stewart, 1901). Male members of the family made noted academic and intellectual contributions – in mathematics, astronomy, and
medicine, in particular. John Gregory was born in Aberdeen into this
already very distinguished academic family, on June 3, 1724. Alexander
Fraser Tytler (Lord Woodhouselee), Gregory's first biographer, begins
his "Life" on Gregory on this note:

> Dr. John Gregory, author of the essays contained in these volumes
> [Gregory, 1788], was descended from an ancient family in Aberdeen-
> shire; and had the honour of counting among his ancestors a succes-
> sion of men eminent for their abilities, and of distinguished reputation
> in the annals of science and literature. It is a singular fact, that this
> family has been noted for mathematical genius for the course of two

centuries (Tytler, 1788, pp. 1-2).

This (male) talent for mathematics first appears in Gregory's grandfather, James Gregorie (1638-1675), as they then spelled it; John modernized the spelling of the family name. James was a son of The Reverend John Gregorie of Drumoak and his wife, Janet, a daughter of David Anderson, who "possessed a singular turn for mathematical and mechanical knowledge" (Tytler, 1788, p. 3). One of David Anderson's cousins was a distinguished professor of mathematics in Paris at the beginning of the seventeenth century. Janet Anderson, both Tytler (1788) and Stewart (1901) agree, brought to the Gregories the "mathematical genius [that] was hereditary in the family of Andersons, and from them [it] seems to have been transmitted to their descendants of the name of Gregory" (Tytler, 1788, p. 3).

Janet Gregorie would seem to have enjoyed in goodly measure the talent for mathematics evident in her family, for she undertook the early education of her son, James.

> The mother of James Gregory inherited the genius of her family; and observing in her son, while yet a child, a strong propensity to mathematics, she instructed him herself in the elements of that science (Tytler, 1788, pp. 3-4).

Thus, in the generation of his great-grandparents John Gregory had the role model of a woman of considerable intellectual ability – the first "academic Gregory," in my judgment – both unafraid of such ability and willing to act on that ability by unselfishly sharing it with her son, James, Gregory's grandfather. Gregory comes to champion such women, women of learning, intellectual accomplishment, and virtue (as we shall see in subsequent sections of this chapter).

Grandfather James went on to a mightily distinguished career. In his twenty-third year James Gregorie invented the reflecting telescope, still in use in Tytler and Stewart's times. His subsequent work in mathematics gained him great acclaim and a fellowship in the Royal Society. Leibniz, with Newton, the greatest mathematician – and no mean philosopher – of the age, expressed admiration for James Gregorie's work (Tytler, 1788, p. 8).

David (1625-1720), James' brother, had a "great love of science" (Stewart, 1901, p. 25). Three of David's sons and three of his grandsons entered academic life. One of his sons, David Gregorie, became the Savilian Professor of Astronomy in Oxford. One of the grandsons was

Thomas Reid, a cousin to whom John Gregory became close in early childhood, and so for life.

John Gregory's father, James Gregorie (1674-1733) took his medical degree in 1698 in Rheims, after studying in Edinburgh, Utrecht, and Paris, thus starting the tradition of physicians in the Gregory family. He returned to Aberdeen, "by no means a dull place" (Stewart, 1901, p. 94). The soon-to-be-famous Rob Roy, James' cousin, was in Aberdeen, recruiting "his clansmen into the Stuart camp" (Stewart, 1901, p. 94) as the series of Jacobite uprisings against the Union began in the second decade of the eighteenth century. By then James was the mediciner, Professor of Medicine, at Kings College and he gave Rob Roy the hospitality of his home.

> ... the charm of the Gregorie household so fell upon the big, warm-hearted outlaw, that in a burst of kindness and enthusiasm he offered to take Dr. Gregorie's little son [James (1707-1755), John's half-brother] and 'mak a man o' him.' Rob Roy thought him far too good to waste upon doctoring, and if the sunny child had got his way, he would have followed the cateran in that delicious life of adventure which he painted – a life of hunting and fighting and success (Stewart, 1901, p. 94).

James senior, "much alarmed," argued that "the child was too delicate and would not live through a highland winter" (Stewart, 1901, p. 95). The argument succeeded.

> So, full of compassion one for another the cousins parted, their roads ran far apart; Rob Roy came to his end claymore in hand listening to the dirge 'Cha till mi tuillidh' (we return no more), while for the doctor there was a career of steady success and a peaceful ending in the sweet house in the middle of the herb garden (Stewart, 1901, p. 95).

James senior was widowed early and, as widowers regularly did, remarried, to a daughter of Principal Chalmers of King's College. John was the youngest of their three children (Tytler, 1788, pp. 24-25). James, who succeeded their father as mediciner at Kings, was thus John Gregory's half-brother and surrogate father. He exercised the paternalism of this role in an important way during John Gregory's medical studies. (See section IV, below.)

Gregory's father died in 1733 and young John's paternal rearing passed to Principal Chalmers and John's half-brother James, who by then had succeeded their father in the professorship of medicine at King's College. Thomas Reid also helped with the rearing and education of the youngster.

John Gregory took the course of classical study at Aberdeen Grammar School and moved on to King's College "under the eye of his [maternal] grandfather," Principal Chalmers (Tytler, 1788, p. 26). There he studied Latin, Greek, Ethics, Mathematics, and Natural Philosophy (which encompassed a very broad range of subjects).

> He was a good classical scholar, and entered warmly into the beauties of the ancient authors; thence deriving a faculty of acutely discriminating the excellencies and defects of literary composition, and forming for himself that pure, simply-elegant, and perspicuous style, which is the characteristic of his writings (Tytler, 1788, p. 27).

No surprise, Gregory also did well at mathematics, reflecting the family "genius" (Tytler, 1788, p. 27). His "Notebook" for 1738-1739 evinces an interest in a wide range of mathematical topics, including "addition, subtraction, algebra, plain trigonometry, heights and distances, and surveying" (AUL 2206/22, 1738).

His Regent, who took students through their entire course of study, was Thomas Gordon, who taught at King's College until the end of the eighteenth century, dying in 1797. Gordon became a member of the Aberdeen Philosophical Society two decades later. (See section VII, below.) Gordon remained close to Gregory the rest of his life, outliving Gregory (Lawrence, 1971, Vol., I, p. 149).

Thomas Gordon devoted his life to teaching. John Ramsay describes the professors of King's College before 1745 in admiring terms:

> They lectured mostly in Latin, and drew their stock of knowledge from the ancient fountains of Greece and Rome, availing themselves also of modern discoveries in the arts and sciences. Instead of attempting to write books which might vie with those of their southern neighbours, they contented themselves with the humble task of storing the minds of their scholars with useful and ornamental learning, which could not well be acquired without the aid of a master. He left it to younger and idler men to write ornate English (Ramsay, 1888, pp. 295-296).

Gordon was reputed by his friends to be a "first-rate classical scholar" (Ramsay, 1888, p. 296). He led his students through the whole course of studies and "in discharging that very difficult task [the practice was abandoned at King's later in the century], this good man displayed much ability, while he gained the hearts of the scholars" (Ramsay, 1888, p. 297). He enjoyed the respect of his colleagues in "academic broils" (Ramsay, 1888, p, 297) and practiced moderation in religious matters, including "Church politics" (Ramsay, 1888, p. 297).

In Gordon, I believe, Gregory found a model that he followed later in his own teaching. Gregory was generous with his students, as Tytler notes in the passage from the "Life" that opens this chapters, and as we shall see later.

[Gregory at Edinburgh] was much beloved by his students, who were all apprised of his learning, ingenuity, and zeal to promote their well-being. Without straining to be popular, he had a great sway over their affection, taking every opportunity of directing their studies while at college, and of befriending them afterwards when beginning to practice (Ramsay, 1888, pp. 477-478).

Gregory, I believe, also found in Gordon a model for patient, learned, intellectual discipline that led naturally, almost necessarily, to moderation, avoiding the extremes of excessive rationalism on the one hand, and enthusiasm, unbridled passion, on the other. Moderation also blunts interest, making for an intellectual virtue that paves the way for scientific inquiry and a moral virtue that opens one to the influence of the social principle.

IV. SCHOOL DAYS, 1742-1746:
EDINBURGH AND LEIDEN

Gregory decided to follow his father and half-brother into medicine. He left Aberdeen to undertake his medical studies at the University of Edinburgh in 1742. His mother went with him, "whose solicitude for her son was at the time much increased by the death of his elder brother George Gregory, a young man of the most promising abilities" (Tytler, 1788, p. 28), who died from consumption in 1741 while following his father and half-brother into medicine at the medical school in Amiens.

At the University of Edinburgh medical school Gregory's teachers included Alexander Monro, *primus* (1697-1767), John Rutherford (1695-1779), who also taught from Boerhaave's work and whom Gregory later succeeded (Underwood, 1977, p. 122), Charles Alston (1683-1760), "the strangeness of whose prescriptions makes it possible for us to grasp what an advance [William] Cullen and Gregory accomplished in medicine" (Stewart, 1901, p. 101), and Andrew Plummer (c.1698-1756) (Tytler, 1788, pp. 29-30). In Edinburgh, Gregory took an active part in the student-originated, Royal Medical Society, which held "the most lively debates upon every subject in medicine and philosophy" (Stewart, 1901, pp. 101-102). Such intellectual, student-run clubs were then very common, functioning in effect as a second curriculum.

In 1745, Gregory went to Leiden, to continue his medical studies (Tytler, 1788, pp. 30-31). In the years just previous, about a dozen English-speaking students had matriculated at Leiden for medical studies, a number that had been declining since Boerhaave's death in 1738 (Underwood, 1977, p. 22). Nonetheless, they were drawn by his legacy and the excellent teachers who remained. There Gregory attended, somewhat irregularly according to Stewart (1901, p. 103), the lectures of Jerome (Hieronymus) Gaub(ius) (1705-1780), Bernard Siegfried Albinus (1697-1770), and Adriaan Van Royen (1704-1779). Ashworth Underwood describes Albinus as "probably the greatest living anatomist" (Underwood, 1977, p. 21). Boerhaave's influence still held sway at Leiden, as did Bacon's there and in many places, including Edinburgh, with important consequences for Gregory's intellectual formation, as we shall see just below.

In Leiden, too, Gregory fell into an interesting intellectual company, especially among British students. These included Alexander Carlyle (1722-1805) and others (Tytler, 1788, p. 30; Stewart, 1901, p. 103). Their conversation was "often brilliant" (Stewart, 1901, p. 104), particularly at their supper parties.

> Gregory's great subjects were religion, and the equal, if not superior, talents of women as compared with men. Everybody made fun of him, for 'he could hardly be persuaded to go to church, and there were no women near whom he could wish to flatter'; but he would not change his mind (Stewart, 1901, p. 104).

Gregory's commitment to deism seems plain and he seems also not to participate in any faith community. His deism comes into play when he considers the relationship between medicine and science, on the one hand, and religion, i.e., belief in a transcendent creator, on the other, in both his *Comparative View* and his medical ethics lectures (Gregory, 1765).

Here again we note Gregory's very interesting views about women. We also learn that he pays a price for these views in Leiden – the friendly teasing of his fellow students. Later, the price increases somewhat, with no apparent dampening – perhaps the opposite effect – on Gregory's commitment to his views on women.

The new discoveries and methods of empirical science deeply shaped the intellectual atmosphere of the medical schools of Edinburgh and Leiden. These include both general developments and the particular teachings of his professors. These had considerable impact on Gregory in three areas: Baconian scientific method applied to clinical medicine; a concept of the nature of medicine; and the physiologic principle of sympathy, a key principle of the new science of the physiology of the nervous system.

A. Baconian Scientific Method Applied to Clinical Medicine

Science generally, and medicine in particular, began to undergo significant change in Enlightenment Scotland. Those changes had their roots in the latter half of the seventeenth century, as Lester King has shown (King, 1970). These changes involved what we might call both a *negative* argument, against the methods and approaches in science and clinical medicine that ought to be abandoned, and a *positive* argument, in favor of a new and better method or approach.

Three critiques form the negative argument. The first critique attacks dogmatism (King, 1970). Dogmatism involves unyielding adherence to authorities, so unyielding that such adherence cannot be corrected by observation and experiment. The second critique assails the "metaphysical" character of the old knowledge. Put more precisely, this critique rejects the metaphysical deliverances of pure reason, unaided and uncorrected by experiment and observation, in particular, the Cartesian method of inspection of ideas by reason alone, to determine which ideas are clear and distinct, and therefore true, which, however, leads only to error. The new science, as we shall see, does *not* reject metaphysics and metaphysical commitments; quite the opposite. The third critique attacks systems, i.e., abstract nosologies of diseases, because they verge on the merely *nominal*. Systemic nosologies may be little more than mere words, with often nothing in reality answering to them.

The positive argument proposes and defends a rudimentary form of what we now call scientific method, propelled by such figures as Francis

Bacon (1561-1626), Thomas Sydenham (1624-1689), Hermann Boerhaave (1668-1738), and Friederich Hoffmann (1660-1742), all of whom
Gregory read and took seriously.

Francis Bacon made numerous contributions to the history of science,
as well as to philosophy of science and philosophy of medicine, as it was
then understood (King, 1978). Bacon bases his scientific method on
experience, correctly understood. He puts aside investigation based
simply on what has been written so far, upon which someone "begins to
meditate for himself," and "logical invention," i.e., reliance on syllogisms
(Bacon, 1875b, pp. 80-81).

> There remains simple experience; which, if taken as it comes is called
> accident; if sought for, experiment. But this kind of experience is no
> better than a broom without its band, as the saying is; – a mere grop-
> ing, as of men in the dark, that feel all around them for the chance of
> finding their way; when they had much better wait for daylight, or light
> a candle, and then go. But the true method of experience on the con-
> trary first lights the candle, and then by means of the candle shows the
> way; commencing as it does with experience duly ordered and di-
> gested, not bungling or erratic, and from it educing axioms, and from
> established axioms again new experiments; even as it was not without
> divine order and method that the divine word operated on the created
> mass (Bacon, 1875b, p. 81).

He goes on at length to describe "Learned Experience, or the Hunt of
Pan" (Bacon, 1875c, p. 413). He introduces this topic in the following
way:

> The method of experimenting proceeds principally either by the Varia-
> tion, or the Production, or the Translation, or the Inversion, or the
> Compulsion, or the Application, or the Conjunction, or finally the
> Chances, of experiment. None of these however extend so far as the
> invention of any axiom (Bacon, 1875c, p. 413).

The investigator of nature goes about gathering *experience* in a disciplined way and only then forms axioms, i.e., hypotheses to be tested by
further careful observation and experiment.

This process follows its own intellectual discipline. Syllogism will
serve us only poorly in this process, because syllogism is based on
propositions, which in turn are based on words, which in turn are based
on "notions of the mind" that may or may not be well founded in ob-

served reality (Bacon, 1875d, p. 24). Nor should the investigator of nature proceed from a "few examples" rapidly to "the most general conclusions" (Bacon, 1875b, p. 111). Instead, the investigator proceeds diligently, in a disciplined way to general truths. This way involves experience, as defined above, the preparation of a "natural and experimental history," and then induction (Bacon, 1875b, p. 127).

The mind disciplines itself to discoverable reality and proceeds under strict intellectual discipline to identify the principles of things. The investigator of nature thus establishes "progressive stages of certainty" (Bacon, 1875b, p. 40). These stages should be duly noted, ranging from "certainly true," through "doubtful whether true or not," to "certainly not true" (Bacon, 1875e, p. 260). He underscores the rigor of this process in the following terms:

> ... if in any statement there be anything doubtful or questionable, I would by no means have it surpressed or passed in silence, but plainly and perspicuously set down by way of note or admonition. For I want this primary history to be compiled with a most religious care, as if every particular were stated upon oath; seeing that it is the book of God's works, and (so far as the majesty of heavenly may be compared with the humbleness of earthly things) a kind of second Scripture (Bacon, 1875e, p. 261).

Thomas Syndenham took observation seriously, too. Lester King provides a concise account of this commitment:

> Syndenham set himself the task of observing nature. Often this term was used as a personalized agent that regulated affairs with intelligence. He rejected all such senses and used the term to mean only 'the whole complication of natural causes; causes which in themselves are brute and irrational'. Nature for him, thus, was the totality of events according to a strict impersonal causal nexus (but this totality of events did exhibit an intrinsic rationality that, in accordance with 17th century doctrine, stemmed from God the Creator) (King, 1970, p. 117).

Syndenham, with Bacon (Bacon, 1875c, pp. 415-416), rejects ultimate or first principles as necessary explanations of disease and seeks instead to understand the natural history of disease and which treatments affect diseases for the better. Syndenham also rejects the "empiric," i.e., someone "who had no rationale for his remedies" (King, 1970, p. 133). Experience, careful observation of disease and how its course is altered by various interventions, supplies the foundation for medical knowledge. Syndenham may not always have met the demands of such a methodological commitment, but he provides an influential account of the empirical method applied in medicine (King, 1970, p. 133ff).

Boerhaave, to be sure, championed the mechanistic model of the human body and of disease. He classified diseases mainly according to fevers. More important, however, "[f]rom 1701 until his death in 1738, Boerhaave was the most influential medical teacher in Europe" (Wilson, 1993, p. 398), an influence Underwood (1977) explores in detail. Boerhaave achieved this standing in good measure because he "used hospital beds for university teaching at Leiden, and his students diffused the practice throughout Europe, particularly to Vienna and Edinburgh" (Gelfand, 1993, p. 1129). Of Gregory's teachers in Edinburgh, Monro *primus*, Rutherford, Sinclair, and Plummer had studied under Boerhaave in Leiden (Underwood, 1977, p. 110). When Gregory later joined the medical faculty at the University of Edinburgh he continued the Boerhaavian tradition, along with Cullen. John Pringle (1707-1782), to whom Gregory dedicates the 1772 edition of *Lectures on the Duties and Qualifications of a Physician*, also studied under Boerhaave (Rothschuh, 1973, p. 117; Underwood, 1977, p. 49).

Boerhaave also espoused and taught a Baconian method for medicine, including epistemological rules for scientific inquiry (Boerhaave, 1751a,b). Karl Rothschuh reads Boerhaave as attempting to apply to medicine recent discoveries in physics, chemistry, anatomy, and physiology (Rothschuh, 1973, p. 116). Boerhaave offered a new integration of knowledge, rather than contributing new discoveries (Rothschuh, 1973, p. 116). Boerhaave dismissed the view that treats physic, or medicine, as "wholly conjectural," holding this to be a "false" view (Boerhaave, 1751a, p. 48). "Wholly conjectural" or speculative science, as it was also called, was "science" based on reason unaided and uncorrected by observation and experiment. Thus, while Boerhaave may have accepted Descartes' mechanistic views of the human body, he certainly did not accept the method of discovering true propositions for which Descartes argues in his *Meditations on First Philosophy*. No inspection of a piece of wax by reason alone, using only the clear and distinct idea of matter as extended substance, could produce true propositions about wax.

Instead, science should proceed only on the basis of what can be established by rigorous investigation. In the spirit of Bacon, Boerhaave puts it this way: "What is demonstrated to us by our senses cannot be disproved in any Age, nor opposed by an Authority, unless by that of the Sceptiks" (Boerhaave, 1751a, p. 48). Something is not "demonstrated to us by our senses" from one observation, but from many, repeated, similar observations. Such observation produces proofs, i.e., true propositions. Only a "sceptik," someone who rejects altogether the possibility of true propositions based on observation, can deny the outcome of this observational scientific method. But this is a radical and unserious position and so Boerhaave sets it aside. It has no place in science nor in science put in service of health.

Not every scientific proposition in medicine can lay claim to truth. But many can, because we do not expect them to be falsified by future observations. This incomplete nature of medical knowledge, however, poses no problem for true propositions that have already been established.

> Besides, the Uncertainty of some things in Physic, does not diminish the Evidence of other Propositions, of whose Certainty we are satisfied (Boerhaave, 1751a, pp. 48-49).

Certainty, therefore, does not invoke nor does it require final truth. Rather, certainty is a function of the *stability of observations*. If these do not alter over time and are not expected to alter, they have achieved functional or pragmatic, i.e., clinical, certainty. One can act on such clinically certain knowledge with a robust – but not final – confidence. This knowledge does not possess the certainty claimed for clear and distinct ideas and, later, for synthetic *a priori* propositions. This concept of certainty is rejected, in the first instance, and foreign, in the second instance, to the Baconian scientific method that Gregory inherits and embraces.

This scientific method yields principles. 'Principle' for Boerhaave did not, however, mean "the constituent Parts or Elements of Bodies, but the Means of Demonstration, or Truths; not depending upon others, but by which others are to be established" (Boerhaave, 1751a, p. 51). Principles thus serve, it seems, an explanatory role.

King reads 'explanation' in the context of medicine at this time to mean "simply the process whereby some particular observation or phenomenon is *referred* to something else" (King, 1970, p. 185). The "something else" may be another thing, as in explanation by analogy, some "hypothetical entity" (King, 1970, p. 185), or "to some concept or general principle abstracted from experience" (King, 1970, p. 185). Boerhaave's approach fits the third of these, with 'principle' apparently referring to the most basic of concepts, those which do not presuppose any others.

Friederich Hoffmann, working in the "corpuscular philosophy" (King, 1970, p. 186), asserted that the first principles were matter and motion (Hoffmann, 1971, p. 2; King, 1970, p. 186). Hoffmann rejected metaphysical explanations of phenomena, particularly explanations that appealed to substantial forms (Rothschuh, 1973, p. 114). Substantial forms are static and they cannot explain change or motion (McCullough, 1996). Motion, according to Hoffmann, "was necessary for the preservation of life" (Rothschuh, 1973, p. 114). Hoffmann's explanations invoked "little bodies" that could not be seen and so his concept of explanation fits the second of the three meanings described by King (King, 1970, p. 187).

Of particular interest for present purposes, Hoffmann, along with Georg Stahl (1660-1734), treats human beings not simply as body-machines, but a complex of mind and body.

6. The life [*vita*] of man consists in the uninterrupted communion of mind [*mens*] and body. It depends on those operations in which concur both the movements of the body and the reflection of the mind.

7. Life is achieved by causes which are wholly mechanical. The mind does not bring life to the body, nor is life oriented to the mind, but rather to the body.

8. When a human body dies it is not that the mind that recedes from the body; rather the body recedes from the mind, since the organs of the body have become corrupt, so that the mind can no longer preside over them (Hoffmann, 1971, p. 11).

Motion or regular activity is to be explained in terms of the vital principle.

21. All philosophers are agreed, that in our machine the first principle of motion is the soul [or vital principle (*anima*)], which you may, if you want, designate as nature, or spirit endowed with mechanical powers, or a most subtle ethereal matter acting in an ordered and specific fashion.

22. It is beyond question that the vital principle in man, apart from its mechanical capacities and power of performing ordered movements, is endowed with a more noble power, namely of thinking and reasoning, which the vital principle of brutes does not have; and by virtue of this

power it is called mind [*mens*], an immortal substance stemming from the decree of God himself (Hoffmann, 1971, pp. 12-13).

This interest in motion – regular, ordered change – reflects the challenge to the static metaphysics of the late Scholastics, Descartes' new discoveries concerning motion in physics, and the biological activity of the *animalicula* observed by von Leeuwenhoek through his microscope (Wilson, 1995). These discoveries required explanations of law-governed change and the old static model and its metaphysics of substantial forms could not provide such explanations. The body-as-machine is inert, not automotive. Thus, Leibniz adapts his concept of individuals and the principle of their individuation into a dynamic model, with the dominant monad the automotive principle of activity or change in a body (McCullough, 1996). Hoffmann and Stahl take up a similar challenge, i.e., to explain physiologic habits, e.g., regular excretion, as law-governed activity (King, 1970, p. 191). They sought a principle to explain regular change that appears in its regularity to be law-governed.

Hoffmann's explanation of such change was that the mind possessed an animal spirit that originated the power of originating purposeful movement (King, 1970, p. 191). This parallels very nicely Leibniz's concept of the monad as comprising both perceptions – relations of each entity to all other entities in creation – and appetition – the intrinsic power of a monad to generate its set of perceptions (McCullough, 1996). Animal spirit in the mind, according to Hoffmann, can produce ideas that, in turn, get "'impressed' on the animal spirits that triggered the muscles" (King, 1970, p. 191). Again, this parallels Leibniz's account of the pre-established harmony of the dominant monad, mind, of a body, or collection of monads brought into a collection by that mind (McCullough, 1996). The difference is that for Hoffmann there was a real causal effect of animal spirit on the body. Stahl took the view that the vital force, or anima (of the soul) "controlled the body's activities and directed its vital motions" (Rothschuh, 1973, p. 121). The anima is the "agens" while the body is the "patiens" (Rothschuh, 1973, p. 121).

Seven points summarize the Baconian method that had emerged in the work of leading figures in medicine and medical science (such as it then was) by the time Gregory attended medical school. First, this method promoted a critical attitude toward any proposition of fact. New propositions of fact, or old ones for that matter, were not to be taken as true on the basis of authority, *anyone's* authority – Hippocrates, Galen, or even one's own teachers and colleagues. No proposition is true or false because accepted or rejected, respectively, by authorities.

Second, any proposition must be tested against "experience." Experience involves rigorous observation and careful experiments guided and disciplined by rules for empirical investigation, e.g., by induction or by analogy.

Third, this method is open to anyone. No one can claim exclusive expertise to establish the truth of a proposition. The sources of scientific knowledge are not in any way internal or peculiar to an individual or social institution. Medical knowledge is therefore based on a public, secular method and is not the exclusive province of physicians.

Fourth, refusal to submit to scientific method earns, rightly, disrepute and disrespect. Any power attached to such willful ignorance lacks moral authority and should therefore be opposed on intellectual grounds. If that power results in harm to people, it should be opposed – vigorously – on moral grounds.

Fifth, for the Scottish scientists, explanation occurred along the path of an appeal to really existing 'principles' or causes in things. Scottish philosophy of science and medicine was realist. Baconian scientific method leads to the discovery of such principles, their correct functioning, and their incorrect or defective functioning.

Sixth, the concept of body-as-machine had been largely discarded and was replaced by the mind-body complex governed by discoverable principles of bodily activities, the seat of which was in the mind. The physiology of mind thus became of paramount importance. As a consequence, *mental diseases came to have increased clinical standing* and to be studied just like any somatic disease, with scientific interest and discipline.

Seventh, science was compatible with deism, especially the new animist science. Stahl, for example, "maintained that the human body was merely the temporary casing of a controlling *anima* or immortal soul created by God" (Tansey, 1993, p. 124). Jerome Gaub and Robert Whytt held that "the Hippocratic *impetum faciens* and 'healing power of nature' and the Christian soul were combined to supply a source of motion for the otherwise inert machine of the body [since no machine can be a principle of its own activity or motion]. It no longer operated through faculties nor was it discernible through a structure-function relationship. Instead, it was investigated experimentally" (French, 1993a, p. 96). This secular way of knowing – the new science – did not exclude, nor did it require the rejection of, deism.

Norman Gevitz concisely summarizes the developments covered by the seven points:

> ... there was also a growing effort to eschew theory arrived at by deduction [reason unaided and uncorrected by experience], and instead to ground medicine upon the direct observation and measurement of phenomena, to conduct controlled experiments and to correlate facts (Gevitz, 1993, p. 605).

Empirical, not deductive, theories were indeed possible, because they were being developed and applied clinically both in *therapeutics for patients* and in *experiments on patients*.

These empirical theories – at least those in which Gregory immersed himself as a student – emphasize observable real principles. Thus, there is a metaphysics hard at work here. Each human being has a nature and that nature is the same in each of us. This, of course, is a straightforward Aristotelian doctrine modified by the later Scholastics (McCullough, 1996). With enough persistent, disciplined observation one can identify with certainty – enough for everyone but the radical skeptics or those, one could add, who think that there are clear and distinct ideas or synthetic *a priori* judgments to be had – the principles of regular activity or habit that constitute that nature as an active force or animal spirit. Nature is not analyzed by Cartesian clear and distinct ideas because these have nothing to do with experience. Nature is discovered; essence = nature's principles that are discovered to be the same in all members of a species. Such discovery follows as the product of "experience," disciplined observation and experiment, following strict method. Ideally, and eventually, we will have exhaustive knowledge of the mental as well as physical principles of health and disease, and life and death in human beings. The static metaphysics of substantial forms is thus replaced with an active metaphysics, a metaphysics of activity, concerned with vital and other causal, physiologic principles.

Gregory absorbed these lessons in Edinburgh, as is evident from "A Proposall for a Medicall Society, Written in 1743" (AUL 2206/45, 1743). This nine-page document is truly remarkable. In it Gregory emerges as a full-blooded student of the new science. More than that, he sets out a virtual outline for the intellectual tasks that would occupy him for the rest of his life. The precocity, though not necessarily originality, reflected in this little document is startling.

This document begins with a lament for the present state of medicine, despite the "multitude of volumes wrote on the subject," and recent advances in "Anatomy, Chemistry, Botany, & every branch of naturall philosophy connected with the art of heeling" (AUL 2206/45, 1743, p. 1). Gregory's diagnosis goes, he thinks, to the heart of the matter: "The whole plan laid down by L[ord]. Bacon for prosecuting enquirys into nature has been applied in some measure to many branches of naturall philosophy, tho not with that accuracy and fidelity proposed by its great Author" (AUL 2206/45, 1743, p. 1). The same goes for medicine, and this lamentable state of affairs requires remedy, improvement. Gregory's response to this problem invokes the scientific world I have just described:

> I would therefore propose that a Medicall Society be founded whose proper business may be to presente Medicall Enquirys in the strict method of naturall history & accurate induction from it, & to make an exact separation of those things in the art which are certain & may be depended on, from such as are uncertain or groundless. To make the undertaking more usefull, the Society should prosecute such enquirys only as have a near connexion with the art of heeling or are immediately subservient to the uses of life. In order to this a distinct list should be made out of these Desiderata in medicine q^{ch} [which] seem to be of most importance. The Society should appoint a particular member to make a diligent enquiry into one or more of these Desiderata, & report the same to the Society within a limited time (AUL 2206/45, 1743, p. 1).

The first task will be to identify the "general heads or articles of enquiry to be prosecuted" (AUL 2206/45, 1743, p. 2). Each article should be provided "a simple narration of facts with the Authoritys for them: & where any of them are untested, a particular mark should be affixed" (AUL 2206/45, 1743, p. 2). "Axioms deduced from the preceding Naturall history" should be connected with the supporting observations and the degree of certainty of each axiom should be noted "from perfect, invariable canons or Axioms down to bare probability's" (AUL 2206/45, 1743, p. 2), just as Bacon prescribes. There should then appear a list of desiderata required to complete knowledge of the topic in question, "with the approximation or method of supplying them. Directions for experiments & observations to be made & references to the best books on the subject" should then be provided (AUL 2206/45, 1743, p. 2). This material should be published in a book by the Society. Gregory wants this book to be thoroughly referenced: "An account likewise should be given of the books he [the investigator] had consulted on the subject, that no body may beat the trouble to search them over again" (AUL 2206/45, 1743, p. 2).

He then lists fifteen desiderata. The first is "A history of every disorder incident to the human body" (AUL 2206/45, 1743, p. 2), which should include:

[a] "definition of disease, q^{ch} [which] ought to be only a simple narration of such symptoms as universally accompany the disease & distinguish it from all others," "Diagnostic Signs of the disease," [the] "manifest antecedent causes of the disease," [and the] "Prognostics to be drawn from the symptoms of the disease or other circumstances; with such marks denoting their different degrees of certainty. These Prognostics should be taken from observation & not copied from Systems" (AUL 2206/45, 1743, pp. 2-3).

There should then be information provided about the "method of cure" (AUL 2206/45, 1743, p. 3). This should include the usual method of treatment by the "best Practitioners," "extraordinary & uncommon methods of curing the disease," and "Particular precepts & Cautions with respect to the administration of every remedy" (AUL 2206/45, 1743, p. 3). This desideratum, for which Gregory provides the most detail, concludes with "An accurate account of the dissection of the body" and "Axioms relating to the form, proximate cause, & cure of the disorder, w^t [with] Queries relating to the present practice or any other which may be thought would be more successful" (AUL 2206/45, 1743, p. 3).

Gregory presents the next thirteen desiderata in single sentences. They include information on the "antecedent causes" of disease; "History of the connexion between the mind and the body & their mutuall effect in producing & curing these diseases," including the effect of the imagination "in the production and cure of diseases"; "History of Old Age with the method of preventing the Consequences of it, or the Renovation of youth"; death and its "proximate causes"; "incurable" diseases, i.e., those that result from "a default in y^e [the] Solids as their can be no hope of remedying"; "History of the senses & the method of preserving & improving them"; "History of Evacuations made by bleeding" and other means; the "Operation of internall remedys," "externall" remedies "such as bathing, rubbing, plasters, & c"; "History of Cautions & precepts w^t [with] respect to the administration of every remedy, also relating to the conjunction & mixture of remedys"; "History of Exercise & its effects on the human body" and the "means of increasing strength" (AUL 2206/45, 1743, pp. 3-4).

The fifteenth desideratum concerns "the best method of prosecuting the Study of medicine, with an account of the best books upon every subject" (AUL 2206/45, 1743, p. 5). To this end each member of the Society should record the natural history of his locale, the "constitution and character of the inhabitants," the diseases to which they are subject, the weather, the "quality of the air," the "rise & progress of Epidemic diseases" and methods to prevent them, with all of these data being recorded in "an exact register." He provides a number of requirements that should guide the updating of this register (AUL 2206/45, 1743, p. 5).

After "Four Capitall Enquirys" (see below) Gregory then lists "Generall Desiderata Previous to these Enquirys" (AUL 2206/45, 1743, p. 6). These include "An Enquiry into the Vitall Principle, what it consists in, its form & consequently the form of Death," a history of man "thro[ugh] all the different stages of his existence," the "different constitutions of mankind as depending on their Sex, manner of living," and environmental factors, "A History of the Union betwixt y^e [the] mind & body, & their mutuall influence upon one another Particularly of y^e [the] power of the imagination not only upon the mind or body of the Imaginant, but upon the mind or body of another," the effects of diet and other factors on the body, medicines and their effects on the body, evacuations such as bleeding, "cautions" with respect to remedies, conception, dreams, and other matters (AUL 2206/45, 1743, pp. 6-7). These desiderata and the fifteen "general heads or articles of inquiry" derive largely from Bacon (1875c).

Gregory includes animal science and research. His interest seems to be mainly comparative anatomy: "A Naturall history of such Animalls & such parts of nature as by their anatomy, constitutions, diseases & instincts can throw any light upon the present subject" (AUL 2206/45, 1743, p. 7).

He calls for study of diseases around the world and then goes on to list desiderata "relating to the Cure of Diseases" (AUL 2206/45, 1743, p. 8). This includes "Enquiry into diseases incurable & what are the circumstances that render them infallibly so," diseases "peculiar to the different ages of men," diseases related to "constitution," sex, heredity, those "which attack people only once in their life," ways of living, as well as endemic and epidemic diseases, and then into particular disease entities such as gangrene, glandular diseases, diseases caused by worms, etc. (AUL 2206/45, 1743, pp. 8-9).

Gregory had learned, and learned very well indeed, all of the major lessons of the new, Baconian science of his time. He commits his proposed Society – and more importantly, as it turns out, himself – to the rigorous scientific investigation of just the sort described above. Many of the ideas expressed in his "Proposall" appear later in the sections of *Observations* and *Lectures* devoted to the philosophy of medicine.

B. A Concept of the Nature of Medicine

The "Proposall for a Medicall Society" would come to shape Gregory's intellectual work in philosophy of medicine, as it was then understood (King, 1978). Gregory follows much of its agenda and provides accounts of whether medicine is a science, of the intellectual virtues of the medical scientist, of concepts of health and disease, of the physiologic concept of sympathy, of the capacities of medicine, and of diagnosis and treatment. Indeed, under the heading "The Four Capitall Enquirys" (between fifteen desiderata and the general desiderata) he lists the following:

1. The Preservation of Health.
2. The Retardation of Oldage.
3. The Cure of Diseases.
4. The Improvement of our Nature (AUL 2206/45, 1743, p. 6).

Here Gregory also exhibits a considerable debt to Bacon, who describes the "three offices" of medicine: "the first whereof is the Preservation of Health, the second the Cure of Diseases, and the third the Prolongation of Life" (Bacon, 1875c, p. 383). Bacon elaborates on these "offices," the moral obligations of medicine:

> But this last [prolongation of life] the physicians do not seem to have recognized as the principal part of their art, but to have confounded, ignorantly enough, with the other two. For they imagine that if diseases be repelled before they attack the body, and cured after they have attacked it, prolongation of life necessarily follows. But though there is no doubt of this, yet they have not penetration to see that these two offices pertain only to diseases, and such prolongation of life as is intercepted and cut short by them. But the lengthening of the thread of life itself, and postponement for a time of that death which gradually steals on by the natural dissolution and the decay of age, is a subject

which no physician has handled in proportion to its dignity (Bacon, 1875c, p. 383).

The office of prolongation of life "is new, and deficient; and the most noble of all" (Bacon, 1875c, p. 390). For "[m]edicine's there are many for preserving *Health*, and curing diseases, but few to *prolong life* ... (Bacon, 1977, pp. 130-131). The first two offices require improvement, as well. The requirement to improve the three "offices" of medicine Gregory finds still unsatisfied as the writes his "Proposall," as we saw above. Gregory picks this up in the desideratum relating to "History of Old Age with the method of preventing the consequences of it, or the Renovation of youth" (AUL 2206/45, 1743, p. 4). Here is geriatrics in its nascent form.

Bacon also addresses subsets of each office. His remarks on those under the preservation of health merit attention. The first concerns incurable diseases:

> Therefore I will not hesitate to set down among the *desiderata* a work on the cure of diseases which are held incurable; so that some physicians of eminence and magnanimity may be stirred to take this work (as far as the nature of things permits) upon them; since the pronouncing these diseases incurable gives a legal sanction as it were to neglect and inattention, and exempts ignorance from discredit (Bacon, 1875c, p. 387).

Gregory absorbed this point as a medical student and returns to it forcefully his medical ethics lectures.

Bacon also addresses pain management under the rubric of preserving health:

> Again, to go a little further; I esteem it likewise to be clearly the office of a physician, not only to restore health, but also to mitigate the pains and torments of diseases; and not only when such mitigation of pain, as of a dangerous symptom, helps and conduces to recovery; but also when, all hope of recovery being gone, it serves only to make a fair and easy passage from life ... This part I call the inquiry concerning *outward Euthanasia*, or the easy dying of the body (to distinguish it from that Euthanasia which regards the preparation of the soul); and set it down among the desiderata (Bacon, 1875c, p. 387).

Bacon also faults physicians for not following a regular use of medications, to learn better about their effects.

Boerhaave takes a similar view on the nature of medicine:

... the whole design of the Art is to keep off and remove Pain, Sickness and Death, and therefore, preserve present and restore lost health; so that every thing necessary to be known by a physician, is reducible to these two heads (Boerhaave, 1751a, p. 51).

Medicine, Boerhaave says, is a science:

Physic [medicine] therefore is the *Science* or knowledge of *those things*, by whose Application and Effects Health may be preserved when present, and restored when lost, by the Cure of Diseases (Boerhaave, 1751a, p. 54).

Under the four rubrics of "Capitall Enquirys," Gregory takes up many of the topics listed in the "Proposall" in his *Comparative View*, his clinical lectures, and in his lectures on medical ethics and philosophy of medicine. Item four becomes the project of *Comparative View*, while the first three items become Gregory's definition of medicine in his lectures on medical ethics and philosophy of medicine.

The second of the first three items appears with altered wording in his medical ethics lectures but not, I think, with altered meaning, which, as we shall see, owes a great debt to Bacon. He also explores this meaning as a medical student. In his "Medical Notes" of 1743 – in which his "Proposall for a Medicall Society" is the first entry – he has entries under "progress of aging" (AUL 2206/45, 1743, p. 99-100), "History of the best method of Preserving and improving the Senses, Memory, Imagination, and the other facultys of Mind" (AUL 2206/45, 1743, p. 706), "History of Protracting Life and Preserving the faculty [of mind] Entire" (AUL 2206/45, 1743, p. 709), and "Enquiry into the Methods of making Death easy" (AUL 2206/45, 1743, p. 713). These four headings are followed by blank pages, but Gregory's familiarity with them indicates that he understands 'preserving life' to mean preserving life when so many died young or in adulthood, children and pregnant and postpartum women particularly, extending life expectancy and the quality of extended life, and providing comfort to the dying. 'Preserving life' does not mean an unlimited obligation to preserve life, for Gregory or for Bacon. Such a view simply parts company with these headings and Gregory's caution in the "Proposall for a Medicall Society" about the harmful effects of medication. The theme of *limits* on medicine's capacities and therefore on the physician's obligations in the exercise of those capacities becomes central to Gregory's medical ethical thinking (from the time he was a medical student) and remains a constant in his thinking. In his "Proposall" Gregory refers repeatedly and thus with emphasis to "precepts and Cautions," i.e., the risks of treatment that therefore should set limits on it.

C. The Physiologic Principle of Sympathy.

Gregory adopts the central concepts and agenda of the new science of the physiology of the nervous system and the science of the mind-body complex. He mentions a vital principle, the mind-body connection, the effect of imagination on the body, and other topics that derive from this new science. The concept of *sympathy* plays a major role in this science, because it helps to explain how the anima can affect the body.

Consider general desideratum 5: "A History of the Union betwixt y^e [the] mind and body, & their mutuall influence upon one another Particularly of ye power of the imagination not only upon the mind or body of the Imaginant, but upon the mind or body of another" (AUL 2206/45, 1743, p. 6). Imagination reflects the power of the animal spirit, as we saw above, to affect the heart and arouse the passions. Imagination can also affect the mind and body of another person.

In another general desideratum, number 11, he calls for "A History of such things as operate on the body at a distance, the Influence of the Celestial bodys, an Enquiry into what are called Sympethys and Antipathys or such things as operate upon the human body by laws of Nature hitherto unknown" (AUL 2206/45, 1743, p. 7).

One principle of the new science of medicine that obviously caught young John's attention at Edinburgh was *sympathy*. In his "Medical Notes" he includes two headings of interest. The first is the following:

History of the effects of Sympathy and Antipathy or any other Principles in Nature not generally attributed to, on the Mind and Body (AUL 2206/45, 1743, p. 703).

The rest of page 703 is blank. The second heading appears earlier in the "Notes" and provides us with a clue as to what Gregory was reading at the time. It reads, simply, "Sympathy," under which the following appears: "See practicall remarks on the Sympathy of the part of the body," followed by an abbreviated reference, which turns out to be to an essay by Dr. James Crawford (EUL E.B. 6104 ED 1, 1744). Crawford also studied under Boerhaave in Leiden and taught at Edinburgh for twenty years, until his death in 1732 (Underwood, 1977, pp. 99-101). The essay to which Gregory makes reference was apparently a reprint of an earlier publication.

Crawford's essay concerns "the influence of each part on the other" in the body (EUL E.B. 6104 ED 1, 1744, p. 480). Crawford describes his general view on this topic – the then current physiology of the mind-body complex:

> An accurate knowledge of the Structure and Œconomy of the several parts of the human Body, and the Influence each particular part hath on another, is the principal Foundation both of Medicine and Surgery ... (EUL E.B. 6104 ED 1, 1744, p. 480).

In reasoning from effects to causes, Crawford identifies three causes for disease in a part, i.e., organ or structure of the body:

> I. That a Part is affected by *Protopathia*, when it is essentially in itself lesed [diseased], and owes not its Origin to any Communication from another Part. Or by *Idiopathia*, when tho' it be essentially lesed [diseased], yet the hurt was at first propagated to it from some other Part. Or lastly, by *Sympathy* or Consent, when the Part in itself is yet whole and sound, and is only affected by the fault of some other Part ... Diseases by Consent are propagated from a Distance, (in which case only I shall consider them) either by long Muscles or Nerves (EUL E.B. 6104 ED 1, 1744, pp. 482-483).

Sympathy is a physiologic principle developed to explain action at a distance – "such things as operate on a body at a distance" (AUL 2206/45, 1743, p. 7). This had posed a deep scientific puzzle for many centuries, in physics as well as in biology (Hesse, 1967). To set Crawford's views – and therefore Gregory's early views – in context, I now turn to a brief history of the development of the concept of sympathy as a physiologic principle.

L.J. Rather explains that in its general sense the concept of sympathy concerned at the time the integrated character of living things, which distinguishes them from non-living entities.

> This is the literal meaning of "sympathy" (*sympatheia*), the term used by the Stoic philosophers to designate a relationship of the kind exemplified by the living body. "Consensus" has a related, if not identical meaning in this context (Rather, 1965, p. 205, n.2).

In the seventeenth century the concept was further developed to explain action at a distance. Sir Kenhelm Digby reports the case of a man with gangrene. In another room where the patient could not see Digby or what he was doing, Digby placed one of the man's garters that had been in contact with the infection in a solution of vitriol and water. The patient, who did not know what Digby was doing, reported improvement. Digby ordered the patient's plasters removed and removed the garter from the solution, whereupon the patient worsened. Digby returned the garter to the solution, left it there, and the patient's gangrene healed in six days (King, 1970, pp. 141-142). Digby posited *sympathy* as the explanatory principle.

Other examples of sympathetic action interested investigators of nature. For example, the imagination can impress itself upon the heart, even though the two are not directly connected. Imagination, by sympathy, can cause a "dilatation of the Heart," arousing particular "Passions" (Digby, 1669, p. 182). There can also be sympathy between a pregnant woman and the infant in her womb, analogous to the vibrating of harp strings when one is struck. Sympathy in this case involves the "harmonious consonance" of atoms in mother and child (Digby, 1669, p. 184).

There can also be "contagion of the imagination" in which the passions of one individual infect others. Melancholy and laughter, for example, can pass from one person to others (Digby, 1669, p. 183). Digby has no explanation for this effect of sympathy.

Hoffmann may have had sympathy in mind as an explanatory principle when he wrote "operations in which concur both the movements of the body and the reflection of the mind (Hoffmann, 1971, p. 11). Jerome Gaub, one of Gregory's teachers at Leiden, provides a somewhat more sophisticated account of the role of sympathy in the mind-body complex:

> The sympathy of the body and mind is ... such that particular affections of the mind will bring on particular disorders of the body, and disorders of the body will in turn affect the mind (Rather, 1965, p. 182).

Crawford picks up on this concept of sympathy and equates it with "consent," which echoes the Stoics on consensus. For Crawford, sympathy functions as a principle to explain how a healthy part of the body becomes diseased in a way like another part of the body from which the first body part is at a distance. In this respect sympathy helps to explain dysfunction in the absence of apparent pathology. He posits "long Muscles or Nerves" as the conduit through which sympathy does its work. (Perhaps here he follows Hoffmann.)

Hume, of course, provides an account – the double relation of impressions and ideas – of sympathy as a principle to explain the "contagion" described by Digby, as well as the distress that we experience when we see others in distress. I have been unable to locate any documentary evidence that Gregory had read Hume's *Treatise* while in medical school in Edinburgh. Hume, however, may have been influenced by views such as Crawford's. As a student at the University of Glasgow Hume was member of the Physiological Library (Steuart, 1725). Crawford was, too, and one of his books was included in this library that students and faculty had assembled for their own use. One of Digby's books is listed in the library, as well (Steuart, 1725). Thus, it is very likely that Hume was aware of the history of the physiologic concept of sympathy that I have just adumbrated. Hume's text make it clear that he treats sympathy as a principle of human psychology, i.e., as a principle of the physiology of the mind that explains the workings of the social principle. Hume can be read as writing a physiology of mind, including moral physiology, with the goal of using experimental method to identify the constitutive active principles of mind. Gregory and Hume were thus "part of a widespread movement from the middle of the eighteenth century which emphasized the primacy of the nervous system" (Bynum, 1993, p. 346).

The interest in sympathy coupled with the concept of the mind-body complex and the effects of body and mind on each other, and of the mind on its subsequent states, led to an interest in mental illness, "nervous diseases." As we shall see (section IX), Robert Whytt becomes the champion of this new clinical diagnostic category. Whytt taught at the University of Edinburgh after Gregory left there as a student, and overlapped for one year Gregory's return to Edinburgh as Professor of Medicine.

The following features of Gregory intellectual development during his medical student days stand out. First, he became steeped in the method of Francis Bacon and other great scientists of natural history and natural philosophy. This method commits Gregory, as it did the Scottish moral sense philosophers, to learning from experience and to experience-based philosophies. Second, he takes a pragmatic approach to epistemological reliability of medical science and judgment: they must have sufficient evidence to be "depended on," i.e., put into practice with confidence that they will benefit the patient. Third, he has a well developed concept of medicine as a science, deeply indebted to Bacon. This concept of medicine identifies three "offices" or capacities of medicine, which together begin to define it as a social institution or practice that transcends individual physicians. This occurs especially when the physician follows scientific method. *These capacities are always limited*, a point that cannot be overemphasized. Fourth, as a physiologist of mind, he accepts the physiologic principle of sympathy, included the interpersonal application of the principle to explain "contagion" of passions. Fifth, he extolls the "superior talents" of women, inspired perhaps by the example of his great grandmother, Janet Gregorie.

Three distinct honors came to Gregory during his school days. On April 18, 1843, while he was yet a student in Edinburgh, the City of Old Aberdeen granted him a "Diploma of the Freedom of the City" and elected him honorary burgess of Old Aberdeen (Lawrence, 1971, Vol. 2, p. 406 and Vol. I, p. 150). To receive such high honors at so young an age added, I suspect, to the academic expectations that his family history had already created.

On March 11, 1746, while Gregory enjoyed student life and the company of English-speaking students in Leiden, King's College awarded him an unsolicited M.D. (Lawrence, 1971, Vol. I, p. 151). Dorothy Johnston (1987) provides a useful account of "irregular" degrees awarded by King's College during this time. Medical degrees were usually granted unearned on the "attestations of two or more doctors, supported by what amounted to a character reference" (Johnston, 1987, p. 138). Gregory's degree was "[s]igned by his half-brother, James Gregory, as Professor and Dean of Medicine, and others" (Lawrence, Vol. 2, p. 406). This would not be the first time that Gregory's career was aided by family connections, a not uncommon phenomenon at that time. That his medical degree was unearned was not without consequence for Gregory's medical ethics, as we shall see in this chapter and the next.

On June 3, 1746 Gregory was elected Professor of Philosophy in King's College, effective upon his return to Aberdeen. Gregory's school days had come to an end and his days as an academic and practicing physician just began.

V. ABERDEEN, 1746-1754:
TEACHING, PRACTICE, MARRIAGE

By mid-century Aberdeen began to experience the growth that typified Scottish cities at that time. There had been "fitful support" for the Jacobites (MacKenzie, 1953, p. 34), even though, as we saw in the story of Rob Roy and Gregory's father, the Gregory family connection was perhaps a bit closer. In any case, "[a]fter 1745 the city settled down to a period of steady expansion" (MacKenzie, 1953, p. 34). Aberdeen expanded commercially, spurred by the growing effects of the Act of Union that had facilitated trade with England. Trade also grew with Northwestern Europe, among traditional markets for Scottish wool, and with New World markets. Aberdeen also began to participate in industrial growth. The economy of the city and region, previously comprising mainly weaving, crafts, farming, and fishing, began to include textile production, shipbuilding, and other industries (MacKenzie, 1953, pp. 35ff.). The Infirmary opened in 1739 and the Poor's Hospital in 1741 (MacKenzie, 1953, p. 41). By 1755, the population exceeded 15,000 and would grow to more than 26,000 by the end of the century (MacKenzie, 1953, p. 87).

King's College still continued the Regent system of teaching when Gregory returned to join his professor, Thomas Gordon, his half-brother James, and Principal Chalmers on the faculty. From 1747 to 1749, Gregory lectured in King's College "on Mathematics, on Experimental Philosophy, and on Moral Philosophy" (Tytler, 1788, p. 32). Stewart quotes from a source that provides an interesting account of academic life at King's:

> Every Professor of Philosophy in this University is also tutor to those who study under him, has the whole direction of their studies, the training of their minds, and the oversight of their manners; and it seems to be generally agreed that it must be detrimental to a student to change his tutor every session ... and though it be allowed that a professor who has only one branch of philosophy for his province, may have more leisure to make improvements in it for the benefit of the learned world, yet it does not seem extravagant to suppose that a professor ought to be sufficiently qualified to teach all that his pupils can learn in philosophy in the course of three sessions (Stewart, 1901, pp. 105-106).

Nothing, however, flowed from Gregory's pen during this time for consumption by the "learned world." That was still to come.[5]

Gregory also maintained a medical practice at the time he was Professor. Medicine drew him more strongly than academic life, however, and in 1749 he resigned his post at King's. His cousin, Thomas Reid, would succeed him in this position in 1751 (Lawrence, 1971, Vol. I, p. 152). Tytler explains this event in Gregory's life:

> In the end of 1749, however, he chose to resign his Professorship of Philosophy, his views being turned chiefly to the Practice of Physic, with which he apprehended the duties of this Professorship, occupying a great portion of his time, too much interfered (Tytler, 1788, p. 32).

Gregory did not go directly into the full-time practice of medicine, however. Tytler continues:

> Previously, however, to his settling as a Physician at Aberdeen, he went for a few months to the Continent; a tour of which the chief motive was probably amusement, though, to a mind like his, certainly not without its profit in the enlargement of ideas, and an increased knowledge of mankind (Tytler, 1788, p. 32).

Gregory entered into medical practice after his grand tour, with all of the advantages of his family name. In taking up this legacy, Gregory participated in the larger social phenomenon in Scotland of "medical families, with one generation of doctors succeeding another ... " (Lochhead, 1948, p. 337).

Marion Lochhead paints a striking portrait of the Scottish doctors of the eighteenth century. They constituted Gregory's competition:

> A good doctor may not have the vices and weaknesses of the flesh, but he may possess as many eccentricities as he will. From 1700 onwards there was a procession of men, many of them pioneers and discoverers, all of them with something in them beyond cleverness – a salt of personality, of wit, even of freakishness (Lochhead, 1948, p. 335).

The reader should recall that all sorts of people practiced medicine at this time – the university degree holder, earned and unearned; the surgeon; the apothecary; the irregulars; the midwives; and others. There was no "external" validation of qualification in terms of medical degrees from accredited medical schools, licensure, or board certification. In short, there was no social institution of medicine, providing a powerful, external, socially sanctioned authority and imprimatur on the physician. The physician was on his own and in hot competition with others (Porter and

Porter, 1989). Eccentricity and other quirks of character made one stand out from the crowd and presumably helped one to gain and hold market share.

Physicians – among the things they did for their patients – wrote prescriptions. Sometimes they did so face-to-face and sometimes they did so by mail. At that time, there were no laws regulating the preparation and dispensing of drugs, including what later became known as "controlled substances," such as opium. Moreover, prescriptions sometimes called for home preparation of remedies; not all prescriptions were meant to be filled by apothecaries, who did fill prescriptions for drugs. Physicians also dispensed drugs themselves. Prescriptions, then, encompassed drugs, home remedies, and changes in diet, e.g., "goat's whey" (Lochhead, 1948, p. 338).

These practices had interesting consequences. Prescriptions did not remain just with those to and for whom they were written; instead, "[p]rescriptions were exchanged like recipes" (Lochhead, 1948, p. 333). No surprise, in such an environment people self-medicated. They also self-diagnosed. "The Scot is often an amateur doctor, and self-physicker," Lochhead summarizes (Lochhead, 1948, p. 339).

No surprise, too, that scientific remedies did not possess the authority they came to have later; and so people also used "folk" or "traditional" forms of healing.

> While scientific remedies and treatment were being discovered and developed [Gregory in his "Medical Notes" comments on how little progress has been made since Bacon], the old homely cures were still in favour, and the good housewife must be able to compound an elixir as well as a pudding. The publication of [William] Buchan's *Domestic Medicine* in 1769 brought science into the home, and for many generations that valuable work ranked not far below *The Pilgrim's Progress* in the opinion of wise householders Lochhead, 1948, p. 339).[6]

I will examine in greater detail the medical practice context in the next section.

After his return from his continental holiday and several years in practice, Gregory courted and wed Elisabeth Forbes, the younger of two daughters who survived Lord William Forbes, the thirteenth of that title (Lawrence, 1971, Vol. I, p. 152). The Forbes traced their ancestry to William the Conqueror and Kind David I of Scotland (Lawrence, 1971, Vol. I, p. 152). Tytler describes Elisabeth Forbes and her marriage to

Gregory in the following way:

> Some time after his return to Scotland, Dr Gregory married, in 1752, Elisabeth daughter of William Lord Forbes, a young lady, who, to the exterior endowments of great beauty and engaging manners, joined a very superior understanding, and an uncommon share of wit. With her he received a handsome addition of fortune [women brought a dowry to marriage]; and during the whole period of their union, which was but for the space of nine years, enjoyed the highest portion of domestic happiness. Of her character it is enough to say, that her husband, in that admired little work, *A Father's Legacy to his Daughters*, the last proof of his affection for them, declares, that "while he endeavors to point out what they should be, he draws but a very faint and imperfect picture of what their mother was" (Tytler, 1788, pp. 33-34).

Stewart reports that "there is a story that her father did not at all approve of the marriage. 'What do you propose to keep her on?' said he [Lord Forbes], and Gregory, getting angry, took his lancet out of his pocket, and said, 'on this'" (Stewart, 1901, p. 107). Gregory, it seems, married the sort of woman whose intellectual abilities and virtues he had extolled during his student days in Leiden.

Obviously, John Gregory practiced medicine in a medical market in which his half-brother enjoyed a good reputation, as did "others of some note in their profession" (Tytler, 1788, p. 34). Perhaps Gregory found the competition awkward, perhaps his Aberdeen practice did not flourish. We do not know.

> ... [Gregory] determined to try his fortunes in London. Thither he accordingly went in 1754; and being already known by reputation as a man of genius, he found an easy introduction to many persons of distinction both in the literary and polite world (Tytler, 1788, p. 34).

VI. LONDON, 1754-1755:
MEDICAL PRACTICE, THE BLUESTOCKING CIRCLE

In London, Gregory experienced an interesting world of practice, in many ways not unlike what he left in Aberdeen. He also made the acquaintance of Elizabeth Montagu, "Queen of the Bluestockings," and her circle of women intellectuals. Both had a lasting influence on him.

A. Medical Practice in Eighteenth-Century England

Medical practice at the time developed, in part, on the basis of concepts of health and disease. Health came to be understood as normal bodily function, where 'normal' was defined in terms of observation, i.e., scientific facts. Observed normal function counted as health. Individuals aimed to maintain their health by attention to the "non-naturals." These included:

> ... environmental agents such as air, food, and drink; natural bodily functions like motion and rest, wakefulness and sleep, and retention and evacuation of nutritive substances; and finally mental activities, including the passions (Risse, 1986, p. 14).

Proper management of one's non-naturals promoted health, while neglect of them contributed to disease. "Among those who subscribed to Enlightenment values, health was a key ideal to strive for, a state of bodily and psychological wellness conducive to a more enjoyable and longer life," and therefore conducive to "human happiness and productivity" (Risse, 1986, pp. 14, 16).

Illness involved the experience of observed abnormal function. Developments in Baconian medical science began to offer pathological explanations for diseases and their symptoms in terms of pathologies of both the body and the mind. With respect to the latter, Robert Whytt deployed the concept of sympathy to explain a variety of psychological disorders, including hysteria and hypochondria (Whytt, 1765), a topic I take up in more detail in section IX, below. Disease resulted in "much personal suffering coupled with financial losses for the afflicted, their families, and ultimately the state" (Risse, 1986, p. 14). Preventive medicine thus grew to increasing prominence.

Individual behavior contributed to disease, as well as environmental factors. Individuals believed that they could alter these behavioral and environmental factors and thus prevent and contribute to the management of disease and illness. As a consequence, people at many social levels began to take a keen interest in health maintenance and remedies for disease, as we saw in the case of the widespread use of Buchan's *Domestic Medicine* (1769). People came to regard disease and illness as avoidable, and health and disease became increasingly individuated, i.e., individuated to the circumstances of each person (Risse, 1986, p. 14).

As we saw in section IV, the ability of medical science to *explain* the causes of diseases and their symptoms began to increase during this period. These scientific explanations competed with each other and changed regularly. Not all explanations or theories satisfied the demands of Baconian rigor, however, as Gregory laments in his "Medical Notes" (AUL 2206/45, 1743, p. 1). Medical knowledge, as we know it – a body of knowledge supported by basic science, clinical trials, and consensus clinical judgment enjoying considerable social support and commitment of money – simply did not exist. Medical theories and practitioners therefore competed in the market place of the increasing commercialized world of eighteenth-century Britain. Andrew Wear succinctly summarizes this context:

> The changing theories of medical practitioners are most appropriately placed within the changing commercial structure of the medial market place ... If political and commercial considerations influenced the type of medical knowledge that practitioners chose, then this is an indication that medical theories of whatever type did not enjoy autonomous existence as 'objective' knowledge independent of society, and that claims for their universal acceptance would fail (Wear, 1992, p. 121).

Because medical knowledge rested on uncertain foundations and because so many kinds of medical knowledge existed in the medical marketplace – allopathy, homeopathy, etc. – the borders between medical and other explanations of disease could be difficult to identify. The marketplace – and the non-existence of a dominant, well-defined social institution called Medicine – led to competing accounts of the nature, origin, and management of disease and illness. In particular, there existed a fluidity between scientific and religious or theological explanations. This had important consequences for patients: "This uncertainty gave a degree of freedom for the individual sufferer who could move from religious to naturalistic explanations or choose to use one or the other" (Wear, 1987, p. 241). This fluidity and uncertainty also had important consequences for the roles of physician and clergy, particularly in reference to the care of dying patients.

Just as health and illness became individuated to each person, medical knowledge to a considerable extent became individuated to each practitioner. Thus, publishing a book on medicine could advance one's career (Porter and Porter, 1989, p. 23). Doing so put one's theories of disease and illness, not to mention one's name, into the marketplace of medical ideas and potential customers. Medical practitioners therefore competed in their explanations and for patients; they had little choice not to, if they wished to live by their knowledge. Failure to protect and promote one's own interest could and did result in loss of patients and income.

This competition was often fierce, sometimes erupting in "fiery rivalry" (Porter and Porter, 1989, p. 22). Physicians competed with physicians, as well as with surgeons, apothecaries, midwives, and they with physicians and each other (Porter and Porter, 1989, p. 19). The Royal College or medical corporation in London strictly limited its membership, excluding physicians such as Gregory with degrees – earned or unearned – from universities in Scotland. The medical corporations defended the use of nostrums and secret remedies, but the corporation did not – and could not hope to – control the medical marketplace. No monopoly on medical practice as we know it, with only physicians with earned university degrees and licenses allowed to practice, existed.

The increasing number of "regular" practitioners, physicians with university training and even degrees, did not drive the irregulars out of the medical marketplace. This came later, in the nineteenth century (Berlant, 1975). This increase of regular practitioners only intensified the competition (Porter and Porter, 1989, p. 24). The distinctions between "reputable, regular practitioners," who might have some university training and an unearned degree – a common pathway into medicine, not just peculiar to Gregory – and "those branded as quacks might seem fluid and elusive" (Porter and Porter, 1989, p. 23). Dorothy and Roy Porter provide a compelling portrait of this medical marketplace:

> Thus a multiplicity of people on pre-modern England claimed to possess some special medical aptitude, and many gained a livelihood, or at least some reward or repute – notoriety even – out of plying their skills (Porter and Porter, 1989, p. 25).

Given the competing medical theories, their lack of scientific warrant in many or even most cases, the variety of practitioners, and the persisting "folk" medicine of the day, it should come as no surprise that eighteenth-century Great Britain experienced an "open world of medicine." Treatment was "hit or miss" (Porter and Porter, 1989, pp. 26, 27), there was no monopoly of physicians on clinical knowledge and practice, and people had money to spend, given the increasing prosperity of the times. Nor could patients, or their physicians, expect to be in control of disease; but they could hope to do their best, a far less ambitious goal.

There thus arose – Risse (1986) notes this as well – an individual responsibility for illness and a strong sense of self-help (Porter and Porter, 1989, p. 53).

Undergoing medication was not a matter of abandoning oneself blindly to professional authority. It involved active decision-making and negotiation, equivalent to buying an estate or selecting an education for one's children (Porter and Porter, 1989, p. 27).

In such a world, the patient possessed considerable power, to which physicians were inattentive at their peril, especially regarding well-to-do patients, the physician's main customers.

Many factors gave pull to the affluent patient. He would do the summoning. He would pay the bill, and so expect to have that say, that sway, which the power of the purse conferred. He might well be of higher social status than the practitioner (Porter and Porter, 1989, pp. 12-13).

Given their economic power and the competition among physicians, patients sought "multiple consultations," regarded by patients and physicians alike as a "perfectly normal" prerogative for the patient to exercise (Porter and Porter, 1989, p. 80). Thus, a physician might find himself called to consult after other practitioners – sometimes without knowing that this was the case – or before other practitioners – again unwittingly. Multiple consultations, of course, could fuel disputes among practitioners and "brought out the disagreements of doctors into broad daylight" (Porter and Porter, 1989, p. 81).

As in Scotland, patients self-medicated, frequently (Lochhead, 1948). Some physicians mounted efforts to condemn and end this practice, without avail (Porter, 1992, p. 105). Instead, "[t]he extension of regular medicine did not curtail self-medication, but rather coopted it" (Porter, 1992, pp. 104-105). Self-help was not "radically opposed to regular professional medicine" but was complementary to it (Porter, 1992, pp. 113-114).

The patient-physician relationship – the Porters rightly emphasize the order of the words – did not enjoy its own, largely autonomous, well-defined social status. Rather, the relationships between patients and their physicians were "... moulded around and modelled upon other familiar and well-established social relations – for example, those obtaining within the family hierarchy, or echoing other ties of service, patronage and employment" (Porter and Porter, 1989, p. 13).

A "professional persona" began to emerge during this period and patients could recognize it (Porter and Porter, 1989, p. 67), but – in the absence of a stable body of scientific knowledge and monopoly definition of who counted as a physician – individual ability and character played a central role.

> Overall, what counted most for patients was character and temperament ... they did seek physicians with a sound judgment. And they themselves were prepared to judge such a man. Patients were concerned to keep ultimate control of the relationship, avoiding placing themselves utterly in the physician's hands (Porter and Porter, 1989, pp. 68-69).

In the absence of a professional identity, which would create a physician-patient relationship – the order would reverse, "[i]t was individual, face-to-face encounters that tipped the balance between distrust and confidence" (Porter and Porter, 1989, p. 69).

Patients – more accurately, *paying* patients – were seen and cared for in their homes; the care of Infirmary patients involved another story, as we shall see in Section IX. Paying patients possessed the means to have their say: they could summon the physician or not, obtain other opinions, choose among them, or even reject them all and "self-physick." Thus, patients "expected to be allowed their say" in the selection of a regimen (Porter and Porter, 1989, p. 87). Patients selected their physicians on the basis of individual trust: "The trustworthy doctor had tact and a certain pliancy; patients hated feeling bullied, and expected at least to have their wishes respected" (Porter and Porter, 1989, p. 95). The physician's ability to succeed economically depended on his ability to be trustworthy and thus exhibit "a capacity to appeal to the individual and community" (Porter and Porter, 1989, p. 132). The Porters summarize the physician's world, the world of medicine Gregory encountered in London:

> Patient power and disease power between them rendered the physician's lot quite precarious. His capacity to cure was often slight; and he could not count on the state and professional protection that nowadays safeguards the incorporated profession of medicine. Relations between individual practitioners were commonly acrimonious, and intra-professional conflicts flared between the physicians and the apothecaries, between general practitioners, druggists and chemists, and between the regulars and the quacks (Porter and Porter, 1989, p. 117).

A physician's trustworthy character provided the patient with an anchor in this social storm: "... practitioners did not dictate to clients, lay people were involved in medical decisions, and worry about social *mores* ... and the need for privacy could affect events powerfully" (Wear, 1987, p. 235). As the eighteenth century progressed, however, the confidence of patients – of anyone – in the ability to identify a physician – or anyone else – as a person of trustworthy character became problematic. Behavior did not emanate from a vacuum; on this the moral sense philosophers and common sense itself were clear. Behavior emanated from one's person and so gave clues to what sort of person one was, to one's character. Social relations were predicated on this common understanding.

The physician who hoped for success thus learned to "ingratiate himself with polite society" (Porter and Porter, 1989, p. 138). This involved the mastery of the expected manners of a gentleman, in the case of the university-educated physician. Manners went further than (at that time) just how to behave at the dinner table without offending one's fellow diners or (in our time) whether to, and who should, open and hold a door for another. Instead,

> ... manners both prescribed correct behavior in certain settings and embodied particular ideas about how people lived in groups, how social structures functioned, and how individual conduct and society overall were connected (Fissell, 1993, p. 19).

Mary Fissell shows how the master-apprentice hierarchy exhibits the relationship between character and behavior. Certain, well-established strictures shaped the relationship between apprentices and their masters. The former conformed to these during training and afterwards, when they themselves became master craftsmen. "It was up to the master to maintain correct and deferential relations with his apprentices" (Fissell, 1993, p. 22). The master's behavior exhibited his character, a character that was expected to be reflected in the apprentices as well, particularly courtesy to customers, ignored by a master at his economic peril (Fissell, 1993, p. 22).

A concern for good manners extended throughout society (Fissell, 1993). A craftsman, for example, dealt with his social betters who were his customers. Failure to meet their expectations could result in economic disaster. The craftsmen also dealt with his apprentices, his inferiors, to whom he should display due paternalistic as well as self-interested regard. The craftsman dealt with his wife and children from the position of superior in the family hierarchy. The craftsman also dealt with fellow craftsman as both peers and competitors. Fissell points out that one can generalize to physicians from such an account:

> In sum, the traditional association between manners and morals, courtesy and virtue, provided a means of discussing behavior towards one's betters and inferiors well-suited to the educational process of early modern medical practitioners. The structures of apprenticeship and household-based professional service were identical to those in other types of work, and advice on manners found many an audience (Fissell, 1993, pp. 25-26).

Fissell illustrates this with the case of Archibald Cleland, trained as a surgeon, who was accused of sexual misconduct in how he had examined three female patients, exacerbated by the fact that these were the patients of other surgeons. Fissell shows how the "key to the dispute lay in character" (Fissell, 1993, p. 29), centering on "issues of behavior, credit, and reputation" (Fissell, 1993, p. 30). The assumption about Cleland's actions was that "[w]hat determined their moral content rested upon character, which could be read through manners" (Fissell, 1993, p. 32).

The crisis in manners occurred, according to Fissell, later in the eighteenth century, when "medical manners and morals became unglued; no longer were codes of conduct based on courtesy functional" (Fissell, 1993, p. 32). People mounted attacks on the "insincerity of manners," because of such factors as "dissimulation," for which the "man of feeling" became infamous, and commodification of manners, which could be bought and sold depending on whether those manners advanced the interests of the individual who might put them on (Fissell, 1993, pp. 33ff.). In the fiercely competitive world of eighteenth-century British medicine, physicians, obviously, could be counterfeit men of feeling or put on manners as suited their interest in gaining and retaining market share. Fissell neatly sums up the crisis of manners that confronted Gregory and to which (as she notes and as we shall see in the next chapter) Gregory mounts a response:

> ... appearance and reality were uncoupled and courtesy [as well as other manners that might ingratiate one to a potential patient] stripped of moral connotation. Manners could be bought and sold, and so could not function as an indicator of virtue (Fissell, 1993, p. 42).

The ability of a potential patient to select a trustworthy physician, the cornerstone of the patient-physician relationship, becomes problematic in this context. One response, to which (as we shall see) Gregory made enormous contributions, involved the creation of "an ethic peculiar to medicine" (Fissell, 1993, p. 42). This crisis of trustworthiness opens up issues of the physician's authority and power, as well as their right use, in the care of patients, power and authority that obviously are open to abuse (Porter, 1992, p. 103).

The rise of man-midwifery illustrates this crisis, as did the case of Cleland, discussed by Fissell (1993). The causes of the increasing involvement of male practitioners in birthing remain something of a scholarly mystery (Loudon, 1993, p. 1051). They appeared as forerunners of obstetricians (Loudon, 1993, p. 1052). This change may have had something to do with the perceived superiority of men in "technical solutions" to medical problems and the introduction of a technology, forceps, in which men were adept (Geyer-Kordesch, 1993b, pp. 903-904). The public increasingly preferred the male practitioner to assist delivery (Loudon, 1993, p. 1051), despite the opposition of female midwives (Porter, 1987). Audrey Eccles claims that pregnant patients had rational justification to prefer man-midwives and the advances that they brought to obstetrics (Eccles, 1982, p. 124).

Roy Porter (1987) examines the problems occasioned by the introduction of man-midwives into clinical obstetric management. Some, Porter says, saw the man-midwife "as a sexual infiltrator, a violator of female modesty." Porter quotes from a tract, written by an offended man, in which such claims are advanced:

> [Man-midwives are allowed] to treat our wives in such a manner, as frequently ends in their destruction, and to have such intercourse with our women, as easily shifts into indecency, from indecency into obscenity, and from obscenity into debauchery (Porter, 1987, p. 217).

The problem that occasioned such an attack was the "manual manipulation of women's private parts," as apparently urged by William Smellie (1697-1763) "as integral to [the man-midwife's] allegedly superior techniques of diagnosis and delivery" (Porter, 1987, p. 217).

Philip Thicknesse, Porter reports, attacked Smellie's practices, taking the view that the man-midwife was "a 'sinister figure'" (Porter, 1987, p. 220), a sexual predator aroused by passion induced from touching female anatomy. Obstetrics, Thicknesse argued, should be returned to the province of women, female midwives, where it had flourished for centuries

(Porter, 1987, p. 217) without, among other things apparently, being such a threat to husbands.

While perhaps at times posing a sexual threat to women – Porter is not able to find much evidence for Thicknesse's accusations, man-midwives did perform useful services, including "colluding in the delivery and subsequent concealment of illegitimate babies" (Porter, 1987, p. 222). Porter summarizes the situation of man-midwives in this way:

> Whether or not there is truth in the allegations made by Thicknesse and others of sexual goings-on between fine ladies and man-midwives, well might the rise of the man-midwife have left husbands feeling dubious about the part they were prepared to play in helping to conceal the fruits of adultery. The enduring role of the doctor as the ambivalent ally of wives in their manoeuvers against their spouses had obviously begun (Porter, 1987, p. 224).

Pregnant women, whether or not they were adulterous, may have wondered whether or not a man-midwife, even of ingratiating manners – what better way to gain access to otherwise forbidden bedchambers? – was trustworthy, as would the woman's husband. A "fine lady" seeking to conceal from her husband, family, social circle, and business associates of her husband the birth of an illegitimate child – no small calamity then – needed a man-midwife whom she could trust, who would help to deceive her husband and everyone not present for her confinement with the dissimulation of trustworthiness. This extreme instability of manners and character illustrates just how deep the crisis of manners could run in the lives of patients and their families.

One cannot omit in describing Gregory's experience in London a consideration of the dying patient and other aspects of medical care that Gregory addresses. Indeed, the Porters report that

> ... traditional medical etiquette required that the dying patient be informed of his likely fate. Then, their part in the proceedings complete, physicians would withdraw, leaving the dying man to compose his mind and will, and make peace with God and his family (Porter and Porter, 1989, p. 146).

If the physician continued to attend to the dying patient, the question of how he should comport himself when clergy were summoned naturally arose. The situation would be not unlike multiple physician consultations and thus fraught with potential for conflict. The existence of competing religious and naturalistic explanations would obviously complicate this matter by prompting further, challenging questions about authority and power of one over the other (Wear, 1987). Gregory (as we saw in Chapter One and will see in detail in Chapter Three) takes it to be a matter of obligation that the physician attend to his dying patients.[7]

There also arose, as the century progressed, a "new ideal" of death: "a peaceful falling asleep." The increasing use of narcotics in many aspects of care, extended to pain relief for the dying may have both enabled and promoted this concept of an ideal death (Porter and Porter, 1989, pp. 147, 148-151). According to Tytler, Gregory died such a death, a "perfect *Euthanasia*" (Tytler, 1788, p. 79).

Drugs generally came to play a larger role and there emerged a phenomenon that we can today easily recognize. "Increasingly, drugs provided the medium of exchange which cemented the implicit contract between patients and practitioners," further coopting "self-physick" (Porter and Porter, 1989 p. 172).

Social class played a major role in defining the relationship between women and their doctors. Gender played less of a role, which may help to explain the rise of man-midwifery (Porter and Porter, 1989, p. 180). Nonetheless, for Gregory the proper care of female patients remains an important topic for medical ethics.

Finally, medical knowledge was open, particularly to persons with an education, because the medical books of the time were not difficult to read. Writers and publishers produced many such books in the eighteenth century. These books did not require considerable scientific and clinical knowledge, as was eventually needed to read the medical books of the latter twentieth century. The Porters summarize the effects of this accessibility of medical knowledge: "Rigid barriers did not exist – they perhaps became more pronounced later – between lay and profession, 'high' and 'low' cultural knowledge" (Porter and Porter, 1989, p. 207).

B. The Bluestocking Circle: Women of Learning and Virtue

In the 1750s the Enlightenment began to develop in earnest, in England, Scotland, and across Europe (Gay, 1966, 1969; Porter, 1990). Recent historical scholarship has called into question the concept of a single, overarching cultural, intellectual, and scientific movement called "*The* Enlightenment." Far more heterogeneity existed than the definite article before 'Enlightenment' warrants. Instead, there existed various national enlightenments, with the emphasis very much on the plural form (Porter and Teich, 1981).

This interpretation should come as no surprise. Students of art history, for example, early on in their studies become keenly aware of national "styles" in art. Thus, for example, the angular drapery of German Gothic statuary distinguishes itself from the more rounded, smooth-flowing drapery of the French Gothic. No one with an ear for music would think that Berlioz was German or Brahms French. The eighteenth century marks no exception to this pattern; I, too, take seriously the concept of national enlightenments. This has a number of important interpretive and philosophical consequences, as we shall see in this and subsequent chapters.

Women participated in different ways and to different degrees in the various national enlightenments. Alexander Murdoch and Richard Sher describe the scene in Scotland:

> Women rarely participated in the world of the Scottish Enlightenment. This state of affairs marks an obvious contrast with the French model, featuring gatherings of *philosophes* at salons run by women of taste and intellect. For all their concern with tolerance and freedom of inquiry, the Scottish literati generally showed little interest in admitting women to their company – let alone their ranks – where intellectual affairs where concerned. Only foreign women of established reputation, such as the English bluestocking, Elizabeth Montagu were accorded real respect as intellects. Yet if women were generally excluded from the formal institutions of the Scottish Enlightenment, their visibility and importance were gradually increasing. Questions dealing with women and marriage were among the most popular topics debated at the Select Society [one of the most prominent intellectual clubs in Edinburgh]. John Millar's *Origin and Distinction of Ranks* (1771) devoted a substantial opening chapter to the changing role of women in different stages of civilisation, and Dr. John Gregory's *A Father's Legacy to His Daughters* (1774) became one of the century's best selling advice books (Murdoch and Sher, 1988, p. 131).

As we saw above, as early as his medical school days, Gregory did not partake in this Scottish view of women. He had the example of his own great-grandmother, a woman of learned – not in the formal, but certainly a real, sense – accomplishment. He married a woman of intellectual ability, as well. And in London, he quickly came into the company of one of the most intellectually accomplished women in that city, Elizabeth Montagu, attended her evening salons, and formed a friendship that lasted through his lifetime and into that of his daughters. Mrs. Montagu became godmother to one of his daughters, Dorothy. Over the years, after his wife's death, Gregory and his daughters were regular visitors to Mrs. Montagu's home in Denton, and Gregory hosted her visits to Scotland. Personally and intellectually Mrs. Montagu had a very deep influence on John Gregory's thinking. She became his exemplar of a woman of learning and virtue, a concept that he championed and made central to his medical ethics. Return for a moment to Gregory during his brief stay in London and to his friendship with the remarkable Elizabeth Montagu.

Here I rely for my account on Elizabeth Montagu and the bluestockings on the magisterial study, *The Bluestocking Circle* (1990), by Sylvia Harcstark Myers, who died before seeing it into print. Elizabeth Robinson, later Montagu, was born in York in 1725, and died in 1800. Probably through her mother, Elizabeth was befriended by Lady Margaret Hartley (later Margaret Bentinck, the Duchess of Portland). The Duchess maintained a country home, Bulstrode, to which Elizabeth Robinson paid regular visits in summer, starting when Elizabeth was in her early twenties. In a time in which formal education for women did not exist, Bulstrode became Elizabeth's academy. As Myers puts it, Elizabeth Robinson's "visits to Bulstrode were an important part of Elizabeth's education, both in terms of worldly experience and of study" (Myers, 1990, p. 35).

Days at Bulstrode began with breakfast and a stroll in the gardens. There might be conversation with the many visitors to the house, including scientists, scholars, and ministers. There was time for formal conversation in the afternoons, as well as for handiwork and reflection (Myers, 1990, p. 35). They read books aloud and discussed them, "investigated nature with a microscope," and debated "sermon topics, especially metaphysics and morality," with visiting clergy (Myers, 1990, p. 40). Here Elizabeth formed friendships with a lasting impact, e.g., with Anne Donnellan (c.1700-1762), who encouraged Elizabeth's continuing education (Myers, 1990, pp. 38-39).

Elizabeth's brothers also supported her educational efforts. Her brother, Matthew, wrote to her in French from Cambridge and "assured her that there were not three people of her sex in England who could write their own language as well as she did French" (Myers, 1990, p. 42). Myers notes that "[h]er oldest brother's compliments and interest must have strengthened her drive towards learning" (Myers, 1990, p. 42). She knew too well the "taboo" against learning for women and "resented it" (Myers, 1990, p. 41). This resentment appears to have become permanent.

Her modest family wealth, such as it was, limited Elizabeth Robinson's marriage prospects, at a time when men and their families expected and negotiated for dowries. She approached marriage with a wary eye:

> Aware that 'fortune', the amount of money a young woman could bring to her husband, was very important in determining whom she could marry, Elizabeth looked with a satiric eye on the manners of those who were courting and marrying (Myers, 1990, p. 92).

Concerned that marriage involved a woman's "bargain and sale," she looked to the Duchess of Portland's marriage as a model, "a union of love and friendship, but an exception to the general state of the other married couples whom Elizabeth observed" (Myers, 1990, p. 93).

She writes to the Duchess, "as for the intrinsick value of a Woman few [men courting for marriage] know it, & no body cares" (Myers, 1990, p. 94). Elizabeth rejected an early offer of marriage. Elizabeth's views on marriage apparently alarmed her mother, who feared that these views would mean she would never marry, a prospect that held, for women of modest family means, economic risk of a high order. Elizabeth writes to her mother:

> I am not at all of your opinion in two things. The first that I shall never marry, & the Second that the wisest and best Men marry for money, I think Sense and virtue in a Man will induce him to chuse prudence & virtue in a Woman, & surely Sense only teaches a Person the true value of things ... but till I can meet with a Deserving Man who rightly thinks the price of a Virtuous Woman above Rubbies I shall take no other obligation or name upon me but that of being ... your most dutiful daughter (Myers, 1990, pp. 94-95).

Soon thereafter Edward Montagu courted her, and they wed in 1742. He was grandson to the Earl of Sandwich, fifty-one years of age, still a bachelor, and possessed of considerable wealth. Even though Elizabeth approached her marriage with some trepidation (Myers, 1990, p. 96), Myers says that "he was the good man for whom Elizabeth was waiting – well off, prudent, and willing to accept her with a small dowry" and she thought of him as "my friend" (Myers, 1990, pp. 96, 97). In a time when women often did not pick their spouses and in which marriage constituted an economic necessity for both women and their families of origin and before the ideal of a marriage based on love and free choice of both partners existed, Mrs. Montagu could count herself fortunate.

Edward was a businessman, an owner of collieries, and a member of Parliament, and moved Elizabeth to his home in Yorkshire, which was "convenient but old and not handsome" (Myers, 1990, p. 97). As was the custom, Elizabeth took a woman companion with her, her sister Sarah. There were few visitors. She bore her first – and, as it turned out, only – child in May of 1743, a son. Afterwards, she suffered various symptoms, including what she called "histeric shaking," as she described it to her mother (Myers, 1990, p. 100). She continued her reading, including Locke's *Two Treatises on Government*, which she recommended to the Duchess of Portland (Myers, 1990, p. 100).

In August, at barely four months of age, baby John, called Punch, died of convulsions, a "devastating experience" for Mrs. Montagu (Myers, 1990, p. 101). The experience of loss – of one's siblings, children, spouse, or parents – was commonplace. Perhaps this phenomenon fed the development of an interest in the social principle and its physiologic mechanism, sympathy. The social principle creates an extended family as a buffer against loss. Too much sympathy for a loved one would make all the more crushing the loss caused by the early death of a loved one. Thus, even in intimate relationships, sympathy needed to be tempered and regulated to protect one from its dangerous excesses.

People commonly feared the death of their children – infant and child mortality rates were high – and the pain that early death would cause them. Elizabeth's mother had expressed just such views to her. Mrs. Montagu grieved for a long time (Myers, 1990, p. 102). Her health continued to be frail and there emerged what Myers thinks was an "unspoken agreement" between the Montagus to have no more children (Myers, 1990, p. 104). Their marriage "remained intact but strained," until Mr. Montagu's death in 1775, through which she attended him and after which "she had the benefits of wealth and independence" (Myers, 1990. pp. 119, 120). She became a business manager of no mean skill and accomplishment, winning the regard of her colliers for her paternalism toward them.

Mrs. Montagu made the acquaintance of other women of intellectual interest and ambition, including Elizabeth Carter and Hester Mulso, later Chapone. These women were visitors in her home and eventually in the 1750s Mrs. Montagu began to host evening salons, to which her women friends and the male literati of London would be invited, during the "season" when the Montagus were at the their home in London. It was at this time that Mrs. Montagu made Gregory's acquaintance and invited him to her salons. She writes to Richard West in 1755 that she found Gregory to be "an ingenious and agreeable man" (Climenson, 1906, Vol. II, p. 73). Gregory also became physician to the Montagu household that year, treating Mrs. Montagu's lady-in-waiting for consumption (Climenson, 1906, Vol. II., p. 76).

Mrs. Montagu believed it to be a scandalous waste of women's intellectual talents to spend so much time at playing cards and attending balls and other festivities. She created her salons accordingly. Guests were to dress plainly, in contrast to the extravagant dress expected at the formal social events her salons were meant to eschew; the grandness of these costumes is evident in portraits from the period. Consider, for example, Thomas Gainsborough's portrait of Sarah, Lady Innes, made about 1757 (Gainsborough, 1757). This work portrays the woman-as-ornament, with a ribbon at her neck, bedecked in a sumptuous gown, holding a freshly picked, about-to-bud rose.

> Lady Innes wears a turquoise-blue silk dress trimmed down the bodice with blue-embroidered lace and at the elbows with white lace ruffles. A black ribbon is tied about her throat, and a white feather ornament is fastened to her brown hair. She holds a pink rosebud of the same color as those on the bush behind her. Her pale face is framed by cool greenish-gray foliage. The misty blue-green landscape melts into a dark, overcast sky (The Frick Collection, 1968, Vol. I, p. 42).

The women of the Bluestocking Circle meant to set themselves apart from such trophies.[8]

At one of Mrs. Montagu's salons, the story has it, one Benjamin Stillingfleet attended with his wool stockings dyed blue, rather than the plainer gray. The name 'Bluestockings' was coined, a word that has come now to have pejorative connotations. Not so in the minds of the women Mrs. Montagu gathered around herself and supported, psychologically and with the generosity of her household for extended visits; nor in Gregory's mind.

Gregory made the acquaintance of Mr. and Mrs. Montagu through their mutual friend, Lord George Lyttleton, who played a key role in seeing Mrs. Montagu's and Gregory's books into print (Tytler, 1788, pp. 34, 35). Gregory frequented Mrs. Montagu's salons during his brief time in London.

Dr. Gregory likewise enjoyed the friendship of the late Edward Montague, Esq; and of his lady, the celebrated champion of the Fame of Shakespeare against the cavils and calumnies of Shakespeare. At her assemblies, or *conversazione*, the resort of Taste and Genius, our author [Gregory] had an opportunity of cultivating an acquaintance with many of the most distinguished literary characters of present times (Tytler, 1788, pp. 35-36).

The Bluestocking evenings included Samuel Johnson, who published works by members of the Bluestocking in his *Gentleman's Magazine* (Myers, 1990), Joshua Reynolds, the great portrait painter, and many others (Stewart, 1901, p. 108). Gregory apparently more than held in his own in this glittering intellectual, cultural, and artistic company.

As the above passage from Tytler's "Life" of Gregory indicates, Mrs. Montagu made contributions to the learned literature of her day. The first appeared anonymously – not uncommon for women writers, to whom there existed no small opposition – as three dialogues in George Lyttleton's *Dialogues of the Dead* (1760). Mrs. Montagu had sent a manuscript version of this to Gregory (by then in Edinburgh) and expressed pleasure with the reception of her "Dialogues" in Scotland (Climenson, 1906, Vol. II., pp. 203-204).

Lyttleton's dialogues cover various virtues, accenting virtue over a life of action. In one of these, Dialogue XVI, between "the Princess of Orange and Lady Sidney," Lyttleton explores the virtue of *tenderness*, one of the cardinal virtues of women. He treats tenderness as a regulated passion for the interests of others, and treats unregulated passion as detrimental, even dangerous to virtue.

Even though anonymous, many of her friends knew that the last three "Dialogues," were Mrs. Montagu's. Lyttleton explains:

> The last three dialogues are written by a different hand, as I am afraid would have appeared but too plainly to the reader without my having told it. If the friend who favoured me with them should write any more, I shall think the Public owes me a great Obligation, for having excited a genius so capable of uniting delight with instruction, and giving to knowledge and virtue those graces, which the wit of the age has too often employed all of its skill to bestow upon folly and vice (Lyttleton, 1760, p. vii).

Mrs. Montagu's dialogues touch upon female virtues, the role of women, the education of woman and other topics. Her first dialogue (number XXVI), between Cadmus and Hercules, includes the following exchange on the effects of liberal learning. Mrs. Montagu, I believe, speaks through Cadmus here.

> Hercules: The true spirit of heroism acts by a sort of inspiration, and wants neither the experience of history, nor the doctrines of philosophers to direct it. But do not the arts and sciences render men effeminate, luxurious, and inactive; and can you deny that wit and learning are often made subservient to very bad purposes?

> Cadmus: ... letters keep a frugal temperate nation from growing ferocious, a rich one from becoming entirely sensual and debauched (Lyttleton, 1760, pp. 298-299).

Liberal learning, then, is not the idle province of the "effeminate," but essential to the proper regulation of the passions. Learning fosters virtue; the two are inseparable.

Her second dialogue (number XXVII) presents a conversation between Mercury and a "modern fine lady." The latter is presented by Mrs. Montagu as the very type of woman the Bluestocking Circle women found anathema. "Mrs. Modish," as Mrs. Montagu dubs her, says that she was "engaged" with her husband and children and that she was also "engaged to the play on Mondays, balls on Tuesdays, the opera on Saturdays, and to card-assemblies the rest of the week; and it would be *the rudest thing in the world* not to keep my appointments" (Lyttleton, 1760, p. 301). The dialogue continues:

> Mrs. Modish: I was much too engaged to think at all: so far indeed my manner in life was agreeable enough. My friends always told me diversions were necessary, and my doctor assured me dissipation was good for my spirits; my husband insisted that it was not, and you know that one loves to oblige one's friends, comply with one's doctor, and contradict one's husband; and besides I was ambitious to be thought *Du Bon ton**.[9] ...

> Mercury: Then, Madam you have wasted your time, faded your beauty, and destroyed your health, for the laudable purposes of contradicting your husband, and being this nothing called the *Bon ton.*

> Mrs. Modish: What would you have had me do?

> Mercury: I would follow your mode of instructing. I would tell you what I would not have had you do. I would not have had you sacrifice

your time, your reason, and your duties to fashion and folly. I would not have had you neglect your husbands happiness, and your childrens education.

Mrs. Modish: As to my daughters education I spared no expense; they had a dancing-master, a music-master, and a drawing-master; and a French governness to teach them behaviour and the French language.

Mercury: So their religion, sentiments and manners were to be learnt from a dancing-master, music-master, and a chamber-maid! Perhaps they might prepare them to catch the *Bon ton.* Your daughters have been so educated as to fit them to be wives without conjugal affection, and mothers without maternal care. I am sorry for the sort of life they are commencing, and for that you have just concluded (Lyttleton, 1760, pp. 301-305).

Mrs. Modish represents everything that the young Elizabeth Robinson resisted and the marriage Mrs. Montagu strived to avoid. Mrs. Montagu, the woman of intellectual ability and ambition, means to distance herself – emphatically – from Mrs. Modish. As we shall see in the latter sections of this chapter, Gregory takes up and extends these views on women.

In the third of her dialogues (number XXVIII), between Plutarch, Charon, and a "modern bookseller," Mrs. Montagu speaks through the bookseller:

Bookseller: The women have far greater obligations to our writers than the men. By the commerce of the world men might learn much of what they get from books; but the poor women, who in their early work are confined and restrained, if it were not for the friendly assistance of books, would remain long in *an insipid purity of mind,* with a *discouraging reserve of behaviour* (Lyttleton, 1760. p. 311).

Women who lack learning – Mrs. Modish's unfortunate daughters spring immediately to mind – will lack virtue as well, and consign themselves to a "dissipated" life. Such a life holds the bleak, unwelcome prospect of no virtue and therefore no worth. Mrs. Montagu means to forge a strong conceptual, as well as practical and lived, bond between learning and virtue.

She also forges a conceptual, practical, and lived bond between these virtues and religion, with reference to the character of Clarrissa in Samuel Richardson's serialized stories, which began in 1747 (Myers, 1990, pp. 140-143).

Bookseller: In the supposed character of Clarissa, (said a clergyman to me a few days before he left the world) one finds the dignity of heroism tempered by the meekness and humility of religion, a perfect purity of mind and sanctity of manners (Lyttleton, 1760, pp. 317-318).

Katherine Rogers describes the impact of the Clarrissa stories in the following terms:

... Richardson's work was enormously helpful to women – affirming the worth of their feelings and their interests, developing a form that women writers could follow (Rogers, 1982, p. 125).

Richardson presents Clarrissa as a model of properly regulated feeling, according to Rogers: "In Richardson, then, sentimentalism is always stiffened by rational emphasis on rights and regulated by practical reality" (Rogers, 1982, p. 133).

The Bluestockings were well aware of this crisis of manners described above: a sundering of appearance – behavior – and reality – character. Hester Chapone, in her *Letters Addressed to a Niece on the Improvement of the Mind* takes dead aim at this problem, when women dissemble in manners. She condemns falsification of tenderness, a paradigm female virtue:

Nothing so effectually defeats it own ends as this kind of affectation [of vanity in the guise of "tenderness and softeness" (Chapone, 1806, p. 58)]: for though warm affections and tender feelings are beyond measure amiable and charming, when perfectly natural, and kept under due control of reason and principle, yet nothing is so truly disgusting as the affectation of them, or even the unbridled indulgence of such as are real (Chapone, 1806, p. 59)

That sort of tenderness, which makes us useless, may indeed be pitied and excused, if owing to natural imbecility; but, if it pretends to loveliness and excellence, it becomes truly contemptible (Chapone, 1806, p. 60).

Chapone provides instructions for her niece's "improvement," i.e., properly regulated behavior, which instructions, we shall see, resonate powerfully with Gregory's views on sympathy in the physician's character:

Remember, my dear, that our feelings were not given us for our ornament, but to spur us to right actions. Compassion, for instance, was not impressed upon the human heart, only to adorn the fair face with tears, and to give an agreeable languor to the eyes; it was designed to excite our utmost endeavors to relieve the sufferer. Yet, how often have I heard that selfish weakness, which flies from the sight of distress, dignified with the name of tenderness! (Chapone, 1806, p. 59)

The same degree of active courage is not to be expected in woman as in man; and, not belonging to her nature, it is not agreeable in her: But passive courage ... patience, and fortitude under sufferings ... presence of mind, and calm resignation in danger ... are surely desirable in every rational creature (Chapone, 1806, p. 60).

In a world in which the appearance and reality of manners had been sundered, the manners of women required some foundation. Religion provides it, a position that Gregory comes to endorse in his *Legacy*.

Shakespeare and his works experienced a revival in the eighteenth century and had been subjected to some criticism by Voltaire, whom Mrs. Montagu read in the French. Lyttleton writes to her in 1760, praising her as a critic (Myers, 1990, p. 185), and in 1764 he urges her to write more:

We will not be satisfied with a fine letter to L[or]d. Bath or Sr. J. Macdonald, which you can write while Mrs. Jenny is combing your Hair; we will have something to be printed, something to be published, something to shew the whole World what a Woman we have among us! (Myers, 1990, p. 186).

In 1769, she publishes her reply to Voltaire, to much acclaim and many subsequent editions into the nineteenth century. She also shared this manuscript with Gregory in 1768 (Myers, 1990, p. 198). Mrs. Montagu offers a critical theory of dramatic poetry and the historical drama that mirrors much of the literature on taste of the time. She then considers several of Shakespeare's plays, including the first and second parts of Henry IV, and Macbeth.

The concept of sympathy plays a central role in her critical theories, indicating an awareness on her part of the development of the concept, including the physiologic concept of the effect of *the imagination* on the heart and passions.

[A drama, tragedy in particular] is addressed to the imagination, through which it opens to itself a communication to the heart, where it is to excite certain passions and affections: each character being personated, and each event exhibited, the attention of the audience is greatly captivated, an the imagination so far aids in the delusion, as to sympathize with the representation (Montagu, 1970, p. 29).

Shakespeare, she shows from various texts, exemplifies mastery of the this sort of dramatic representation. Indeed, he has no equal in this art.

She also uses sympathy in the sense of a "contagion" of affect from the characters on the stage to the audience.

> ... such enthusiasm is to be caught only from the stage, and the effect alone of a strong-working sympathy, and passions agitated by the particular force and activity of the dramatic manner (Montagu, 1970, p. 30).

> The business of drama is to excite sympathy; and its effect on the spectator depends on such a justness of imitation, as shall cause, to a certain degree, the same passions and affections, as if what was exhibited was real (Montagu, 1970, pp. 32-33).

Mrs. Montagu seems to know what method acting involved long before it gained prominence. She offers Macbeth as an example:

> [Shakespeare] has so tempered the constitutional character of Macbeth, by infusing into it the milk of human kindness, and a strong tincture of honour, as to make the most violent perturbation, and pungent remorse, naturally attend on those steps to which he is led by the force of temptation (Montagu, 1970, pp. 164-165).

In her published work Mrs. Montagu articulates a central theme of the bluestocking philosophy: learning and virtue were synergistic and the worthwhile life involved their mutual cultivation. This synergy then creates bonds of friendship, built on sympathetic affection and mutual regard, which has analogies in the dramas of Shakespeare.

> For the bluestockings themselves, learning, virtue, and friendship *were* inextricably linked. In their own eyes, to be a bluestocking meant to be an impeccable member of an intellectual community which included both men and women (Myers, 1990, p. 11).

Gregory experienced this world of women of learning and virtue in Mrs. Montagu's *conversazione* in London, and in his life-long friendship with her and her regard for and interest in the education of his daughters.

Gregory met with a "feminist consciousness" (Myers, 1990) in Mrs. Montagu and her bluestocking friends. This *feminism* did not build on the feminism of the previous century but found its own way (Myers, 1990, p.

121). Myers summarizes their feminism, using the work of Gerda Lerner (1986) as her template:

> They were becoming aware of wrongs [against women] and were developing a sense of sisterhood – of sharing common problems with other women ... They did have goals – better opportunities for education and a more equal partnership with men in social and intellectual life. They hoped to change the attitudes of both men and women by advocating the cause of women. But they had no feeling for the political and legal changes which would eventually be necessary. As for the future, they did not really think in terms of long-term improvements for life on earth, but looked forward to eternal life in the hereafter, when they hoped all inequities would be redressed, and they would continue in pure friendships with those to whom they had been close on earth (Myers, 1990, p. 121).

The link between learning and virtue forms the core of this Bluestocking feminism. Learning and virtue go hand in hand for everyone. Myers quotes from a letter of Hester Chapone:

> ... 'tis certain that every accession of understanding, whether in man or woman, in its natural tendency, leads to the improvement of the heart (Myers, 1990, p. 145).

This feminism reflects, Myers thinks, the "growing emphasis on female virtue as the key to public acceptance of women who studied or published" (Myers, 1990, p. 128). This was an unusual, perhaps even threatening – to men and women alike (think of Mrs. Modish) – thing to do and women engaged in it needed to earn the public trust. Individual virtue provided the key to that process, just as it did for the acceptance and trust that physicians had to win from potential patients. Gregory, we shall see in Chapter Three, made this connection.

We should not attempt to read or understand the Bluestockings as feminist "in the contemporary sense" (Rogers, 1982, p. 3)

> Surely feminism should not be limited to single-minded, systematic compaigning for women's rights, but should include particular sensitivity to their needs, awareness of their problems, and concern for their situation (Rogers, 1982, p. 4.)

Mrs. Montagu herself expresses awareness of the limits of what form her feminism could take. This resentment appears to have become longstanding. In 1760 Mrs. Montagu writes to Lyttleton:

> Your Lordship's account of Miss Lyttleton rejoiced me on your ac-
> count and hers. I congratulate you that your son is fit to grace public
> life, your daughter to bliss domestic. Your Lordship's forming care
> will polish her virtues, till they are smooth and soft, and never idly
> wish to make them bright and dazzling. Extraordinary talents make a
> woman admired, but they will never make her happy. Talents put a
> man above the world, and in a condition to be feared and worshipped;
> a woman that possesses them must always be courting the world, and
> asking pardon, for uncommon excellence (Montagu, 1809, Vol. 4, p.
> 311).

The Bluestockings, as Myers notes, were not influenced by their prede-
cessor feminists. Instead, they came to their views individually: "women
were entitled to independence in both their intellectual and social lives"
(Laurence, 1992, p. 163) but had to accommodate this ambitious goal to
the realities of the time, whether regarding marriage or public display of
their abilities. Anne Laurence summarizes their resultant strategy:

> The Bluestockings obviously were more than simply patrons and dillet-
> tantes; they made the case for women's intellectual independence in a
> socially acceptable form (Laurence, 1992, p. 163).

As Mrs. Montagu's 1760 letter to Montagu indicates, however, they did
not do so altogether happily.

The Bluestockings did avail themselves – sometimes with help from
men, as in Mrs. Montagu three "Dialogues – of the "only profession that
actually developed for them," writing for publication. Mrs. Montagu's
and others' salons also played a role (Rogers, 1982, p. 21).

> The Bluestocking parties created an atmosphere in which men could
> appreciate women's society in a nonsexual way and demonstrated that
> women could sustain the interest of intelligent men without the aid of
> cards or dancing (Rogers, 1982, p. 31).

Gregory experienced this world directly and, I believe, took from it
fortification for the view that "women are to be regarded as intellectual
beings, as capable as men of strength and self-discipline, and that they
owe it to themselves to be virtuous and reasonable" (Rogers, 1982, p.
217).

Later in the eighteenth century, women of learning and virtue came to be seen as moral exemplars – here, Gregory was ahead of the social trends long before he moved to London. "Gradually the idea that women are actually superior to men in spiritual qualities such as capacity for friendship and love gained ground" (Myers, 1990, p. 128). Gregory takes this one step further: women of learning and virtue and friendship based on these exemplify the sympathetic bond that should form the basis of the professional physician-patient relationship. These themes emerge in the course of the correspondence between Gregory and Mrs. Montagu, that continued until just before his death.[10]

In 1754, Gregory added to his honors when he was made a Fellow of the Royal Society "and daily advancing in the public esteem, it is not to be doubted, that, had he continued his residence in that metropolis, his professional talents would have found their reward in a very extensive practice" (Tytler, 1788, p. 36). In November 1755, however, James Gregory died in Aberdeen, "occasioning a vacancy in the Professorship of Physic in King's College Aberdeen" (Tytler, 1788, p. 37). King's offered the Professorship to John Gregory and he accepted, returning to Scotland in 1756, "and took upon him the duties of that office to which he had been elected in his absence" (Tytler, 1788, p. 37).

VII. ABERDEEN, 1755-1764:
SCIENCE OF MAN, ABERDEEN PHILOSOPHICAL SOCIETY, SYMPATHY, AND LAYING MEDICINE OPEN

King's College still had no medical school when Gregory returned to *alma mater*. With the encouragement of his physician-colleague, David Skene, Gregory offered lectures on medicine, in the hope of stimulating enrollment and developing a medical curriculum and school.

> But the students did not attend. It was an indignity to the university, keenly felt by these professors, that an Aberdeen degree should be the laughing stock of all the other universities; but without an Infirmary [i.e., access to the then existing Infirmary?] it was impossible to teach the Practice of Physic, and the attempt had to be given up for the time (Stewart, 1901, p. 109).

Gregory had no teaching duties, his professorship therefore amounting to a sinecure, of which Gregory took full intellectual advantage. Aberdeen may have been a provincial town, but it was home.

To [Gregory] the society of Aberdeen had many attractions. It was there he had spent the most delightful period of his life, the season when the heart is alive to its warmest affections; and there, of consequence, he had formed his most cordial intimacies. These had been contracted chiefly with a few persons of distinguished abilities and learning, whom it was now his fortune to find attached to the same place, and engaged in pursuits similar to his own. The animosities and mean jealousies, which so often disgrace the characters of literary men, were unknown to the friends of Gregory, who, educated in one school, professing no opposite tenets, or contending principles, seem to have united themselves, as in a common cause, the defence of virtue, of religion, and of truth (Tytler, 1788, pp. 37-38).

The "literary men" included Thomas Reid, George Campbell, Alexander Gerard, and James Beattie. Their pursuits similar to his own focused on the Scottish Enlightenment science of man and of morals. They undertook this and other pursuits under the auspices of the Aberdeen Philosophical Society.

Gregory returned to King's College and Aberdeen at a time when both were changing. The Regent system, which Gregory had experienced as Thomas Gordon's student at King's, was beginning to be challenged; at nearby rival Marischal College it was replaced by a more open and specialized curriculum (Ulman, 1990, p. 22). Aberdeen began to participate in the Scottish Enlightenment, which had been well under way in Edinburgh and Glasgow. Faculty members at King's College and Marischal College played a key role in this development, as did their academic colleagues in the larger cities to the south.

As I noted above, the view propounded by Gay and others of *one* Enlightenment that embraced Europe, the British Isles, Ireland, and North America has come in for sustained criticism in recent historical scholarship (Porter and Teich, 1981; Porter, 1990). Gay's view continues to influence recent work in the history of philosophy, however, particularly that of Alasdair MacIntyre (1981, 1988, 1990). MacIntyre believes that there existed in the eighteenth century something called *The* Enlightenment "project" in moral philosophy. Readers familiar with MacIntyre's work know that he argues for the view that the Enlightenment project was bound to fail. I shall challenge MacIntyre's interpretation on both counts, as it applies to the Scottish Enlightenment. (See Chapter Four.) At present, I undertake the task of characterizing the Scottish Enlightenment, identifying its project in moral philosophy and the contributions of the

Aberdonians to the Scottish Enlightenment project in moral philosophy.

A. The Scottish Enlightenment: Science of Man, Science of Morals.

The Scottish Enlightenment almost certainly did not come *de novo* into existence. Adnand Chitnis (1987) argues that the drive to improvement – which, we shall see shortly, is central to the Scottish Enlightenment – had deep roots in agricultural, religious, social, and intellectual developments in the seventeenth century. How, then, should we understand the Scottish Enlightenment? This turns out to be an interesting and challenging question (Withrington, 1987).

Roy Porter provides a concise account of the leitmotivs of the Enlightenment phenomena in different countries, an account that can be usefully employed as a template for beginning to understand the Scottish Enlightenment. One central leitmotiv involved the development of a "true 'science of man'" (Porter, 1990, p. 12). The science of man developed in different ways in different countries. Porter summarizes the main elements of the science of man (which should be recognizable in light of the discussion in the preceding sections of this chapter):

> Systematic doubt (as advocated by Descartes), experimentation, reliance upon first-hand experience rather than second-hand authority, and confidence in the regular order of nature – these procedures would reveal the laws of man's existence as a conscious being in society, much as they had demonstrated how gravity governed the motions of the planets in the solar system (Porter, 1990, p. 17).

The new science of man in Scotland took its intellectual and moral authority from a secular method, namely, *experience or the experimental method*. Experience, as we saw, involves Baconian scientific method: disciplined observation; critical evaluation of authorities; the careful formulation of hypotheses that might explain observed facts; rigorous testing of hypotheses by experiment and observation; and subsequent labeling of the probability-value of the tested hypothesis. The secularity of the Baconian method had important implications:

> The new Enlightenment approaches to human nature ... dismissed the idea of innate 'sinfulness' as unscientific and without foundation, arguing instead that passions such as love, desire, pride and ambition were not inevitably evil or destructive; properly channelled, they could serve as aids to human advancement (Porter, 1990, p. 19).

The new science of man in Scotland has a normative function that justifies undertaking the science: the improvement – by seeking relief from adversities – of the human condition in all its aspects, from agriculture to education, medicine, and taste in the musical, visual, dramatic, and poetic arts. Indeed, *improvement* becomes the core value of the Scottish Enlightenment, reflecting its strong pragmatic orientation.

> Francis Bacon in an earlier age had defined the aim of knowledge as 'the relief of man's estate', and this well defines the aims of the men of the Scottish Enlightenment, who in addition to (and often in conjunction with) their interest in what Hume called the 'science of man', were also interested in the improvement of agricultural techniques, improved civic planning and the creation of model villages, improvement in the design of prisons and hospitals, and improvement in taste and what they called 'refinement' (Daiches, 1986, p. 5).

We have already noted the Scottish Enlightenment ethos of improvement in Gregory's medical-student concept of medicine and its improvement and the need for a "Medicall Society" devoted to this high purpose. Developing the capacity to cure disease relieves man's estate by treating abnormal function and restoring normal function. Developing the capacity to preserve health relieves man's estate by preventing abnormal and thus maintaining normal function. Developing the capacity to prolong life relieves man's estate by increasing life expectancy and the quality of life – i.e., cure of diseases of aging and preserving health into advanced years – of each individual's extended life span. Gregory also emphasized the *limits* that constrain the relief of man's estate via the improvement of medicine: the risks of iatrogenic disease, injury, pain, and suffering that are unavoidably built into efforts to develop further the capacities of medicine. The improvement of medicine must take account of these limits and press against them only with the utmost scientific and moral care. Physicians should avoid both the excessive caution of following authority – no improvement would then occur – and the excessive pursuit of knowledge, a dangerous form of enthusiasm, with no limits – too much harm would then occur. In my view, Gregory's experience with the "hit or miss" character of medicine, the uncertain state of medical knowledge, competition of ideas and of practitioners, abuse of patients, self-physicking, and the host of other social dimensions of medical practice at once spurred his interest in the improvement of medicine and reminded him of the importance of limits in undertaking such an ambitious project.

Disciplined, moderated progress counts as improvement, not an unconstrained, go-for-broke attempt to overcome the limits of medicine's always-limited, though not unchangeable, capacities.

Gregory had a keen interest in agricultural improvement; he was involved in the Gordon's Mills Society for agricultural improvement in Aberdeen (Ulman, 1990, p. 16). Agricultural improvement took on urgency at a time punctuated by crop failures and the need to feed a growing population. Crop failures could and did result from excessive caution, on the one hand, or enthusiasm for untested techniques, on the other. There were limits to agricultural improvement that one ignored at one's – and the country's – peril. This applies, as well, to medicine and every other human endeavor. Scottish relief of man's estate, at least in Gregory's case, aims at progress in a disciplined fashion, attentive to the disciplined management of risk. Failure to achieve this discipline would only compound adversity and thus offer no relief or improvement.

The new science of man in Scotland, while its secularity surely challenged religious authority and belief, did not displace a creator God, i.e., deism. The secularity of the new science of man might, however, displace particular religious and theological beliefs and traditions, i.e., theism. As Porter puts it, "[v]ery few intellectuals wanted to replace religion with out-and-out unbelief" (Porter, 1990, p. 35). Eighteenth-century Scottish thinkers could not and, with perhaps the exception of Hume, were not willing to part with deism, illustrating Chitnis' thesis that religion "impregnated" many aspects of Scottish life (Chitnis, 1987, p. 79).

Gregory, as we shall see, strives for a secular science of man that does not, as a matter of logical necessity, exclude deism. In this respect, Gregory participates in the mainstream of the Scottish Enlightenment: "the fear of scepticism and the desire to found a Science of Man that would serve the interests of Christians as well as of unbelievers were among the hallmarks of Scottish learning in the age of Enlightenment" (Phillipson, 1981, p. 20).

Porter also takes the view that the "Enlightenment helped to free man from his past" (Porter, 1990, p. 75). I wish to challenge this interpretation's application to the Scottish Enlightenment. The social principle takes an essential part of its meaning from the medieval, feudal institution of paternalism. Given that almost half of the population at mid-century still resided in the Highlands, this should come as no surprise. Paternalism, we shall see, played an essential role in the Aberdonian concept of sympathy and a less cynical one than that which Porter (1981, pp. 11-12) assigns to it. Thus, Gregory and other sympathy, moral sense theorists in Aberdeen do not make a decisive break with the past. Far from it; they retain a key, medieval, feudal concept at the core of their moral philosophy. In this crucial respect they are pre-modern, conservative thinkers, as were so many so-called "modern" philosophers (McCullough, 1996).

Porter also sees the Enlightenment as involving, for some, the twin views that "man was his own maker" and a "secular vision of limitless human drive towards economic growth, scientific innovation and human progress, which the Enlightenment developed" (Porter, 1990, pp. 20, 21). This does not apply to Gregory's science of man. Deeply influenced, I think, by his early and life-long Baconian scientific methodological commitments and growing clinical experience, Gregory's science of man emphasizes limits and limited, moderate – and, therefore, steady – relief of man's estate. Gregory's Enlightenment agenda does not include the perfection of man (Porter, 1990), i.e., the limitless, cost-free pursuit of human betterment, simply because there is no such thing. Sooner or later medical progress imposes unacceptable clinical costs and so we must always observe the "cautions" of medical progress (Gregory, 2206/45, 1743).

Gregory's commitment to deism reinforces his disciplined Enlightenment ambitions for improvement. There is a law-governed order of nature, created by God and ordained by God to human good, if we understand, respect, and only carefully push against the discoverable limits imposed on creation by God. To think and do otherwise is to elevate man to the status of being a god, which status deism shows to be impossible and therefore false and empty – a dangerous enthusiasm, indeed.

Scottish moral sense philosophers, devoted to the science of man in the form of a science of morals, regularly contrast their method with that of metaphysics, just as Bacon – their exemplar – did. We should read such remarks with great care. The attack of the Scottish Enlightenment on metaphysics is an attack on *the metaphysics of pure reason*, one variety – but not the whole – of metaphysics. Descartes epitomizes this false metaphysical method: we cannot do metaphysics or ontology on the basis of clear and distinct ideas and *a priori* reason uninformed by experience as the whole of metaphysical method. Experience neither informs nor corrects Cartesian metaphysical method; this defect cripples Cartesian method. *Au contraire*, experience "corrupts" Cartesian method, its adherents would quickly say in reply. This the Scottish philosophers do indeed attack as false and pernicious metaphysical methodology.

The Scottish Enlightenment philosophers do not, however, attack or deny the possibility of metaphysics itself. They do not attack or deny the possibility of discovering the most general truths about reality, the task of metaphysics bequeathed to them by the Ancient Greek philosophers. The Scottish philosophers do indeed do metaphysics; they just don't do it in a fashion similar to Descartes or Kant or the English moral realists. They do it in Bacon's way instead.

The Scottish philosophers also reject Descartes' static model of reality, which he had inherited from Aristotelian, medieval, and later Scholastic philosophy. The Cartesian-medieval metaphysics invokes the concept of unchanging substantial forms, i.e., essences. Nature or essence is fixed, unchanging, and unchangeable. This metaphysics involves a descriptive anatomy of the most fundamental structures of static reality. This began to change with the development in the seventeenth century of physics and its distinctive mathematics, the calculus of regular, law-governed change invented by Leibniz and Newton. Activity in the form of *regulated motion* or *law-governed change* becomes the most fundamental category of reality, not static structures such as individual substance, substantial form, nature, and essence. The old "anatomical" metaphysics is supplanted by a new "physiologic" metaphysics, in a return to central Aristotelian metaphysical themes.

This shift began also in the seventeenth century with the new science of microbiology. Von Leeuwenhoek describes the regular motion or activity of animalicula, the microbes that he observed through his microscope. The idea of the mind as an active force, "connected" by the sympathy of nerves and muscles also developed during this period. Physiology undertakes the scientific description of law-governed, regular biological activity or change, supplanting anatomy as the primary basic science of medicine.

Instead of a geometry of space, expressed in the Aristotelian astronomy of interlocking spheres or the Copernican astronomy of planar, elliptical orbits, there is now a physics, a physiology of space. Regular, law-governed change is explained by an active force, operating at a distance, namely, gravity. Gravity is a real physical/physiologic "principle" in entities possesses of mass. Gravity explains the natural and mutual attraction of inert, mindless matter.

Instead of an anatomy of mind, explained by the anatomy of individual, non-material substances = thinking things and their ideas = individual accidents inhering in those substances, there is now a physiology of the mind as active force. The social principle is the chief moral principle of this new physiology of man (and mind), because the social principle explains the mutual regard and concern of human beings that *acts at a distance* to draw them together into nations with distinct national identities, as Millar (1990) and Hume (1987a) explain. The social principle becomes the first principle of moral – and political – physiology.

We considered the initial development in biological physiology of the principle of sympathy. Hume, in his *Treatise of Human Nature* (Hume, 1978), builds on these developments and takes them to a level of sophistication that results in *the first complete account of the mental physiology of the social principle*. Scottish moral sense philosophers before Hume, Francis Hutcheson in particular, used the term 'sympathy' but they did not have a developed account of it as a moral and therefore social and physiologic principle of considerable normative significance and content. Scottish philosophers after Hume, Reid in particular, take up Hume's account essentially unchanged, although in altered linguistic dress – perhaps to put themselves at arm's length from Hume's apparent irreligiosity.

Hume, as a scientist of man, held commitments typical of the new metaphysics of activity. There is something real called "human nature" that is active, not static. We can discover the constituent operating or physiologic principles of human nature by following experience but never by following *a priori* or pure reason. 'Experience' functions as a shorthand for Baconian method, a shorthand that Hume's contemporaries in Scotland would readily recognize because they used it all the time. So, too, they would understand 'principle' to mean a real, constituent physiologic function of human nature, not, as John Biro (1993) mistakenly thinks, an anatomical structure. The physiologic principles of human nature, sympathy included, exhibit observable normal and abnormal ranges of function. Principles such as sympathy can be improved in that their normal function can be preserved and refined when intact, as well as restored when lost.

The normal range of function of the principle of sympathy operating at the micro level of face-to-face relationships among individual human beings – or the social principle writ small – is normative for such relationships, because *sympathy* naturally binds us together and thus makes us other-regarding, *moral* beings. Sympathy is therefore is an inherently normative principle. The principle of sympathy operating at the macro level of a community defined by common, Humean moral causes – the social principle writ large – is normative for the Scottish nation because sympathy is an inherently normative principle. In other words, the method of the science of man leads to the descriptive discovery of the inherently normative principle that guides and allows us to judge human action. Pure reason does not contribute to this discovery; it is helpless to this crucial moral, social, and political task of discovery of the normative. The metaphysics of Scottish moral sense philosophy thus encompasses a rich terrain of topical and methodological commitments.

Let us turn to Hume's *Treatise*. Although Hume felt that it fell stillborn from the press, this was not its fate in Aberdeen.

Hume explicitly denies that there was ever a "state of nature," in which humans existed as isolated, independent, social atoms, a concept that must be regarded as a "mere fiction" (Hume, 1978, p. 493). Human nature is inherently active and inherently social, i.e., other regarding; this is another defining leitmotiv of the Scottish Enlightenment. To act merely on self-interest therefore involves wrongdoing:

> Of all crimes that human creatures are capable of committing, the most horrid and unnatural is ingratitude, especially when it is committed against parents, and appears in the more flagrant instances of wounds and death. This is acknowledg'd by all mankind, philosophers as well as the people; the question only arises among philosophers, whether the guilt or moral deformity of this action be discover'd by demonstrative reasoning, or be felt by an internal sense, and by means of some sentiment, which the reflecting on such action naturally occasions (Hume, 1978, p. 466).

Observation supports the fact of a normative judgment and experience in response to ingratitude, i.e., turning against those to whom one is bound by kinship and acting only for oneself. The source of this normative judgment lies in human nature itself, either in reason or the passions.

> Human nature being compos'd of two principal parts, which are requisite in all its actions, the affections and understanding; 'tis certain, that the blind motions of the former, without the direction of the latter, incapacitate men for society: And it may be allow'd us to consider separately the effects, that result from the separate operations of these two component parts of the mind. The same liberty may be permitted to moral, which is allow'd to natural philosophers; and 'tis very usual with the latter to consider any motion as compounded and consisting of two parts separate from each other, tho' at the same time they ac-

knowledge it to be in itself uncompounded and inseparable (Hume, 1978, p. 493).

This, of course, is the metaphysical language – in use for centuries before Hume – of distinct inseparables. This involves the medieval idea that there are realities that cannot exist apart as separate things but which can still be distinguished from each other conceptually. Any motion or human nature is individual, but it comprises parts or elements that are real and thus ground a distinction between the parts of elements that together inseparably make up the motion or human nature. The distinct inseparables of human nature are the affections or the passions, on the one hand, and understanding or reason on the other.

These are the functional parts of human nature. The anatomy or corpuscles, as it were, upon which they operate comprises impressions and ideas, the two, exhaustive types of perceptions. The difference between the two is discoverable upon disciplined observation of mental physiology.

> The difference betwixt these consists in the degrees of force and liveliness with which they strike upon the mind, and make their way into our thought or consciousness. Those perceptions, which enter with most force and violence, we may name *impressions*; and under this name I comprehend all our sensations, passions and emotions, as they make their first appearance in the soul. By *ideas* I mean the faint images of these in thinking and reasoning ... I believe it will not be very necessary to employ many words in explaining this distinction. Every one of himself will readily perceive the difference betwixt feeling and thinking (Hume, 1978, pp. 1-2).

As students of Hume know very well, he goes on to discover the many rules that govern the activity of the soul or mind with respect to impressions and ideas, e.g., rules for the proper and useful association of ideas. These principles of activities, physiologic principles of mind, constitute "the science of human nature" (Hume, 1978, p. 7). These principles govern the associations of ideas with ideas, impressions with impressions, impressions with ideas, and ideas with impressions.

Morality, obviously, has to do with actions and the mental activity of judging actions, behaviors, and character in oneself and others. The explanation of morality therefore requires an "active principle" (Hume, 1978, p. 457). Reason or understanding, however, is "inactive in itself [and] it must remain so in all its shapes and appearances, whether it exerts itself in natural or moral subjects ..." (Hume, 1978, p. 457). Hume's argument proceeds as one would now expect:

> Since morals, therefore, have an influence on the actions and affections, it follows, that they cannot be deriv'd from reason; and that because reason alone, as we have already prov'd, can never have any such influence. Morals excite passions, and produce or prevent actions. Reason of itself is utterly impotent in this particular. The rules of morality, therefore, are not the conclusions of our reason (Hume, 1978, p. 457).

These rules can, therefore, be a function of the only other distinct inseparable constitutive element of human nature, the *passions* or *affections* and the principles that govern the activity of the passions or affections with respect to impressions and ideas.

Sympathy is the chief of these constitutive principles of the generation and therefore explanation of normative behavior and judgment, or morals.

> No quality of human nature is more remarkable, both in itself and in its consequences, than that propensity we have to sympathize with others, and to receive by communication their inclinations and sentiments, however different from or even contrary to our own (Hume, 1978, p. 316).

Sympathy "communicates" or acts at a distance, just as does the inter-organ physiologic principle of sympathy discussed earlier. Sympathy causes us to "enter so deep into the opinions and affections of others, whenever we discover them" (Hume, 1978, p. 319). Sympathy is the principle that explains the social principle. We can observe sympathy in children, in people who share national identity, and in myriad other circumstances (Hume, 1978, pp. 316ff).

> We may begin with considering a-new the nature and force of *sympathy*. The minds of all men are similar in their feelings and operations [i.e., there is one mental physiology], nor can any one be actuated by any affection, of which all others are not, in some degree, susceptible (Hume, 1978, pp. 575-576).

How does sympathy work? Hume's answer involves a "double relation" of impressions and ideas. When one sees another human being in pain, for example, on the basis of this impression of suffering one forms an idea of *oneself* being in pain. This idea leads naturally and automati-

cally to the impression of that same pain in one. When properly functioning, i.e., not deformed or defective, as in the case of ingratitude, the
double-relation process of sympathy operates automatically. As a result,
one has the same impression – or direct, vivid experience – of pain that
the individual observed has.

> This idea is presently converted into an impression, and acquires such
> a degree of force and vivacity, as to become the very passion itself, and
> produce an equal emotion, as any original affection (Hume, 1978, p.
> 317).

Various factors can affect the operation of the principle of sympathy.
Sympathy works with perfect strangers. That is, no other individual is a
stranger to one. We are all built the same way and function the same way,
unless we have fallen into abnormal function or moral deformity, a
condition that be improved by relieving one of moral deformity. When we
resemble others, sympathy will produce impressions of greater vivacity,
as will contiguity.

> When I see the *effects* of passion in the voice and gesture of any per-
> son, my mind immediately passes from these effects to their causes,
> and forms such a lively idea of the passion, as is presently converted
> into the passion itself. In like manner, when I perceive the *causes* of
> any emotion, my mind is convey'd to the effects, and is actuated with a
> like emotion. Were I present at any of the more terrible operations of
> surgery [namely, amputations], 'tis certain, that even before it begun,
> the preparation of the instruments, the laying of the bandages in order,
> the heating of the irons, with all the signs of anxiety and concern in the
> patient and assistants, wou'd have a great effect upon my mind, and
> excite the strongest sentiments of pity and terror. No passion of another
> discovers itself immediately to the mind. We are only sensible of its
> causes or effects. From *these* we infer the passion: And consequently
> *these* give rise to our sympathy (Hume, 1978, p. 576).

According to Hume, simply being in someone's presence who is in pain
powerfully gives one the idea of that individual's pain and automatically
an impression of his or her pain. This experience activates sympathy,
putting one into action to relieve that individual's pain and thus obliterat-
ing all social and other differences, even national identity, that might
otherwise make us moral strangers. There can be, in short, no moral
strangers in Hume's ethics. Hume on sympathy or humanity – he uses the

two words interchangeably – reflects precisely Reid's claim, considered earlier: "Homo sum & nihil humanum a me alienum puto" (Reid, 1991, p. 139).[11]

Sympathy moves one to action to respond to both present and future needs of others.

> 'Tis certain, that sympathy is not always limited to the present moment, but that we often feel by communication the pains and pleasures of others, which are not in being, and which we only anticipate by the force of imagination. For supposing I saw a person perfectly unknown to me, who, while asleep in the fields, was in danger of being trod under foot by horses, I shou'd immediately run to his assistance; and in this I shou'd be actuated by the same principle of sympathy, which makes me concern'd for the present sorrows of a stranger. The bare mention of this is sufficient. Sympathy being nothing but a lively idea converted into an impression, 'tis evident, that, in considering the future possible or probable condition of any person, we may enter into it with so vivid a conception as to make it our own concern; and by that means be sensible of pains and pleasures, which neither belong to ourselves, nor at the present instant have any real existence (Hume, 1978, pp. 385-386).

Sympathy governs morality because it leads to one's having the same pain or sorrow or distress or pleasure as another, and being moved by this impression to appropriate action. This is not empathy, wherein one has an idea of another's plight, links it to one in some imaginative fashion, and then suffers for or with that unfortunate individual. Humean sympathy causes one to suffer *just as another suffers*, to find that unpleasant – to be made uneasy Hume says in many places – and moved to action to resolve the uneasiness.

Sympathy creates an unease that can lead to both negative and positive sympathetic responses to a situation.

> I have mention'd two different causes, from which a transition of passion may arise, *viz.* a double relation of ideas and impressions, and what is similar to it, a conformity in the tendency and direction of any two desires, which arise from different principles. Now I assert, that when a sympathy with uneasiness is weak, it produces hatred or contempt by the former cause; when strong it produces love or tenderness by the latter (Hume, 1978, p. 385).

Sympathy that moves one to respond positively to another grounds the virtue of tenderness. Eighteenth-century British writers used 'tenderness' to denote a female virtue, exemplified above in the *Letters* of Hester Chapone. Hume thus genders positive sympathy female, with no small consequences for Gregory's medical ethics. Gregory understands sympathy, properly developed, to generate tenderness: an asexual virtue that moves us to enter into the suffering and distress of others appropriately, with women of learning and virtue the exemplars that show us what 'appropriately' means.

In Hume's *Treatise* sympathy is a real, constitutive, active, moral, mental physiologic principle of human nature that explains the "contagion" of behavior and passions (Hume, 1978, p. 317). More to the point, sympathy explains the natural, built-in, evident, observable, and fundamental other-regardingness of human beings that Scottish moral philosophers took to be an empirically established fact. Sympathy is real, a *constitutive, normative element of human nature*, distinct yet inseparable from the other principles of human nature and from the corpuscles, as it were, of human nature, impressions and ideas through which sympathy acts. Hume summarizes:

> Thus it appears, *that* sympathy is a very powerful principle in human nature, *that* it has a great influence on our taste of beauty, and *that* it produces our sentiment of morals in all the artificial virtues. From thence we may presume, that it also gives rise to many of the other virtues; and that qualities acquire our approbation, because of their tendency to the good of mankind (Hume, 1978, pp. 577-578).

In summary, Hume takes a thoroughgoing, Baconian approach to the science of man with a particular view to generating a *science of morals*. He thus makes good on the promise of the *Treatise's* subtitle: "Being an Attempt to Introduce the Experimental Method of Reasoning into Moral Subjects" (Hume, 1978). Hume picks up on the main themes of the new physiology, with which he was familiar since his student days at the University of Glasgow and his membership in the "Physiological Library" (Steuart, 1725; Barfoot, 1990). With the new physiologists Hume holds that *the mind is an active power and sympathy one of its main causal principles*.

The experimental method leads Hume to distinguish reason from the passions and to observe with certainty (in the meaning of that term described earlier: not likely to be falsified by any future experience) that reason is inactive and the passions active. The possibility of reason actively forming synthetic *a priori* judgments appears in Kant, of course, but not in Hume. For Kant, reason undertakes under its own rules to add with necessity to the concept named by the subject term of a proposition content from the concept named by the predicate term of that proposition. These synthetic *a priori* judgments represent a distinct *activity* of pure reason, leading Kant to a physiology of pure reason as active. The German Enlightenment, obviously, took a direction strikingly different from the Scottish, at least in physiology (philosophy) of mind.

Mind as active can influence things at a distance; any one mind can communicate its passions to other minds. Thus, Hume provides an explanation for the then widely noticed and remarked upon phenomenon of a *contagion of mood*, which we considered above. The physiologists before Hume provided no explanation of how this contagion occurred; Hume did. Sympathy, as it were, exercised in the social principle, functions analogously to the long nerves and muscles that, according to the physiologists, connected the active mind to the inert body. The double relation of impressions and ideas can be seen a moral and social nervous system, naturally binding everyone together. This binding varies in degree, depending on contiguity, national identity and other moral causes, but the binding together *always* occurs to *some* degree. There are no moral strangers, i.e., individuals in response to whom sympathy is never activated. A crucial normative dimension of sympathy is that one can always take account to some extent the interests of others – across social, class, gender, national, language, race and other differences that might separate us.

Impressions and ideas, as it were, function analogously to the corpuscles or atoms of the body that are activated by mind. For Hume, impressions and ideas are activated in the observer and, unless the observer is morally deformed or defective, they function automatically, just as do the corpuscles of the heart when activated by hypochondria in the mind.

Moral evil is thus not a function of ignorance, as Plato had it. Rather, moral evil results from failure to maintain the active powers of mind, the passions, in their proper working order, individually and in relationship to each other. Hume thus devotes considerable attention to the topic of the types and interrelationships of passion, as do Reid and other Scottish moral sense theorists. Moral evil, in other words, results from sloth or willful perversion of one's natural faculties, sympathy included.

Consider the following examples. If one dissembles and exhibits a false sympathy – e.g., to ingratiate one's way as a physician into the social circle of those trusted by a potential patient – then one acts deceptively in a way that, by repetition, produces a defect of character: false sympathy, false because merely nominal or verbal. If one views sympathy as dissipating, as the languid response of women of ornament, then, as Chapone puts it, one is "contemptible" (Chapone, 1806, p. 60). Sympathy should not lead to dissipation, i.e., tears in response to suffering, but no action. If one rejects sympathy because tenderness is a female virtue unfit for a man of vigor and action, then one makes oneself deformed. Sympathy's exemplar is found in the female virtues, properly developed. Men of commercial and city life become hardened by self-interest, the corrosive and therefore dangerous opposite of sympathy. The Scottish response to moral evil, of course, is improvement: the maintenance of sympathy in its properly modulated, moderate, regulated working order and the relief (restoration) of this normal function when it is lost through deformity or defect.

In other words, Hume's science of man discovers sympathy and thus generates a virtue-based ethics *par excellence*. Virtues are traits of character that routinely dispose and motivate one to blunt self-interest, incline one to focus on the interests of others as one's primary concern, and move one to act to protect and promote the interests of others when those interests are harmed or threatened with harm (McCullough and Chervenak, 1994). Sympathy does all three.

Phillipson provides a useful general summary of the Humean and Scottish science of man:

> Of what did this Scottish Science of Man consist? Technically, it was founded on a desire to study scientifically what we should call the contents of the mind and what contemporaries called 'ideas' or 'beliefs'. These ideas made intelligible the external world, God and even the self, and to understand their origins was the key to understanding the principles of morality, justice, politics, and philosophy. The Scots thought that it was unscientific to trace the origins of ideas back to abstract conceptions like reason, however convenient that might be theologically. The scientific study of the mind involved an empirical investigation of its operations and of the process of socialization (Phillipson, 1981, p. 20).

Hume's science of man and the account of sympathy that this science produces in Hume's hands fit these particulars. The first, experimental

Baconian method, is entirely embraced and the second, principles of morality, become the subject of intense discussion at the Aberdeen Philosophical Society, along with other topics.

B. The Aberdeen Philosophical Society.

Intellectual societies or clubs provide another distinguishing features of the Scottish Enlightenment. Phillipson (1981) points out that a concern for virtue was central to the Scottish Enlightenment. Hume sees sympathy as the basis of the virtues and virtues play a central role in Gregory's writings, as we shall see. Virtue possessed an evident importance for the Bluestocking Circle, as is plain from the texts of Montagu and Chapone. This feature probably helped to attract Gregory to the Bluestocking women of learning and virtue as exemplars. The joining of learning and virtue reflects, as well, the Scottish idea of the "close relationship between intellectual and social development," a theme that looms large in Gregory's medical ethics (Withrington, 1987, p. 16).

Virtue did not flourish in the world of commerce, dominated as it was by the pursuit of self-interest, although commerce did indeed contribute to the improvement of man's estate, an ambivalence that appears in Gregory's thoughts on the subject.

It was better to seek a life of virtue within commercial society itself but away from the world of business, politics and fashion. The key lay in the *salons*, coffee-houses and taverns of modern cities. Here men and women met each other as friends and equals and were able to enjoy the sense of ease that good conversation could bring. Addison and Steele saw coffee-house conversation as a form of social interaction that taught men tolerance, moderation and the pleasures of consensus (Phillipson, 1981, pp. 26-27).

Phillipson sees the coffee-houses and salons as the sites from which developed the intellectual societies and clubs (Phillipson, 1981, p. 27). Allan Ramsey, the great Scottish painter, for example, founded the Select Society in Edinburgh in 1754, of which Gregory's fellow student in Leiden, Alexander Carlyle, was a member. The Select Society and other clubs included academics, and professional people – including physicians, lawyers, and businessmen. The latter might include, as in the case of the Select Society, according to Carlyle, "Robert Alexander, a wine merchant, a very worthy man, but a bad speaker [who] entertained us all with warm suppers and excellent claret" as payment of a kind for the intellectual "constipation" to which his speeches subjected the other members (Carlyle in Daiches, 1986, p. 23). Conversation thus went with good food and drink, and there were rules about such matters included in the minutes of the Aberdeen Philosophical Society: "Any member may take a Glass at a By table while the President is in the Chair, but no Healths shall be drunk during that time" (Ulman, 1990, p. 76). Perhaps such rules and the activity to which they make reference led to the Society also being known in a not altogether complimentary fashion as the "Wise Club."

Aberdeen participated in this club phenomenon. Some of the societies were practical, e.g., the Gordon's Mill Farming Club (Ulman, 1990, pp. 15-16). We saw earlier in this chapter the important role that agricultural improvement played in eighteenth-century Scotland. Given the periodic crop failures and the difficult farming conditions of the North East, the Gordon's Mill Farming Club occupied a place of importance in Aberdeenshire. Societies and clubs also formed to advance the arts. This was the case for the Aberdeen Musical Society (Ulman, 1990, p. 15). Gregory belonged to this Society, as well as the Gordon Mills Farming Club. Questions relating to agricultural science and methods, as well as music, were taken up by the Aberdeen Philosophical Society. These societies in Aberdeen reflected the spirit of the Scottish Enlightenment and made significant contributions to it.

> The thinkers of the Scottish Enlightenment often turned their thoughts and efforts to practical matters, for they believed that scientific or philosophical inquiry should benefit society by improving practical arts such as manufacturing, agriculture, medicine, and oratory (Ulman, 1990, p. 14).

Gregory came to Aberdeen with his interest, since his medical student days, in the improvement of medicine. As we shall see, his thoughts on this subject continue to advance during his time in Aberdeen.

In Aberdeen, this interest in improvement appears to have combined pragmatism and idealism, for which the professors at the two colleges there were known (Carter and Pittock, 1987b). Indeed, there existed a "characteristic Aberdeen concern with the teacher's responsibilities for and to his students" (Carter and Pittock, 1987b, p. 3). We saw this in the case of Thomas Gordon, Gregory's teacher at King's, and we will see it in Gregory's treatment of his medical students in Edinburgh.

H. Lewis Ulman concisely describes the context created by these intellectual societies and clubs in eighteenth-century Scotland: "a tradition of organized debate, discussion, correspondence, and cooperative inquiry that characterizes eighteenth-century Scottish intellectual culture" (Ulman, 1990, p. 9). This was the context in which this Society began in 1758, at the initiative of Gregory, George Campbell (1719-1796) and Robert Traill (1720-1775), who were Presbyterian ministers, John Stewart (ca.1708-1766) and Thomas Reid (1710-1796), professors (Reid had been a minister for fourteen years until 1751, when he was appointed to King's faculty), and David Skene (1731-1770), a physician (Ulman, 1990, pp. 24-34). Nine additional members were soon elected, with academics predominant, including Gregory's protegé, James Beattie (1735-1803), and Gregory's teacher, Thomas Gordon (Ulman, 1990, pp. 34-43). Gordon took extensive notes of meetings, with the result that his "papers provide the richest and most varied collection of materials concerning the Philosophical Society" (Ulman, 1990, pp. 36-37).

Ulman rightly claims that the presentations and discussions at the Society meetings led to "the compositions of some of the most influential philosophical works published in Scotland during the latter half of the eighteenth century" (Ulman, 1990, p. 12).

Thomas Reid's *An Inquiry into the Human Mind, On the Principles of Common Sense* (1764) established the Scottish Philosophy of Common Sense, influenced the shapers of the American republic, and made a lasting contribution to philosophical inquiry. James Beattie's *An Essay on the Nature and Immutability of Truth* (1770) also championed the virtues of "common sense" and earned a wide following in its day, but its reputation has since suffered because it is more polemical and far less rigorous than Reid's work. Alexander Gerard's *Essay on Genius* (1774) is one of the most significant contributions to the "science of human nature" written during the period, and George Campbell's *Philosophy of Rhetoric* (1776) ranks among the most important rhetorical treatises in the Western tradition. Two volumes of somewhat lesser note, John Gregory's *A Comparative View of the State and Faculties of Man with Those of the Animal World* (1765) and James Dunbar's *Essays on the History of Mankind in Rude and Cultivated Ages* (1780), complete the catalog of books that grew directly out of the discourses read to the Philosophical Society (Ulman, 1990, p. 12).

Ideas formulated by Gregory in his writings for the Society also found their way into the *Lectures*.

The Society adopted a set of rules. These concerned venue, officers, dues, fines, accounts, nominations and elections, record-keeping, rules for discussion (aiming at "cooperative inquiry by encouraging the full attention and participation of all the members" (Ulman, 1990, p. 46)), changes to the rules, and elections of new members (Ulman, 1990, pp. 45-50).

The intellectual work of the Society took the form of Questions, which were addressed orally or in writing and then read to the group, and Discourses, which were usually written out ahead of time and then read to the group. The seventeenth and final rule sets the parameters for questions and discourses and reflects many of the themes of the Enlightenment science of man:

> The Subject of the Discourses and Questions shall be Philosophical, all Grammatical Historical and Philological Discussions being conceived to be forreign to the Design of this Society. And Philosophical Matters are understood to comprehend, Every Principle of Science which may be deduced by Just and Lawfull Induction from the Phænomena either of the human Mind or of the material World; All Observations & Experiments that may furnish Materials for such Induction; The Examination of False Schemes of Philosophy & false Methods of Philosophizing; The Subservience of Philosophy to Arts, the Principles they borrow from it and the Mean of carrying them to their Perfection. If any Dispute should arise whether a Subject of a Discourse or Question falls within the Meaning and Intendment of this Article it shall be determined by a Majority of the Members present (Ulman, 1990, p. 78).

Henry Sefton claims that "Hume was probably in the founders' minds when they drew up Rule 17, which states that one aim is 'the examination of false schemes of Philosophy and false methods of Philosophizing'" (Sefton, 1987, p. 125). I think that this reading is mistaken; the false methods of philosophy were those based on pure reason and/or uncritical reliance on authorities. Hume announces his *Treatise* to be committed to the "Experimental Method of Reasoning" (Hume, 1978), as we saw above, and so Hume's method falls precisely within the methodological boundaries defined by Rule 17.

As it turned out, Hume's *Treatise* was indeed much on the minds of the members of the Aberdeen Philosophical Society. Hume, in bringing to moral philosophy the Baconian method of natural philosophy, probably attracted men whose "own education taught them that natural and moral philosophy were integrally related and interdependent" (Ulman, 1990, p. 59. n. 74). Reid, surely a critic of Hume, nonetheless held to the "experimental method" in his own work.

Hume's skepticism did trouble some of the members of the Society, especially when it came to religion or, more properly, deism. This reflected what Ulman sees as the earned reputation of the Society – "representing the conservative, religious opposition to Hume's skeptical philosophy, a sometimes cordial, sometimes contentious confrontation ..." (Ulman, 1990, p. 54). On the other hand, members seem to have taken strongly to Hume's method and the account of sympathy based on that method. Thus, they do not reject Hume's skeptical method in a moderate form, i.e., one consistent with deism, but they do not accept that skepticism when it leads to claims that experimental method excludes God's existence as creator. Hume's attack on evidence for a creator in the *Treatise* represented, I believe, the Humean skepticism they reject, which is, decidedly, the whole neither of Hume's method nor of its fruits. The intellectual relationship of Society members to Hume, therefore, exhibited an interesting ambivalence.

Hume was aware of the interest in his work by the Aberdeen Philosophical Society through Hugh Blair, a friend of John Farquhar, one of the first elected members of the Society. Farquhar introduced Blair to Campbell and Reid, and Blair sent drafts of works by them to Hume for his comments. There developed a correspondence between Blair and Hume in reference to the Society's discussion of Hume's work. In 1763, Hume writes to Blair: "I beg my compliments to my friendly adversaries, Dr. Campbell, and Dr. [Alexander] Gerard [both ministers who opposed Hume's irreligiosity], and also to Dr. Gregory, whom I suspect to be of the same disposition, though he has not openly declared himself such" (Burton, 1846, p. 154). We shall consider whether Hume should suspect Gregory to be a possible opponent. Blair passed Hume's remarks on to Reid, who replied:

> Your friendly adversaries, Drs. Campbell and Gerard, as well as Dr. Gregory, return their compliments to you respectfully. A little philosophical society here, of which the three are members, is much indebted to you for its entertainment. Your company would, although we are good Christians, be more acceptable than that of St. Athanasius; and since we cannot have you upon the bench, you are brought oftener than any other man to the bar, accused and defended with great zeal, but without bitterness. If you write no more on morals, politics, or

metaphysics, I am afraid we will be at a loss for subjects (Burton, 1846, p. 154).

This letter exhibits a number of interesting features. First, Reid, Gregory's cousin and intimate, does not seem to include Gregory in the category of "friendly adversaries," only Campbell and Gerard are included. An alternate reading is that all three are *friendly* adversaries, not implacable foes. Second, Reid expresses his debt to Hume for the "entertainment" of the Society, i.e., for offering it a work that required serious intellectual engagement, and in this sense of the term the Society was "entertained" in the Questions and Discourses of the Society. As we shall see, sympathy provides a great deal of "entertainment" to the Society's members. Third, Hume is both accused *and* defended and, in both cases – especially the first – "without bitterness." In other words, the Aberdeen Philosophical Society took Hume's work seriously, with the diffidence required of intellectually disciplined philosophers.

Fourth, Hume's skepticism regarding religion is not mentioned, an interesting omission. After all, Hume writes in the *Treatise*: "Generally speaking, the errors in religion are dangerous; those in philosophy only ridiculous" (Hume, 1978, p. 272). In this respect Reid's reference to Athanasius is interesting. Athanasius, Church Father and Patriarch of Alexandria, condemned the Arian heresy, which was anti-Trinitarian in its denial that Jesus was one in substance with God the Father. The Athanasian Creed, attributed to Athanasius, was then a lively subject of debate, given the emergence of Arianism, which dissented from the view that Jesus was divine. Reid's comment can be read as the expression of the Society's preference for a skeptic who argues, rather than an unquestioning, unreflective, perhaps enthusiastic believer. 'Enthusiastic' here means a person who does not regulate his or her passions and so they are deformed, by being let to go out of control into excess. On this reading, which is at least plausible, Reid's comment counts as a subtle compliment to Hume's skepticism, perhaps even in some of its applications to religion, e.g., to the religious enthusiasm that could then be found in many churches in Scotland.

We saw in Chapter One, and will consider in the next chapter, Gregory's skepticism about enthusiastic ministers being involved with dying patients. Skepticism properly understood plays a central, indispensable role in the experimental method in Bacon – which Gregory wholly embraced in his medical student days, and which Rule 17 of the Society endorses. Only hypotheses based on observation and experiment are worthy of serious consideration and even their truth value may be less than certain. Thus, Baconian method makes one a skeptic regarding both untested and proven claims. This skepticism, as we will see, plays a central role in Gregory's concept of scientific method in medicine, and the obligation of the physician to be "open to conviction."

Gregory addressed seven questions before the Society.

Question 1, February 8, 1758: "Whether the greatest Part of the Matter that composes Bodies of Vegetables and Animals is not Air? Or some Substance that is mixed with Air and Floats in it?" (Ulman, 1990, p. 189);

Question 20, February 11, 1759: "What are the Plants that enrich a Soil, and what are those that impoverish it and what are the Causes of their enriching & impoverishing it?" (Ulman, 1990, p. 190);

Question 30, January 8, 1760: "Whether the Socratic method of Instruction or that of Prelection is preferable?" (Ulman, 1990, p. 191);

Question 39, January 13, 1761: "What are the Natural Consequences of high national Debt & whether upon the whole it be a benefite to the Nation or not?" (Ulman, 1990, p. 192);

Question 50, February 9, 1762: "What are the good & bad effects of the provisions for the poor by poor rates, infirmarys, hospitals, and the like?" (Ulman, 1990, p. 193);

Question 59, July 12, 1763: "Whether the art of Medicine as it has been usually practised, has contributed to the advantage of mankind?" (Ulman, 1990, p. 193); and

Question 70, May 8, 1764: "What are the distinguishing characteristicks of wit and humour" (Ulman, 1990, p. 194).

Manuscript material survives for Question 59, which we consider in this section. With slight modifications, virtually the entire content of Question 59 reappears in the sections of *Observations* and *Lectures* that address the philosophy of medicine.

Gregory delivered six Discourses to the Society.

October 11, 1758: "The State of Man compared with that of the lower Creation" (Ulman, 1990, Table A-4);

August 28, 1759: "An enquiry into those faculties which distinguish man from the rest of the animal creation ... (To continue his former subject)" (Ulman, 1990, Table A04);

March 31, 1761: "Dr Gregory continued" (Ulman, 1990, Table A-4);

January 11, 1763: "The influence of religion upon human nature" (Ulman, 1990, Table A-4);

August 9, 1763: "The foundation of taste in Music" (Ulman, 1990, Table A-4); and

August 14, 1764: "The prolongation of human life & retardation of old age" (Ulman, 1990, Table A-4).

I will also consider manuscript sources for some of these Discourses. The first five become the backbone of *Comparative View*, the general aim of which Gregory had already set out in his fourth "Capitall Enquiry" in 1743, "The Improvement of our Nature" (AUL 2206/45, 1743, p. 6). The title of the sixth indicates that Gregory was giving further thought to the second of his "Capitall Enquirys" from his medical student days, "The Retardation of Oldage" (AUL 2206/45, 1743, p. 6). In all, Gregory attended ninety-five meetings of the Society between its founding in 1858 and the end of 1764, before he left for Edinburgh (Ulman, 1990, Table A-2).

The Society's Questions and Discourses ranged over many subjects. These included slavery, which "debases human nature in the person of the slave, and excites ferocity and a savage turn of mind in the master" (AUL 3107/2/4, 1761, p. 119), thus morally deforming the slave *and* master, a repellant outcome. The Society also took up the topic of just wars by a nation "in its own defense, when an army of invaders attacks their lives or liberty .. " (AUL 3107/2/6, 1764, p. 214). Three subjects of the Society's deliberations concern us here: sympathy; skepticism and the experimental method of natural philosophy; and medicine. We will consider the manuscript sources for each in turn, to chart the intellectual ferment and center of activity that the Society represented and to trace its influence on Gregory's intellectual development as it occurred in conjunction with the activities of the Aberdeen Philosophical Society.

1. Sympathy
The term 'sympathy' was used regularly by the moral sense philosophers. Francis Hutcheson, whose work the members of the Aberdeen Philosophical Society knew well, predates Hume. While Hutcheson does not

provide a developed account of sympathy, as does Hume, Hutcheson does anticipate much of Hume's account (Beauchamp, 1993).

Hutcheson notes that we have certain perceptions or impressions made on us by others. His debt to the physiologists, who described and attempted to account for the same phenomena, seems plain.

> We have multitudes of perceptions which have no relation to any external sensation, if by it we mean perceptions, occasioned by Motions, or impressions made upon our bodies, such as the Ideas of ... *the pains of remorse, shame, sympathy* ... (Hutcheson, 1730. p. x).

In other words, we have non-sensory perceptions that result from action at a distance on us of events that impress on us certain ideas. These excite our passions, an active, affective response to the ideas impressed on us.

Hutcheson attacks those who deny that such passions exist. Some oppose a *"natural affection, friendship, love of country or community"* (Hutcheson, 1730, p. 14), but they are mistaken. Hutcheson can be read as developing the social principle to explain the concept of community and nation, in response to the crisis of Scottish national identity discussed at the beginning of this chapter. The *social principle* operates at three levels: interpersonally, in forming friendships; socially, in the formation of national identity; and at the intermediate level of formation of community – larger than a family or clan but smaller than a nation. All three levels display kinship-like affinities. Hutcheson calls this "natural affection" sympathy:

> That this sympathy with others is the effect of the constitution of our nature, and not brought upon ourselves by any choice, with view to any selfish advantage, they [opponents of the "natural affection"] must own; whatever advantage there may be in sympathy with the *fortunate*, none can be alleged in sympathy with the distressed. And every one feels that his *publick sense* will not leave his heart, upon a change in the fortunes of his child or friend; nor does it depend on a man's choice, whether he be affected with their fortunes or not (Hutcheson, 1730, p. 14).

Sympathy is part of human nature, not something that we put on artificially. Hutcheson's metaphysical commitments seem plain: human nature is the essence of our species and sympathy is definitive of that nature. Sympathy functions automatically, not by will, just as other physiologic principles function. Sympathy does not function on behalf of self-interest, for no one would choose to be in distress at the distress of a child or friend. Sympathy is by nature *other-regarding*.

These reflections lead Hutcheson to the concept of the moral sense, a "calm universal benevolence" (Hutcheson, 1730, p. 31), an other-regarding habit of mind and behavior that encompasses sympathy and other passions. The moral sense explains why we act morally, i.e., in an other-regarding fashion and not just for "selfish advantage," and also provides the basis for moral judgment (Hutcheson, 1730, p. 14):

> An action is *good, in a moral sense*, when it flows from benevolent affection, or intention of absolute good to others. Men of such reflection may actually intend *universal absolute good*; but with the common rate of men their virtue consists in intending and pursuing *particular absolute good*, not inconsistent with universal good (Hutcheson, 1730, p. 37).

To be useful and defensible, the concept of "particular absolute good" requires specification. Hume takes the view that pain and distress are obvious harms and their absence obvious goods, as experience teaches. Gregory, as we shall see, adds some further specification for the goods relevant to medicine.

Henry Home, Lord Kames, also employs a concept of sympathy. His work, too, was well known to the members of the Aberdeen Philosophical Society and his work on morality was in Gregory's library. Kames follows Hutcheson and Hume in this account:

> It may be gathered from what is above laid down, that nature, which designed us for society, has connected us strongly together, by participation in the joys and miseries of our fellow creatures. We have a strong sympathy with them; we partake of their affections; we grieve with them and for them; and, in many instances, their misfortunes affect us equally with our own (Kames, 1751, pp. 15-16).

Sympathy, contra its critics, is not either "vitious or faulty" (Kames, 1751, p. 16). One needn't worry about dissipation and lack of manliness in exercising sympathy, because "mutual sympathy must greatly promote the security and happiness of mankind" (Kames, 1751, p. 17). Sympathy is expressed in the female virtue of tenderness, which is all to the good:

> When we consider our own character and actions in a reflex view, we cannot help approving this tenderness and sympathy in our nature; we are pleased with ourselves for being so constituted, we are conscious

of inward merit; and this is a continual source of satisfaction (Kames, 1751, p. 17).

Kames, however, takes exception to Hume in a key respect:

According to this author [Hume], there is no more in morality but approving or disapproving of an action, after we discover by reflection that it tends to the good or hurt of society. This would be far too faint a principle to control our irregular appetites and passions. It would scarce be sufficient to restrain us from encroaching upon our friends and neighbors; and, with regard to strangers, would be the weakest of all restraints (Kames, 1751, pp. 57-58).

Kames does not reject sympathy; rather, he fortifies it with the theological concept of "merited punishment, and dread of its being inflicted upon us" (Kames, 1751, p. 64) in the afterlife. Gregory does not accept this criticism, though he is concerned with the fostering of sympathy, as would any secular moral sense theorist.

Against the background of Hutcheson, Kames, and Hume, the members of the Aberdeen Philosophical Society take up and endorse the concept of sympathy within their consideration of physiology (philosophy) of mind. Skene, for example, lauds Hume's contributions : "... we so seldom meet with a Locke, a Hume or such a genius as is able to throw light on the Gloom, with which the object [the mind as an object of "natural knowledge"] is obscured" (AUL 37, 1758a, p. 170v-171r). Hume and Locke have lifted the gloom on "various branches of natural knowledge" (AUL 37, 1758a, p. 170v). Skene then goes on to defend sympathy as a principle of mind against its critics:

I shall conclude this part of the discourse with one argument for the steady prosecution of the philosophy of the mind, drawn from an objection that has been warmly urged against it. I mean the disagreeable and pernicious tendency of some of its pretended discoverers. I am far from thinking the authors of these are bad men. As nothing is more common than to see a man strenuously endeavoring to prove that sympathy and compassion are weaknesses which man ought not to have or justice a political contrivance, which convenience may dispense with, while perhaps he lives in the uniform practice of both (AUL 37, 1758a, p. 174r-174v).

Elsewhere Skene makes the same criticism of "pretended discoverers" of principles of the mind (AUL 3107/1/3, 1758a, p. 42). He means to establish that sympathy is anything but a weakness. True, women exemplify tenderness, but they are *superior* to men in this respect:

> It is a fact which holds almost universally, that the women have more of this knowledge [of "sentiments and manners"] than the men. For one man, who can enter into the spirit of a particular character with readiness and propriety, we shall find several women (AUL 3107/1/6, 1760, p. 20r)

Men and women alike share human nature and its principle, sympathy, because human nature and its principles constitute their essence as human beings. They are built this way: sympathy is a function of our "constitution," a word frequently used by eighteenth-century moral sense philosophers. Sympathy must be properly cultivated and regulated, if it is to serve us well as the basis of moral actions and judgment – without the theological adjunct Kames wants. Women show the way to more hard-hearted men, especially those who view sympathy or tenderness to be a weakness to be strenuously avoided. Gregory, I think, found fortification in such remarks for his own views on women as moral exemplars.

Sympathy, in other words, must be properly cultivated and regulated, just as any human capacity must. The agricultural metaphor is apt; a properly cultivated field regularly yields crops while a poorly cultivated field either dies – i.e., dissipates – or grows wild – i.e., becomes enthusiastic. In the history of Western philosophy this idea of discipline in moral capacity and its virtues goes back to the Stoics and, before them, to Plato and Aristotle.

This view of sympathy finds a counterpart in the discussions in the Society of the nature of "genius" – "the faculty of invention; by means of which a man is qualified to make new discoveries in science, or to produce original works of art," as Gerard puts it (AUL 3107/1/3, 1758, p. 3). "Caprice" and other undisciplined activities of the imagination do not lead to invention, Gerard argues (AUL 3107/1/3, 1758, p. 22). Judgment serves the requisite regulatory, disciplining function:

> Tho genius consists properly in a comprehensive regular & active imagination, yet it can never attain perfection, or assert itself successfully on any subject except it be united with a sound and vigorous judgment. The strength of imagination carries us forward to invention: but understanding must always conduct it and regulate its motions (AUL 3107/1/3, 1758, p. 23).

So, too, understanding and judgment must regulate sympathy, which "carries us forward" to relieve those in distress. Unregulated, untutored sympathy could impel us pell-mell. In clinical practice this would wreak havoc as often – probably more often – than good for patients. Gregory, who makes sympathy the cornerstone of his moral-sense-based medical ethics must find the counterpart for sympathy to the regulation of imagination by understanding. As we shall see in the next chapter, he does so.

The Society also considered Adam Smith's theory of sympathy (Smith, 1976). Smith's moral sense theory varies slightly – but crucially for present purposes – from Hume's. In a commentary on the concept of sympathy, particularly in Hume's philosophy, Phillip Mercer draws the distinction between Hume's and Smith's understandings of sympathy:

> Although, like Hume, he [Adam Smith] thinks that to sympathize with another is to have the same feelings as this other person has, Smith does not conceive the process by which this happens in Hume's mechanical terms. Whereas Hume held that sympathy consists in the idea of an emotion being converted into the emotion itself through the enlivening association of the impression with the impression of self, according to Smith sympathy involves imagining oneself in the other person's situation and thus, in one's imagination, going through all of the emotional experiences he would be going through. We change places 'in fancy with the sufferer' (Mercer, 1972, p. 85).

The difference between Hume and Smith on sympathy can be put in the following terms. For Hume, for one individual, A, to sympathize with another, B, is for A to *experience directly* the same emotion, E, that B experiences. We saw this in Hume's examples of the terror invoked by the preparation for surgery or the sight of someone in a field in danger of being trampled by a horse. By contrast, for Smith, when A sympathizes with B, A *imagines* E. For Hume, A and B share the same emotion, whereas for Smith there need be no emotion common to A and B, because A is at a remove from E by an act of the imagination. For Smith, sympathizing is more *cognitive*, while for Hume it is *visceral*, an automatic, instinctual, physiologic response.

Stewart takes up Smith's theory and describes it as follows:

> Sympathy acts by substituting ourselves first in the place of the Agent who confers a Benefit upon another, & then in the place of the Person who receives the Benefit and if the same or similar Emotions are excited in Us as in the Agent and if by entering into the Gratitude of the

Person benefitted, we observe the Conduct of the Agent to be agreeable to the general rules according to which those two Sympathies act, we approve of the Action as virtuous (AUL 3107/1/10, 1765, p. 5).

Stewart finds this latter process something that he does not understand – although he gives no further explanation for his lack of understanding – and expresses favor for Hutcheson, whose views are more understandable, because they can be captured by "conscience" (AUL 3107/1/10, 1765, p. 6). Stewart finds this more attractive because of its *theological* content.

Campbell endorses Hume's theory of sympathy. His discussion of pity mirrors Hume's exactly, I think: "... pity is only a participation by sympathy in the woes of others, or whatever kind they be, their fears as well as their sorrows" (AUL 3107/2/1, 1758, p. 9v). He later says:

Nothing, says Mr Hume, "endears so much a friend as sorrow for his death. The pleasure of his company has not so powerful an influence." Distress to the pitying eye diminishes every fault, & sets off every good quality, in the brightest colours. Nor is it a less powerful advocate for the mistress than the friend: often will this single circumstance of misfortune make a weak and languid passion flame out with a violence which it is impossible any longer to withstand. There is then in pity 1st commiseration purely; 2dly – benevolence, or a desire of the relief & happiness of the object pitied, a passion, as was observed above, of the intermediate kind; 3dly Love, a word which denotes one of the noblest and most exquisite pleasures of which the soul is susceptible, & which is itself in most cases sufficient to give a counterpoise of pleasure to the whole (AUL 3107/2/1, 1758, p. 10r).

Sympathy thus supports the virtue of self-effacement as one its immediate effects. It moves to one side the differences that might otherwise count in separating two individuals, e.g., another's "faults." Features that attract one to another, that person's "every good quality" of which sympathy naturally makes us aware, draw us to that person.

Reid does not equate virtue with sympathy: "Virtue does not consist in any one affection such as benevolence or sympathy" (AUL 2131/6/1/12, 1760, p. 2); he argues instead for a multi-faceted moral sense which he calls "conscience" (in many places in his presentations to the Society). Later, in his *Essays on the Active Powers of the Mind*, Reid identifies a number of "benevolent affections" (1863). The third of these is "*Pity and Compassion toward the Distressed*:"

In man, and in some other animals, there are signs of distress, which nature both taught them to use, and taught all men to understand without any interpreter. These natural signs are more eloquent than language; they move our hearts, and produce a sympathy, and a desire to give relief (Reid, 1863, p. 562).

There is, for example, "compassion with bodily pain" (Reid, 1863, p. 563). There is also a social affection:

The *last* benevolent affection I shall mention is, what we commonly call *Public Spirit*, that is, *an affection to any community to which we belong* (Reid, 1863, p. 564).

I read Reid to hold to Hume's theory of sympathy as causing us to enter into the distress of others and thus to be moved by compassion for them to relieve their distress. Reid also holds the social-principle version of sympathy, because his description of "Public Spirit" fits it well. He wants to deny the primacy of sympathy as a benevolent affection. Reid, as it were, employs Hume's concept of sympathy, but in the company of other affections to explain and judge human behavior.

Gregory also contributes to the Society's deliberations on sympathy.[12] He begins with an essentialist account of the principles (in the sense of built-in, causal functions) of human nature, directly in the tradition of the metaphysics of the new experimental method of natural and moral philosophy, and of their maintenance and improvement:

Human nature consists of the same principles every where. In some people one principle is naturally stronger than it is in others. But exercise & proper culture will do much to supply the deficiency. The inhabitants of cold climates having less natural warmth & sensibility of heart enter but little into those refinements of the social principle which those of a different temper delight in. But if those refinements are capable of giving the mind innocent & substantial pleasure, it should be the business of Philosophy to search into the proper methods of cultivating & improving them (AUL 3107/1/4, 1759, p. 8).

The "Philosophy of taste & manners" (AUL 3107/1/4, 1759, p. 8) has been neglected in Britain, Gregory claims. He is well aware of what is at stake for patients and physicians in this lack of cultivation of manners and the deficiency of false manners, i.e., good manners put on but not *internalized* as virtues, just as Fissell (1993) describes it.

> A physician, when he sets out in life, quickly perceives that the knowledge most necessary to procure him a subsistence, is not the knowledge of his profession, what he finds most essential to that purpose are the arts of deceiving mankind, into an opinion of his understanding by an appearance of solemnity & importance in his whole deportment, and the various arts of flattery & dissimilitude, views very different from the pursuit of genius and science – He can expect no patrons of his real merit, because none are judges of it, but a few of his profession, whose interest it is to keep it concealed (AUL 37, 1762, p. 163r)

Physicians, as the Porters (1989) point out, entered a competitive world of practice in eighteenth-century Britain. There was no established body of science and no certification of a physician's "genius" and qualifications. Patients, perforce, chose their doctors based on trust. This became a perilous enterprise, as illustrated by the problem of men-midwives. Genuine sympathy – sympathy properly developed and regulated, "cultivated" and "improved" – could provide an antidote, Gregory will come to argue in his ethics lectures.

The two principles of human nature are instinct and reason.

> It has been imagined that man was designed to be guided by the superior principle of reason entirely. But a little attention will shew that instinct is a principle common to us & the whole animal world & that as far as it extends it is a sure and infallible guide, tho' the depraved unnatural state into which mankind are plunged [by commerce and city life, which promote self-interest] often stifles its voice, or makes it impossible to distinguish it from other impulses which are adventitious & forreign to our nature. Reason is but a weak principle in respect of instinct & we greatly overrate it when we substitute it in the others place. Reason's business is to investigate causes discover consequences point out the best means, & to be a check upon our instincts our passions & our tastes. But there must still be the governing & impelling principles, & without these life would not only be joyless & insipid, but quickly stagnate & be at an end (AUL 3107/1/3, 1758b, p. 30).

Gregory is completely committed to the new experimental method, which he had laid out in his "Medical Notes" in 1743, and Hume's results of its application to physiology (philosophy) of mind. With Gerard he has a developed notion of reason, functioning in the role of judgments to properly regulate passions such as sympathy, that both govern and impel judgment and action, so that they can better achieve the end of human happiness.

> Nature has made man to be governed by affections, by passion, by taste, by temper & by imagination. These may be considered as analogous to instincts (AUL 3107/1/3, 1758b, p. 33).

More precisely, these are instincts properly regulated. In his medical ethics lectures Gregory will address how the passion of sympathy is tempered by steadiness, an expression of judgment about the proper limits of sympathy.

Gregory explicitly addresses sympathy, again in a way that Hume would recognize, because the text makes it evident that Gregory takes his account from Hume. Sympathy is a built-in principle, affected by "relations" such as contiguity and "circumstances" such as national identity. Sympathy produces pleasure when one acts on it and it acts independently of understanding. Sympathy must be cultivated and regulated.

> The next distinguishing principle of mankind [after reason], which was mentioned, is that which unites them into societies, & attaches them to one another by sympathy & affection, which operate with different degrees of strength according to the variety of relations & circumstances in life. This principle is the source of the most heartfelt pleasure, which we ever taste. It does not appear to have any natural connexion with the understanding. It was even observed formerly, that persons of the best understanding possessed it frequently in a very inferior degree to the rest of mankind; but it was at the same time noticed that this did not proceed from less natural sensibility of heart but was entirely owing to the social principle languishing from want of being exercised. It must be acknowledged that the idle, the dissipated and debauched draw most pleasure from this source ... (AUL 3107/1/4, 1759, p. 7).

With Hester Chapone and Mrs. Montagu, Gregory does not equate virtue with dissipation and lassitude; these are deformities of sympathy that need to be relieved, and corrected, to improve human beings so that sympathy functions normally more often.

Sympathy properly regulated leads to a consideration of alcohol consumption in the "cold climate" of Scotland and the relationship between drinking and the cultivation of the social principle:

> Even drinking, if not carried to excess is found favorable to friendship, especially in our northern climates, where the affections are naturally

cold as it produces an artificial warmth of temper, opens & enlarges the heart, & dispells that reserve, natural perhaps to wise men, but incompatible with friendship, which is entirely a connexion of the heart (AUL 3107/1/4, 1759, pp. 7-8).

Thus, the rules governing drinking in the Aberdeen Philosophical Society – its members met in a local tavern, the Red Lion – promoted friendship and the social principle. The social virtues promote intellectual virtues, just as the former promote the latter. There is more to this delightful passage: properly developed sympathy creates a genuine warmth of temper, opens and enlarges the heart and so improves character by making one routinely, by nature, other-regarding, open to the impressions of the "motions" of others on one's heart, and dispells the natural reserve to intrude into the distress and privacy of patients, especially when one is meeting that person for the first time and may be bringing bad news. In other words, properly cultivated sympathy makes one worthy of trust, because properly cultivated sympathy makes one a person of "real merit" when that sympathy is joined to genius and science in medicine. Thus, the intellectual and social virtues form a synergy at the heart of Gregory's medical ethics.

Years later, James Beattie, who was a member of the Society, captures the spirit of Hume and Gregory on sympathy:

Nothing is more odious, than that insensibility, which wraps a man up in himself and his own concerns, and prevents him from being moved with either the joys or the sorrows of another. This inhuman temper ... results from habits of levity, selfishness and pride (Beattie, 1976, p. 176).

Sympathy with distress is thought so essential to human nature, that the want of it has been called *inhumanity* (Beattie, 1976, p. 180).

Gregory takes this view, which comes straight from Hume. Gregory adopts Hume's concept of sympathy as is plain in the above-noted texts. Sometimes, however, in his medical ethics he uses language closer to Reid's, whose views essentially coincide with Hume's on the notion of sympathy. Lack of sympathy is inhuman because it perverts our naturally other-regarding nature. Hence, the conceptual interchangeability of 'sympathy' and 'humanity'.

2. Skepticism.

As we have seen in Rule 17 and in the documentary record of the Society considered so far, the members of the Aberdeen Philosophical Society committed themselves in varying degrees to the new experimental method in natural philosophy and morals. Skepticism in one of its meanings plays a central role in this method. Gregory's "Medical Notes" from 1743 reveals that he appreciates this fact and takes the concept of skepticism from Bacon. One should be skeptical in not taking authorities as true simply because they have become established or are regarded as authorities. One should be skeptical by assigning the appropriate degree of truth-value to both observations and tested hypotheses. Finally, one should be a skeptic toward one's own knowledge based on the new method: Openness to new observations and revision in the degree of truth-value of hypotheses constitutes the central intellectual virtue of the new method. Gregory calls this "openness to conviction" in his ethics lectures, and the concept plays a central role in the *synergy of intellectual and social virtues* that drives his medical ethics.

The members of the Aberdeen Philosophical Society took up the topic of skepticism in this context. Skene, for example, expresses the following view: "Nothing is more disagreeable to the mind than suspense of action or scepticism in science" (AUL 3107/1/3, 1758a, p. 40). Skene's objection, I take it, is to the inaction of paralyzing skepticism. This view comes as no surprise from a thinker who was a physician and participated in a movement committed to the practical goal of *improvement*, not just the intellectual goal of *understanding*. When understanding or knowledge is disconnected from practical goals, inaction or indefinite suspension of belief poses no problem. The opposite obtains in understanding devoted to relief of man's estate, e.g., by reducing frightful infant and childhood mortality rates. So, the members of the Aberdeen Philosophical Society rejected – and they should have – skepticism that paralyzes action and thought. Doing so did not require a rejection of Baconian skepticism, however.

Reid objects to Hume, as I read the objection, on the grounds that he did not follow his own method in the *Treatise*, a method that holds itself open to correction from common sense.

> Presumption in D[avid].H.[ume] to pretend to give an entire system – probably he some parts of it have escaped him, and others have been mangled and distorted to ply to his system – A prejudice against the Philosopher that [there] are so many things in his plan which shock the

common sense of mankind. A Philosopher can never be matched with a harder adversary than common sense, who never fails at last to triumph over those who wage war with her (AUL 3107/1/1, 1758, p. 18).

Hume, Reid claims, has in the *Treatise* erected a "system," a word with all the negative connotations it has in the new physiology. The reason, it appears, that Hume built a system detached from common sense was that he employed a *dogmatic* skepticism that does not question its own limits. Dogmatic skepticism has no more use than dogmatic reason or authorities.

> But what if these profound disquisitions into the first principles of human nature, do always necessarily lead a man into such determined and dogmatic skepticism. Tho we have some examples of this in great works I am not apt to believe that it must necessarily be so. – I have always found Philosophy an agreeable companion, a kind and benevolent guide, a friend to common sense and to the happiness of mankind (AUL 3107/1/1, 1758, pp. 18-19).

Skepticism *per se* cannot be objectionable, only when it is "determined and dogmatic," i.e., unregulated by understanding and judgment formed from common sense, Reid holds. In effect, Reid, in effect, accuses Hume of being an enthusiastic skeptic who does not follow the experimental method in morals. Skepticism does indeed play a role in that method; Hume gets it wrong, according to Reid.

Gregory endorses the inductive method in medical science and medical education (AUL 37, 1762, p. 162v). He mounts a fierce attack on the then present state of medical education:

> Medicine as taught in colleges is digested into a regular and compleat system – In this view it is beheld by the young student who embraces the theories with the same facility & unsupported confidence that he would do facts – He understands the causes of all diseases & the operation of all medicines – His mind is at ease in having allways sure and fixed principles to rest upon – In the mean time the art can receive no improvement from him, as he does not imagine that it stands in need of any – If a patient dies he is quite satisfied everything was done for him, that art would do – It is difficult for men to give up favourite opinions, the children of our youth, to sink from a state of security & confidence, into one of suspense and scepticism – Accordingly, few physicians change their principles and very seldom their practice (AUL

37, 1762, p. 162v).

In the absence of skepticism – a conscious and deliberate suspension of belief that impels one to seek new knowledge and therefore not treat patients as incurable because that is what one has been taught or what the authorities say – patients die unnecessarily. Note Gregory's choice of words: "suspense," not paralysis. He thus agrees with his cousin, Reid, that pragmatic skepticism is permitted and with Skene that paralyzing skepticism is a problem.

Gregory thus endorses skepticism in the form of diffidence, a central intellectual virtue of the account of the compleat physician provided by his philosophy of medicine:

> If we can enquire now into the effects produced on the mind by acquiring knowledge in the slow method of observation and induction from experiments, we will find him very different – The mind here gains a close & accurate attention to facts (having nothing else to trust to), slow in forming principles from these facts & diffident of them when formed, instead of being assuming & dogmatical, is modest & sceptical – A physician of this stamp never loses a patient but he secretly laments his ignorance of the proper means of having saved him, which he blames rather than the disease being incurable in itself – This naturally stimulates to the improvement of knowledge, both from a love to science & from humanity, and a principle of conscience (AUL 37, 1762, p. 163r).

A disciplined skepticism in the form of diffidence, even toward one's favored hypothesis and the rejection of "incurable" as *the* explanation of first resort for a patient's death, forms one of the central maxims of Gregory's philosophy of medicine, in Aberdeen and thereafter. He maintains this commitment for the rest of his life.

Gregory also claims that skepticism as an intellectual virtue is not only consistent with religious belief; it improves religious belief:

> Surely their education in their own profession does not dispose them to infidelity – A physician who does not see the weakness & futility of the best of a physical system is certainly well prepared to embrace any system of religious belief that can be offered to him (AUL 37, 1762, p. 163r).

For Gregory, then, diffidence – the properly developed form of skepticism – instructs one to question and test religious belief, for some forms of it will not be good for the proper formation, maintenance, and cultivation of one's character. Those that foster unregulated passion, the enthusiastic religions, will not pass the test. Thus, as a scientist and moral philosopher, Gregory creates a secular apparatus that allows him to test the worth of particular religious beliefs and practices. The standpoint for doing so is non-theological and so is neutral to all religious belief; this forms, I believe, the sources of the intellectual and moral authority of the standpoint for Gregory. This is just the sort of skepticism that Hume exhibits about miracles and superstition and so it seems that Gregory – though not Reid – would accept these results. But, he might add, Hume has only shown that some theological matters, theism in particular, cannot hold up well under skeptical examination. Gregory would not, however, accept Hume's skepticism on whether there was a creator for the world we see around us. Gregory remains a committed deist, although he nowhere addresses Hume's attack on the argument for God's existence based on the need for an author or originator of the evidently ordered natural world.

3. Medicine
The members of the Aberdeen Philosophical Society addressed topics in the philosophy of medicine. They saw medicine, as they saw every practical art, as related to and serving human happiness. This is just what we would expect from practical thinkers concerned to relieve man's estate. They say little about the content of human happiness, because they know what is involves: steady, incremental relief of man's estate. Their concern is more with the necessary and sufficient material conditions of relief than with some perfect end-state.

Skene, for example, takes the view that happiness has three requisites: health, employment, and "such a constitution of mind as may enable us to make a proper use of these advantages," a constitution that includes "good humour" (AUL 37, 1758, pp. 180v-181r). Medicine, obviously, contributes to the preservation of health, Gregory could surely add. Moreover, the developing scientific and clinical interest in sympathy leads at this time to a new clinical category, *nervous diseases*, as we shall see in Section IX. Thus, medicine has something to contribute to maintaining a "constitution of mind" that will, indeed, "enable us to make a proper use" of health and employment.

We have seen already that Gregory thinks that the new scientific medicine and its proper skepticism, diffidence, have a central role to play in the improvement of medicine and its contributions to physical and mental health and thus to human happiness. Gregory adds to his understanding of medicine through his contributions to the Society.

He expresses a concept of disease as a "disorder" in anatomical structure or function that is accompanied by "distress" (AUL 37, 1762, p. 162r). Properly speaking, the first provides a concept of disease and the second a concept of illness, a distinction that writers of this period do not address systematically. Friederich Hoffmann, with whose works Gregory was familiar, uses *morbus* or "disease" and *aegrotus* or "the sick" but he does not make this distinction (Hoffmann, 1749). Diseases in humans cannot be left to run their natural course, as they do in the animal world. Nature does have its correctives to abnormal structure and function, but these correctives should not be left to themselves, because they do not always work.

> Some times we cannot safely allow nature her course in carrying off a disease as in the case of an internal infection which nature would generally carry off by a supposition satisfactory to the patient – The efforts of nature are sometimes so violent as to require a check & sometimes so feeble as to stand in need of cordials & an additional stimulus – In some case nature makes no sensible effort for her relief – These facts which cannot be disguised lay the foundation for the art of medicine (AUL 37, 1762, p. 162r).

Baconian method teaches us that nature does not always serve human health. Medicine should use its capacities of preserving health, curing diseases, and prolonging life to supplant or correct nature as necessary. The assumption here about medicine and human good is minimal:

> No people would be so stupid or obstinate as to leave a dislocated arm to nature – The only question then can be to settle the proper limits between the art and the operation of nature in the cure of disease, a question which requires the deepest knowledge of both to resolve (AUL 37, 1762, p. 162r).

The content of this medical morality is that it is only safe to assume that virtually everyone values restoration of observed normal anatomy and physiology, except for the "stupid" and "obstinate." The first do not appreciate that medicine can make a difference in a very painful injury, such as a dislocated shoulder, and its consequent disease processes. The second don't mind the pain or so fear medicine that they would rather live with the pain and consequent disease processes of a dislocated shoulder. This is hard to imagine and nearly impossible to take seriously, especially since sympathy is impelling the physician to relieve evident agony by attempting to relocate the shoulder in its proper anatomy and restore its proper function. The "stupid" and "obstinate" need not be coerced; left alone for a time, they will be instructed by nature in the form of the experience of unrelenting pain and dysfunction to change their minds.

Gregory addresses other aspects of medicine in a way that sets the stage for his medical ethics lectures. Medicine well practiced, Gregory maintains, benefits mankind. But under what conditions is medicine practiced well?

> ... let us enquire whether it is most beneficial to have it [medicine] confined to & practiced as it usually is by a sett of men who live by it as a profession or to leave it large to be studied, like all other branches of natural philosophy by those who love science, but have a Fortune or some other profession to support them (AUL 37, 1762, p. 162r).

The argument in favor of the first view, Gregory says, turns on the claim that medicine not practiced by those who earn a living from it would "rather decline than be improved" and that quacks would prosper, "lessening the confidence and implicit faith in the physician which is convenient for the patients own sake as well as the doctors" (AUL 37, 1762, p. 162v). Gregory doubts the latter, on the grounds that medical science and education, still dominated by systems, does more harm than good, as we saw in the passage above where he discussed the case of explaining mortality from so-called incurable diseases.

We also saw that Gregory's antidote to medical systems is Bacon's: the diffidence of rigorous, experience-based skepticism. We considered already the passage in which Gregory endorses diffidence as an intellectual virtue. He takes a distinctly pragmatic, clinically-oriented approach to skepticism and its proper limits:

> We own that the phylosophical mind, if not carefully attended to, may become very detrimental in the practice of physic & indeed in all practical parts of life, by making a physician timid and irresolute when he must act – But true phylosophy leads [the physician] to be diffident and cautious in forming opinions, but when there is occasion to act, to be quick in forming a resolution, & resolute & fearless in putting it into execution (AUL 37, 1762, p. 163r).

The problem is that the competitive world of medicine and its rival theories, which we considered earlier, would lead more often to attacks on the results of diffidence than support for them, because science should be able to address the competition, sorting out reliable from unreliable theories and practices – a deep threat to unscientific physicians (then and now). In effect, diffidence becomes the cornerstone of the authoritative body of science that could legitimately earn public confidence as *science*.

In other words, Gregory's concept of diffidence as a crucial intellectual virtue of physicians contributes to the formation of the concept of medicine as a fiduciary profession based on science. The science of medicine would be developed in a disinterested way; hence it should no longer be practiced as a trade. The science of medicine would be practiced for human good, not the physician's interest in reputation, market share, and income – for which physicians were mightily scrambling at that time. Self-interest would lose its power; indeed, diffidence and sympathy – intellectual and moral virtues in a perfect synergy – remove the influence of self-interest from thought, motivation, and action. Thus, the virtues alone are adequate to the task of medical ethics in forming the concept and practice of a genuine medical profession; punishment in this world or the next is not required for men of professionally formed intellect and moral character.

> There is a consequent dignity that generaly accompanies genius whenever found, which renders the possessor equally impervious to the feelings of envy, & all the low arts of dissimulation (AUL 37, 1762, p. 163v).

Physicians and scientists of these qualities will improve medicine, despite the opposition to them of the "faculty," i.e., the practicing community and the royal corporations. Their opposition arises from interest and in the long run such moral deformity, Gregory implicitly assumes, will fail to improve medicine. Thus, society has an interest in the creation of a genuine intellectual and moral profession of medicine, so that man's estate can be relieved and thus improved, which forms one of the primary purposes for the existence of human society in the first place.

> These observations make it probable that little improvement can be expected in medicine while it is monopolised by a sett of men who have a separate and distinct interest from that of the art, & that if it were cultivated by men of sense & knowledge who have no other view but the love of science & the benefits of method, it would make ad-

vances correspondent to the other branches of natural knowledge (AUL 37, p. 1762, 163v).

The alternative is calamity for patients, going to the "nearest physician" (AUL, 1762, p. 163r) or "a physician, whom accident throws in their way" (AUL 37, 1762, p. 163v). This is not only not improvement; this, the present state of medicine-not-yet-a-profession, is patent decline and undermines health and thus the prospects for human happiness.

Medicine's capacities need to be put in proper working scientific and clinical order. This requires the creation of a body of science in the sense of an authoritatively established body of scientific and clinical knowledge, authoritative because it has been established on the basis of the correct, i.e., Baconian, method. Creating such a body of science and then giving it over to those of diffidence, dignity, and self-sacrifice will create a profession of medicine in its intelletcual and moral senses: a group of men (in those days women could not enter medical school) who live a life of obligation to their patients in which self-interest is systematically blunted and removed to a secondary status, even removed altogether in the ideal case. Universities have a major role to play in this process; hence Gregory's commitment to university life as a professor of medicine. Universities of liberal spirit foster the new science, and the medical faculty have an independent basis of income that frees them to practice medicine with diffidence in regard to their own monetary interests.

While he was in Aberdeen, Gregory desired to be in a larger urban setting. To this end, Mrs. Gregory writes to Mrs. Montagu on June 28, 1756, in an attempt to advance her husband's cause:

> I have been informed by mama of the many obliding proofs of friendship you have shown her Dr Gregory & me since we left London – allow me Dear Mrs Montagu to assure you of the just sense both Dr Gregory and I have of all those and all former Instances of friendship in England – your friendship like your self is superior to the rest of the world – how greatly I think my self Honoured by and indebted to it. would require Mrs Montagu's Elegance and Politeness to express. whether or not Dr Gregory succeed if the expected vacancy in St George's Hospital should happen. .. it would still add to our obligations if Dear Mrs Montagu would be so good as to favour Dr Gregory with her advice what steps she judges proper for him to take at this junction (HL MO 1063, 1756, p. 1).

Elisabeth Gregory's mother, Baroness Forbes, played a role in trying to obtain a position for Gregory at the University of Edinburgh. On December 20, 1760, she writes the following to Mr. Montagu:

> Mrs Gregory, has wrote twice to Mrs Montagu to intreat her assistance in procuring the Professorship of Botany, in the Colledge of Edin^{br} for Dr Gregory. Since she wrote we have learnt, that it is a very fine thing (HL MO 949, 1760, p. 1).

She goes on to suggest the strategy of Mr. Montagu communicating with the Duke of Newcastle to advocate for Gregory, because Cullen, who serves as the physician to the Duke of Argyle, has sought this professorship. She concludes, "the only chance we have is that the Duke of Newcastle can give it, without asking the D[uke] of A[rgyle's] leave" (HL MO 949, 1760, p. 2).

VIII. THE DEATH OF GREGORY'S WIFE

On September 29, 1761, Elisabeth Gregory died at the age of 33 in confinement with her sixth pregnancy (Lawrence, 1971, Vol. I, p. 157, p. 167). Like many women of the day she spent about half her married life pregnant and died young from complications of pregnancy before which medicine was powerless. Stewart wrongly reports the year of her death as 1763, though she does provide a concise account of the effect of this event:

> It was the greatest sorrow of his life, and afterwards when high honours came to him in his profession [in Edinburgh], and when the world praised him [as the author of *Comparative View*], he never ceased to think with longing of the early joyous days of his love. Elizabeth Gregory was very happy, and even in her memory there is something tender and simple, something to make one smile, and feel the better of it (Stewart, 1901, p. 111).

Gregory thus experienced the loss of a loved one that then so commonly occurred to parents, spouses, and children. He did not marry again.

Among the Montagu manuscripts and other materials at the Huntington Library can be found an undated letter, which is most likely from late 1761, from Gregory to Mrs. Montagu in which he expressed his grief and loss. I find this letter to be completely in the spirit of Gregory the moral sense philosopher and member of the Aberdeen Philosophical Society.

Madam

My Friends used to say that I was Proud of my Wife and my children. I
believe I was so. It has now pleased Almighty God to humble that
Pride in the Dust. Every One Laments my losing the best of Wives, &
my children losing the best of Mothers. But Alas, Madam, I have lost
what the World knows not nor can know. I have lost my Friend, my
Mistress, the Partner of every Joy & every Sorrow I ever felt since we
were United. I was ever sensible of her Superior Genius & Capacity. I
felt this but was not hurt by it because she never seemed Conscious of
it Herself. I was & I saw with every Sentiment which Love & Gratitude
could inspire a Heart of the greatest Sensibility, thoroughly detached
from every Pleasure which Life affords & entirely devoted to me & my
Children. I perfectly revered her Piety. Religion never appeared in a
more amiable form. Most of the time she could spare from the Dutys of
her Family & the Dutys of Humanity & Charity were spent in her Private Devotions, & yet she had the most uninterrupted chearfulness &
sweetness of temper & the most uniform Vivacity of Spirit I have ever
known. These Quality's attached my Heart to Her in the tenderest
manner, They engaged my Confidence to so great a Degree that I gave
myself & my whole affairs entirely up to her Direction. Our Sentiments, our Tastes, Our Views of Life were the same, We talked of our
Childrens Education & all our little matters together but I left the Execution of Every thing to her, so that for these nine years past I have
lived the life of a Child in my own House, & excepting in my own Profession, I gave myself trouble about Nothing. The only Merit I ever
took to myself was that my Heart felt & was gratefull for all her Goodness. – The Sensibility of your own Heart will easily represent to you
how desolate & forlorn is my present Situation. At a time when I require the greatest Vigor of mind to perform many nameless Dutys
which are entirely strange to me I feel my Spirit broke & incapable of
any Exertion. The Sight of the Children perfectly unmans me. O,
Madam, they were a remarkable Family. This is not the suggestion of
Vanity. I think I am now Dead to every Sentiment of Vanity. but indeed they were spoke of all over the Country & considering the attention given to their Education it was no wonder that they were distinguished. Alas I wander. Dear Mrs Montagu, forgive the overflowings
of a heart oppressed and feeble. The common Esteem & Affection my
Dear Betsy had for you, makes me think my self entitled to pour out

my Griefs before you. It is a Melancholy Pleasure, but still it is a very great Pleasure. I know it has never been in the power of Posterity & an affluent Fortune to harden your Heart against the tender feelings of Humanity. I know it possessed of a Delicate Sensibility, & what is still more uncommon, of a sense of Piety, seldom preserved in the tumult of Public Life, & more generally acquired in retirement or the School of Afliction. Your letter was a new proof of that Goodness which I have so often experienced & shall ever gratefully remember. It suggested the greatest, the only motive I can at present feel for exerting myself, the Consideration of my Infants. This I trust to the Goodness of God, will have its effect, tho the thousand tender melancholy Ideas which every surrounding Object recalls to my Memory, may render it very distant. I always thought cool reason & Philosophy very insufficient Aids to support the Mind under certain Pressures & aflictions. Little did I think how sensible a Proof I was to have (HL MO 1064, n.d., after September, 1761).

In this letter Gregory does what he writes about: He exercises his humanity in a regulated, earnest fashion, with the depth of his grief coming from a heartfelt sentiment made powerful by its regulated expression. As one should, if one takes sympathy seriously, Gregory takes a "very great Pleasure" in the "Melancholy Pleasure" of pouring out his "Griefs" in writing. Sympathy, I think, made Gregory "grateful for all her Goodnesses." Moreover, Gregory invokes the concept of sympathy in his characterization of the response he expects from Mrs. Montagu to his letter, for her "Heart will easily represent to you how desolate and forlorn is my present Situation." Mrs. Montagu will sympathize with Gregory as she would a character in a Shakespeare play, and experience the emotions of profound loss that Gregory describes in his letter. This is the Humean double relation of impressions and ideas at work in the context of mutual grief. We all know it too well, except those among us who are young or heartless.

We also find in this letter a portrait of a woman whom Gregory not only loved, but whom he plainly admired for her intellectual capacities and moral character. They had for their short nine years together a marriage he describes in the most admired terms of the day: "Friend, Mistress, and Partner," an equal. Elisabeth Gregory also possessed not just intellectual ability, but "*Superior* Genius and Capacity" (emphasis added), not unlike another "academic" Gregory, Janet, of generations before. Elisabeth had cultivated the best sentiments of heart, as well as piety, and "chearfulness & sweetness of temper & the most uniform Vivacity of Spirit I have ever known." Obviously, for Gregory, female virtues of such exquisite development occupied a high plane, so high that his confidence in his wife was complete. His family was "remarkable," he says in effect, because of her unflagging efforts, including the education of their children, girls and boys alike.

Gregory goes on to put Mrs. Montagu on the same plane with his wife. Mrs. Montagu lives a life unlike that lived by Mrs. Gregory, namely a "public life" in her salons and writings. Her heart had not been hardened "against the tender feelings of Humanity" by city life and the high-powered intellectual circles in which she moved. She is an exemplar of *tenderness*, the very word Hume uses to characterize sympathy, a term interchangeable, we saw, with 'humanity'. Mrs. Montagu, too, had suffered great loss with the death in infancy of her only child.

Finally, Mrs. Montagu and Mrs. Gregory were distantly related. One of her kinfolk had died.[13] Mrs. Montagu and Elisabeth Gregory were women of learning and virtue, exemplifying properly regulated passion. Properly regulated passion, even though its exemplars are female, does not "unman" because regulated passion involves the properly cultivated expression of passion and thus human virtue. Only vice or weakness can unman someone. Moreover, women of learning and virtue can remain so in public life, in the arena of contesting self-interest.

Religion and piety occupy an important place in both women's lives, Gregory notes. We shall see in his *Comparative View* and *Father's Legacy* how important he thinks religion is, particularly for women and for his daughters.

Gregory closes his letter with a reflection on the power of philosophy to see him through loss. As we saw above and will see shortly in a closer examination of his views in *Comparative View*, Gregory does not think much of reason; it is a "weak principle" of mind, as we saw above. Hence, in this letter Gregory characterizes reason as "cool," not "warm," the distinguishing mark of the heart and passions, the other principle of mind. Now Gregory has "sensible Proof," in his own experience of desolation, of the truth of his philosophical commitments.

The letter itself displays the antidote to grief that the instinct of passion provides: regulating it, properly expressing it, and thus bringing it into the correct place in one's life, as one of its permanent features. Too little humanity, too little grief, would indicate a defect of sensibility and per-haps, as well, an excessive and misplaced reliance on reason. One does not think through grief, one lives with discipline in it. Excessive grief lacks this discipline and so it is too warm, enthusiastic, and thus dangerous because self-consuming and therefore self-destructive. Excessive grief "unmans," because it is a simplistic, self-serving outpouring of feeling. It serves its own small purposes and not the larger end of improving oneself in response to profound loss. Excessive grief is undisciplined; hence, it dissipates, it unmans by weakening one.

Gregory struggles in this letter, I believe, to live by his philosophical commitments and his closing comments indicate that he takes himself to have succeeded, at least initially. Occasionally, he is indeed "unmanned" by the "Sight of the Children." To be unmanned, we now know, means to act in the weak and despicable way in which Mrs. Modish had had her daughters instructed and against which Hester Chapone strongly admonishes her niece. Women of dissipation lack self-discipline in the expression of heart, and become in this letter the exemplars of the "unmanned." This concept, it turns out, plays a crucial role in the case that Gregory makes for sympathy in his lectures on medical ethics, as we shall see in the next chapter.

IX. EDINBURGH 1764-1773:
PROFESSOR OF MEDICINE, THE ROYAL INFIRMARY NERVOUS
DISEASES, THE BEATTIE-HUME CONTROVERSY, AND
GREGORY'S CORRESPONDENCE WITH MRS. MONTAGU

A. Appointment in the University of Edinburgh

The Minutes of the Aberdeen Philosophical Society record Gregory's last attendance at a meeting on November 22, 1764 (Ulman, 1990, p. 126). Shortly thereafter Gregory moved to Edinburgh. Tytler's version of this decision runs as follows:

> Dr. Gregory remained at Aberdeen till the end of the year 1764, when, urged by a very laudable ambition, and presuming on the reputation he had acquired as affording a reasonable prospect of success in a more extended field of practice, he changed his place of residence to Edinburgh. His friends in that metropolis had represented to him the situation of the College of Medicine as favourable to his views of filling a Professorial chair in that University, which accordingly he obtained in 1766 on the resignation of Dr Rutherford, Professor of the Practice of

Physic. In the same year he had the honour of being appointed first
Physician to his Majesty for Scotland, on the death of Dr. Whytt
(Tytler, 1788, pp. 53-54).

Stewart's version provides some addition insights into the circumstance
of this change in Gregory professional, academic, and personal life.

Some years after his wife's death Dr Gregory was invited to go to Edinburgh. Professor Rutherford, who held the chair of the Practice of
Physic, wished to retire, but he would not resign his place to [William]
Cullen, whom he held a heretic in medicine. So the old professor ar-
ranged that John Gregory should be asked to come from Aberdeen, and
set up practice in Edinburgh. At another time Professor Gregory would
have hesitated, but in his distress and despondency he thought of what
a benefit it would be to himself to leave the sad associations of Aberdeen and allay his sorrows in the fulness of work which he knew
would await him. His university did not ask him to resign his chair at
King's College, but in 1765 Sir Alexander Gordon of Lesmore was
appointed as joint-professor (Stewart, 1901, pp. 111-112).

Gregory bought a home at 15 High Street, which no longer exists, and
which was near Lord Monboddo's, one of Edinburgh's literati. He was
granted the high honor of Freedom of the City of Edinburgh on February
12, 1766 (Lawrence, 1971, Vol. II, p. 397). He received his license from
the Royal College of Physicians of Edinburgh on March 5, 1765, and was
made a Fellow of the College on August 6, 1765 (Royal College of
Physicians of Edinburgh, 1882, p. 4).

His practice began to grow and *Comparative View* appeared, in which
publishing project Lyttleton assisted, as he had Mrs. Montagu. *Comparative View* appeared in its first editions anonymously, part, it would seem,
of a concerted effort by Gregory and his friends to win him the coveted
opportunity to succeed Rutherford.

London read the book, Aberdeen read the book, and so did Edinburgh,
and Gregory was made at once a member of that literary Edinburgh as
he had in his youth been received by Mrs Montagu and her friends in
London (Stewart, 1901, p. 112).

Controversy surrounded Gregory's appointment to the medical faculty
in the University of Edinburgh. Emily Climenson (1906, Vol. II, p. 226)
reports on the 1760 letter of Baroness Forbes that we considered earlier. It
seems that Gregory had an interest in a position in the University as early

as 1760 and used family connections to advance his cause. Gregory's interest in a professorial position in Edinburgh thus predates by four years events described by Tytler and Stewart. As in the case of his unsolicited M.D. degree from Aberdeen, family connections apparently played a role in Gregory's eventual success in gaining a post at the University of Edinburgh. Stewart continues:

> Professor Rutherford watched with growing satisfaction the success of the Aberdeen doctor, whom he regarded as a protege of his own. It was unfortunate for Gregory that he stood as it were a rival of Cullen, for whom he had throughout his life the profoundest regard. But nevertheless this was the case. In 1766 matters came to a climax in the appointment of Gregory to the Chair of the Practice of Physic, made vacant by the retirement of Professor Rutherford. There was an immediate and furious outcry against this election, which was known to be mostly due to family influence. Gregory was a great man, and proved himself a brilliant teacher, but he was at this time absolutely untried, whereas Cullen had already made himself a name as one of the greatest teachers of the day (Stewart, 1901, p. 113).

As was then the custom, the appointment to the chair was made by the Edinburgh Town Council, which made Gregory "teacher of the practice of medicine" on March 6, 1766, to succeed Rutherford (Lawrence, 1971, Vol. II, p. 407). When Whytt died in 1766 a vacancy occurred in the chair for the Theory of Medicine. At the time Cullen was Professor of Chemistry.

As was also then the custom, students entered into this decision-making process about what to do next, with an "Address of the Students of Medicine to the Right Hon. the Lord Provost, Magistrates, and Town-Council of the City of Edinburgh" (AUL G404a, 1766), a printed broadside dated April 28, 1766, and signed by one hundred and fifty-six students.

The students opened with the following: "The following address is presented with the sole view of promoting the good of the University and the City of Edinburgh" (AUL 404a, 1766, p. 3). About the latter the concern was literal. D.B. Horn reports that arguments about appointments in those days turned on such considerations as keeping students at home to study, and on the economic impact on a city like Edinburgh of doing so. "One tourist calculated that the Edinburgh students spent in the town during a session of seven or eight months not less than 30,000 pounds"

(Horn, 1967. p. 72). The students made the following proposal:

> ... we are thoroughly convinced, that Dr. GREGORY'S relinquishing
> the professorship of the Practice for that of the Theory of Medicine,
> that the appointment of Dr. CULLEN to the practical chair, and elec-
> tion of Dr. BLACK as Professor of Chemistry, will add to the reputa-
> tion of the University and Magistrates, and contribute highly to the
> good of the city and the Students of Medicine (AUL 404a, 1766, p. 7).

The students argued for Cullen's appointment on the basis of Cullen's
enormous improvements in Chemistry, which had come to rival anatomy
for the number of students engaged in study of it, and the quality of
Cullen's clinical teaching at the Royal Infirmary (about which, more
below). They clearly esteemed Cullen as a superior clinical teacher in the
Infirmary.

> There, from the justness and importance of his observations, from the
> accuracy and chastity of his reasoning, from his extensive knowledge
> of the practice of other countries, from the wideness of his views, from
> the perspicuity of his method, from the ease and elegance of his deliv-
> ery, every hearer has obtained the most thorough conviction of his
> abilities as a Practical Professor, and conceived the most ardent desire
> of hearing his instructions from that Chair (AUL 404a, 1766, p. 4).

Benjamin Rush, who was then a medical student in Edinburgh, expressed the negative view of his fellow students of Gregory, as not a very
good clinical teacher (EUL mic. m. 28, 1766). This, perhaps, explains the
way the students' made their case for Gregory.

> Nor is our opinion of Dr. CULLEN meant, in the least, to detract from
> the merits of Dr. GREGORY. On the contrary, a principal motive to
> our expressing the sentiments we do, on this occasion, is the high
> opinion we entertain of that Gentleman's capacity. By a very late and
> ingenious performance, by every body attributed to him, we imagine it
> is evident what advantages the University must reap from Lectures on
> the Theory of Medicine, delivered by a thinker so just and original, and
> so universally acquainted with human nature [a reference, it would
> seem, to *Comparative View*]. With pleasure too we reflect, that his
> character is not less respectable as a man, than as a philosopher; we
> therefore cannot suppose, that, were the publick emolument to be ob-
> tained, even at the expense of his private interest, he would not rejoice
> to make the honourable sacrifice, far less, that he would, in the least,

hesitate to favour a scheme for promoting the publick utility, when his private advantage is consistent with it (AUL 404a, 1766, pp. 4-5).

The Town Council was not moved by this petition and Cullen was given the chair of the Institutes of Medicine, to succeed Whytt (Lawrence, 1971, Vol. I, p. 161), and Black the chair in Chemistry. One hundred and fifty-one of the students responded with the proposal that Gregory and Cullen alternate, "which, however expedient, has hitherto been unexperienced in this University" (AUL 404a, 1766, p. 7). This was, however, implemented sometime later, in 1769, an arrangement that worked well (Lawrence, 1971, Vol. I, p. 161). Four of the students signed an addendum to this broadside:

A report having been malignantly propagated, that the framing and presentation of this address were suggested and conducted, not by the Students of Medicine, but by one of the Professors of Medicine; we, the committee appointed for presenting it, who have had access to be acquainted with the particulars relative to these matters, hereby declare, in the most solemn manner, that such report is entirely without foundation (AUL 404a, 1766, p. 8).

This may be a reference to Rutherford's involvement in this process, which apparently did occur (Grant, 1884, Vol. 2, p. 403). It might also be a reference to possible efforts by Cullen on his own behalf. Stewart comments that "[t]his can hardly have been a pleasant reading for Gregory" (Stewart 1901, p. 114). Gregory includes his ethics lectures at the start of his lectures on the practice of physic as early as 1767, for which year a dated manuscript version of student notes exists (WIHM 2618, 1767; RCPSG 1/9/5, 1767). Perhaps he began his ethics lectures to increase his appeal as a clinical teacher.

While the above events transpired, Gregory was also appointed to succeed Dr. Whytt as First Physician to His Majesty in Scotland, no small honor and not without impact, we can suppose, on his standing and reputation among potential patients. This appointment was commissioned "under the Privy Seal of Scotland" on May 12, 1766, and "entered into Exchequer" on June 26, 1766 (Lawrence, 1971, Vol. II, p. 407). James Coutts, partner in a powerful banking house and son of John Coutts, the former Lord Provost of Edinburgh, writes on May 13, 1766, to Provost J. Stewart of Edinburgh about this appointment. After a cryptic reference to private letters that he had directed to Provost Stewart, Coutts indicates that there had been three candidates, among whom he had selected Gregory and had already recommended him to the King who made the appointment before Coutts learned of the Provost's "private wishes for Dr. Cullen" (EUL La II. 647/98, 1766, p. 1v). Coutts explains his decision:

> ... but Dr Gregory being earnestly recommended to me, by several of my warm friends, and understanding that heretofore no Precedency has been given either to the Seniority or the Rank of the Professor [a reference to Cullen's seniority in the University, apparently]. I from that motive *alone* with great pleasure named D[r] Gregory, who was named to the King, and the Commission made out for his signing ... (EUL La.II. 647/98, 1766, pp. 1r-1v).

This appointment came "with a salary of 100 sterling per annum, during the King's pleasure" (Lawrence, 1971, Vol. I, p. 161), and Coutts makes reference to "dividing the salary, amongst his present Brethren" (EUL La. II. 647/98, 1766, p. 1v). With whatever survived of his wife's dowry, with his own inheritance, with the annual income from the King, and with his salary from the University, Gregory became the sort of physician who ought, as he argued in Aberdeen, to be in medicine: those "who love science, but have a Fortune or some other profession to support them" (AUL 37, 1762, p. 162r).

B. *The Royal Infirmary of Edinburgh*

Gregory lectured on the theory and practice of medicine in the University and taught clinical medicine in the Royal Infirmary of Edinburgh. He addresses problems at the Infirmary in his medical ethics and other lectures. The Royal Infirmary was established in 1739 as a small house with just a few beds. It grew by the 1760s into a large, multistory facility of over two hundred beds. The growth of the Royal Infirmary in many ways paralleled and supported that of the medical school at the University of Edinburgh, and rose to international prominence by the time that Gregory joined its distinguished ranks in 1766 (Risse, 1986, p. 2). Rutherford had begun a teaching ward in the Royal Infirmary in 1750s and Cullen and Gregory continued to attend to patients and teach in this ward (Risse, 1986).

In his masterful scholarly study of the Royal Infirmary, Guenter Risse provides a concise summary of the context that gave rise to the Royal infirmaries that developed in major British cities during the eighteenth century.

British infirmaries were part of a comprehensive program to institutionalize the poor under the banner of a vigorous philanthropic movement fueled by religious and humanistic concerns. Whatever the expense involved, erecting such hospitals also seemed to make economic sense to eighteenth-century leaders, who believed that medical efforts could restore sizable numbers of workers to their previous health and productivity, thereby decreasing welfare costs and even expanding the population (Risse, 1986, p. 1).

The Royal Infirmary of Edinburgh had its impact on the city it served.

During the second half of the eighteenth century, the Edinburgh infirmary therefore became part of the city's network of influence, patronage, and power. Appointments to its governing board and medical staff conferred prestige and furthered social and professional status. Among the managers and subscribers were key figures of Edinburgh's governmental and intellectual circles (Risse, 1986, p. 3).

The Royal Infirmary was thus a complex institution, like all human institutions. Annual gifts of subscription from wealthy individuals supported its operating budget. While these gifts were surely motivated in many cases by charity, they also were motivated by self-interest, as Risse notes above.

Sickness among the working poor occasioned economic calamity for families, whether it was the male or female head of household who became so ill that seeking hospitalization sometimes became a necessary alternative (Risse, 1986, p.8., p. 17, p. 26). Recall, that in a society of "self-physickers" one would probably have to be very ill indeed to seek admission to the Royal Infirmary. Sickness among the working poor deprived one of workers or soldiers; commerce and warfare required a healthy, sturdy population. The Infirmary's charity extended thus to the worthy poor, those who had earned some claim on the resources of the well-to-do – not some euphemistic "community." The Royal Infirmary thus served as an instrument of economic and political power in the newly emerging industrial society of Scotland and empire of Great Britain.

Physicians, especially the University professors, quickly saw the opportunities for teaching and experimentation that the Royal Infirmary afforded. In its wards were to be found patients living in a structured, regulated environment ideal for the study of the natural history of disease and the effects of medical and surgical interventions on that history. Physicians, with the aid of the permanent staff (and the many rules of the house that the staff enforced), had "considerable authority and control over their hospitalized patients" (Risse, 1986, p. 21). As we saw in section VI, patients had and used their power and control in the out-patient setting; this situation did not obtain in the royal infirmaries. The Royal Infirmary, in Edinburgh and elsewhere in Britain, made a crucial contribution to the social transformation of the *patient-physician* relationship into the *physician-patient relationship*. Gregory thus confronted the task of writing medical ethics for both paying out-patients and Infirmary in-patients and in response to two different kinds of relationships, the patient-physician relationship and the incipient physician-patient relationship.

The institutional context with its rules and regulations and the marked social-class differences between physicians and patients created power for physicians that did not exist in the out-patient setting. This change occurred in concert with the growing intellectual authority of scientific medical knowledge. It is unlikely that the capital required to build and the income required to maintain the Royal Infirmary in Edinburgh would have been forthcoming had not their donors believed that the enterprise was worth supporting. By supporting the scientific medical enterprise these donors made it more successful. They put in place, without realizing it, the institutional basis to which scientific knowledge could be wed to create a profession of medicine in the ethical sense, i.e., as a fiduciary profession, *if and only if* the power and authority thus created could be directed and regulated with ethical justification. If directed toward self-interest, scientific knowledge and institutional power would not result in a fiduciary profession. If power and authority were directed to the interests of patient, a fiduciary profession would indeed result. Gregory's medical ethics thus forms the bridge from growing scientific knowledge and institutional power to a fiduciary profession, as we shall soon see.

Individuals who became ill and wanted admission to the Royal Infirmary had to obtain a ticket of admission from one of the Infirmary's benefactors (Risse, 1986, p. 19). These authorizations soon took standardized printed form, with the prospective patient's name simply filled in. Eventually, the forms were printed in the newspapers and widely distributed through the churches (Risse, 1986, p. 82). By 1770, the total number of admissions was 1,170 annually (Risse, 1986, pp. 46-47), with about 180 beds occupied in the house on any given day, this roughly 80% census being mainly a function of available funds (Risse, 1986, p. 84).

The Infirmary included a variety of wards, e.g., for seamen, soldiers, those with venereal disease, and a lying-in ward. The professional staff of the Infirmary included physicians appointed by a variety of titles and functions, surgeons, apothecaries, nurses, and lay staff. The latter exercised no small amount of power, since it fell to them to enforce the rules.

The teaching ward operated each year from November 1 through April 30 (Risse, 1986, p. 90), coincident with the University "session." This ward produced about 15% of the Infirmary's income (Risse, 1986, p. 36), primarily through the sale to students of tickets to attend ward-round and other forms of clinical teaching (Risse, 1986, p. 34). The Infirmary strictly regulated student contact with patients (Risse, 1986, p. 247). Diseases seen, treated, and taught about in the teaching ward included the following categories: infectious diseases (25% of total); respiratory disease (16% of total); diseases of the digestive system (11% of total); and neurological or mental diseases (10% of total), to name the four most common (Risse, 1986, pp. 256-257). These numbers reflect the newly emerging status of neurologic and mental illnesses, about which, more below.

The teaching ward afforded the students direct clinical observation under the supervision of their professors, exposure – "most of them for the first time" given their social backgrounds – to "the plight of lower-class patients as they interrogated and examined them," and education in the "skepticism of William Cullen" (Risse, 1986, p. 277), in the sense of 'skepticism' explained above, namely, as an essential element in the newly-emerging science of clinical reasoning in diagnosis and treatment.

Disease, we saw in our consideration of Gregory's development, involved departures from the observable norm of function. 'Cure' thus came to mean a "return to health" (Risse, 1986, p. 177). The strategic vagueness of this concept of cure resulted in a loose use of the term 'cured' in patient records.

> British hospital physicians applied the term "cured" with great liberality to any patient who appeared to be on the mend, even if the recovery was partial or temporary (Risse, 1986, p. 230).

In two volumes of Gregory's cases, "D^r cured" is by far the most common entry that Gregory made (RCSE C12, 1771-1772).

This analysis of the meaning of 'cure' provides a context for Gregory's concern about the label 'incurable' and its misuses. Gregory's position amounts to the claim that usually there is something that might be done to improve the patient's condition, even if that improvement falls short of complete recovery of normal function. The good that medicine can achieve for patients therefore is not some robust concept of the good, but an incremental, cautious concept of medical good as simply improving function to whatever degree the patient's disease permits, following, of course, cautions about the risks of doing so.

Drugs played a prominent role in medical therapeutics in the Royal Infirmary, along with many other modalities of treatment (Risse, 1986, pp. 189ff). Alcoholic beverages had their place, as well. Because of their cost, alcoholic beverages came under "strict professional control" (Risse, 1986. p. 225). We see here an instance of the power of managers, who insisted on such control and accountability. While physicians gained power in the institutional setting, they also came under institutional power and regulation.

> For the opportunity of closely observing large numbers of patients and becoming more proficient in clinical skills, hospital practitioners gave up some of the freedom and power they enjoyed in private practice. At the Edinburgh infirmary they owed their appointments to lay managers who demanded an oath of allegiance and promises to follow the institution's statutes and regulations. All budgetary decisions and authority to hire ancillary personnel were out of the physicians' hands. Fiscal constraints even affected treatment plans, including the quantity and quality of the diet, the need to prescribe in-house remedies [from the in-house apothecary], and limits on the use of beer and wine. Such bureaucratic rules tended to streamline and standardize medical treatments, in direct opposition to the practitioners' desire to assert their prerogatives in individualized care. The resulting tensions between hospital authorities and medical staff could never be totally avoided and were the source of innumerable conflicts (Risse, 1986, p. 185).

Gregory did not address these matters, although Thomas Percival did, thus writing the first secular, professional ethics of medical institutions (Percival, 1803). (See Chapter Four.) We can also see here the embryonic form of the set of ethical issues that have emerged in our time with the advent of managed care, in which authority shifts from the individual physician to the corporate managers.

The Infirmary experienced a mortality rate of about 3.5%. In part, this reflects a selection bias on the part of the Infirmary patrons and managers against serious diseases with high mortality rates, e.g., fever (Risse, 1986). The patrons of the Infirmary wanted their new institution to be successful and so they segmented their market. A voluntary, secular eighteenth-century institution would seem to have invented market segmentation based on institutional self-interest, not, as some might think, twentieth-century for-profit institutions! The teaching ward experienced the highest mortality rate of the wards, more than double the average mortality and well above its nearest rival wards for this dubious honor (Risse, 1986, p. 235).

Risse reports that critics found fault with the infirmary as an institution.

> ... some eighteenth-century critics pointed to the inherent inhumanity of the voluntary hospital system. A prime target was the avarice and insensibility of the staff, from the penny-pinching behavior of the administrators and the despotism of "Lady Matron" and "Lord Steward" to the greed and brutality of the nurses or other servants who dispensed their services only to those willing to bribe them (Risse, 1986, p. 24).

Charges of "cruelty" were brought against physicians and surgeons.

> "Gentlemen who arrived to the highest pitch of eminence in their profession use exceedingly harsh language and apparently unfeeling treatment to their patients," asserted one author. Another, writing under the pseudonym "Mac Flogg'em," sarcastically suggested to physicians that "you must manifest your authority by being as rigidly severe and contumacious as possible and exert every endeavor to render the situation of the afflicted poor as irksome and miserable as you can." The proper "pomp and Hauteur" could only benefit the institution by driving from the hospital burdensome patients "who had rather leave it half cured than to submit to your insolent and barbarous behavior" (Risse, 1986, pp. 24-25).

Patients came to the Infirmary with a combination of "fear and hope," Risse claims: fear that they were gravely ill and hope that the Infirmary could help them. They had to tell a convincing story in order to be admitted. They were sometimes met with "arrogance and condescension," as well as "exploitation and harassment" as patients and as research subjects (Risse, 1986, pp. 182-183, 184-185).

Risse reads Gregory to be using sympathy in his medical ethics lectures in response to these problems.

> John Gregory ... emphasized the importance of "sympathy," a human quality that allowed people to identify with each other's emotions. In

his view patient and physicians needed to become friends. The advantages accruing to both parties and to the therapeutic process in general was clearly recognized (Risse, 1986, p. 185).

Sympathy, as we saw above, involves more than identifying with another's emotions; sympathy creates an active moral bond of regard and response from the physician to the distressed patient. Sympathy, as we shall see in the next chapter, creates the physician-patient relationship as a *professional relationship*, modeled on friendship and the paternalism of friends toward each other.

> John Gregory's recommendations for the establishment of a bond of "sympathy" between patient and practitioner were part of the protective paternalistic attitude that transcended self-serving efforts toward improved relations with the public (Risse, 1986, p. 283).

Lindsay Granshaw provides a useful summary of the type of institution in which Gregory taught and of some of the problems to which he responded in his medical ethics:

> These hospitals, like their medieval predecessors, were set in the context of charity: they were for the poor, or more particularly, the 'deserving' poor. Social, not medical, criteria were to determine admission, and donors were given rights proportionate to donation to admit patients. Donors, as hospital governors, had far greater power over affairs in the hospital than the doctors whom they appointed. Patients received treatment free of charge, but first had to secure an admission ticket from a benefactor. The benefactors sought to exclude the socially 'undeserving' such as prostitutes or drunks, but also a range of other 'inappropriate' cases. The hospital, if it was to use its charity well, was intended to help as many of the poor as possible. From this, it was concluded that incurables should not be admitted, nor should fever cases, lest fever spread to other patients. Once in hospital, patients had to obey a strict set of rules; and if disobedient, they would be ejected from the institution. Convalescent patients had to help their sicker neighbours, mend sheets, clean the ward, or tend the fire (Granshaw, 1993, pp. 1185-1186).

C. Nervous Diseases

As noted above, the Royal Infirmary admitted and treated many patients with neurological and mental illness, what were often together called "nervous diseases." Robert Whytt, who overlapped with Gregory one year at the University of Edinburgh and whom Gregory succeeded as First Physician to His Majesty in Scotland, played a leading role in giving legitimacy to nervous diseases.

Whytt focused on diseases that should properly speaking be classified as nervous. He notes that most "complaints" could be labeled "nervous." He wants a sharper clinical and scientific focus.

> The design, however, of the following Observations is far different. In them, it is only proposed to treat of these disorders, which in a *peculiar* sense deserve the name of *nervous*, in so far as they are, in great measure, owing to an uncommon delicacy or unnatural sensibility of the nerves, and are therefore observed chiefly to affect persons of such a constitution (Whytt, 1765, p. iv).

Nervous diseases, in turn, are the result of abnormalities of "that sympathy which obtains between the various parts of the body" (Whytt, 1765, p. v).

Whytt's understanding of sympathy and "sympathy of the nerves" (Whytt, 1765, p. v) builds on the work of the physiologists of sympathy (whose work we considered earlier in this chapter). Whytt lays down his principles of sympathy.

> Our bodies are, by means of the nerves, not only endowed with feeling, and a power of motion, but with a remarkable sympathy, which is either general or extended through the whole system or confined, in a great measure, to certain parts (Whytt, 1765, p. 9).

Whytt then adduces ample observational evidence in support of this systemic sympathy.

Nerves – not to be mistaken for how we now think of them – communicate sympathy in both sound and diseased states. Nerves also communicate the passions to the body. Nerves, as part of the body, fall prey to diseases.

> When the feeling of the nerves is too acute; disagreeable or painful sensations will be excited in the body ... In such a condition of the nervous system, the passions of the mind, errors in diet, and changes in

heat and cold, or the weight and humidity of the atmosphere, will be apt to produce morbid symptoms; so that there will be no firm or long continued state of health, but almost a constant succession of greater or less complaints (Whytt, 1765, pp. 88-89).

Nervous ailments are thus real, a claim the support for which Whytt believes he has ample and convincing experimental and clinical observations, which he reports. Thus, a physician should always include symptoms such as "PALPITATIONS or trembling of the heart" (Whytt, 1765, p. 101), "giddiness" (Whytt, 1765, p. 100), and the like (in their differential diagnosis, we would now say). Whytt goes on to claim that hysteria in women and hypochondria in men are not two diseases, but one. The difference between the two involves only the fact that women have a uterus from which sympathetic changes can emanate and that men do not. This source of nervous illness in women can sometimes lead to more severe symptoms than those observed in men.

William Cullen continued and expanded Whytt's concept of nervous diseases. Nervous diseases became part of Cullen's nosology, "by making all diseases the general product of excess or deficient 'nervous force' and by incorporating specific neurological disorders as one of the four fundamental categories of disease" (Brown, 1993, p. 447).

Whytt also takes note of interpersonal sympathy, as did others who studied the physiology of sympathy before him, what he calls "a still more wonderful sympathy between the nervous systems of different persons, whence various motions and morbid symptoms are often transferred from one to another, without any corporeal contact or infection" (Whytt, 1765, p. 219). This phenomenon is owed to the effect on the "*sensorium commune*" by "different impressions," although Whytt admits that he cannot explain the phenomena he reports, e.g., reddening face accompanying a sense of shame (Whytt, 1765, p. 220).

Hume, of course, does have the needed explanation: the double relation of impressions and ideas that we examined above. Whytt can be read as adding to Hume the reminder that interpersonal sympathy is not just ideational, it is physical, involving the nervous system as the mediator of sympathy. It is in this physiologic sense that Gregory understands sympathy, as did Hume. It is clear that Gregory takes from Whytt a commitment to the new science of neurology and to the clinical reality of nervous diseases. He takes up the ethical challenges posed to physicians by patients with nervous diseases, as we shall see in the next chapter.

D. The Beattie-Hume Controversy

We have seen so far the considerable debt that Gregory owes to Hume concerning philosophical method and the principle of sympathy. Gregory also knew and socialized with Hume. As Tytler puts it, "Dr Gregory lived in great intimacy with most of the Scottish literati of his time," including Hume (Tytler, 1788, p. 84). Gregory did not accept Hume's attacks on religion, more specifically, on a creator God. This will become clearer by consideration of what I shall call the Beattie-Hume controversy.

James Beattie was in his time an important figure, but his stature has waned in the history of philosophy. Beattie attacks Hume in 1770 for irreligiosity (Beattie, 1976). Breaking his normal pattern of silence in response to his critics, Hume responded publicly, leading to a crossfire in which Gregory was caught as Beattie's friend and mentor, and as Hume's friend and student. How Gregory sorted out matters in this controversy provides an interesting insight into his relationship with Hume and Hume's philosophy.

E.C. Mossner, Hume's biographer, notes that Beattie's attack "was chiefly responsible for disturbing the philosopher's tranquility" (Mossner, 1980, p. 577). Mossner portrays Beattie's attack in stark terms: " ... it was Beattie's intention to arouse the emotional prejudices of his readers" (Mossner, 1980, p. 577).

Mossner quotes from a letter from Gregory to Mrs. Montagu on June 3, 1770 about Beattie's attack:

... but the Authors [Beattie's] Zeal for his Cause has made him [Beattie] treat Mr Hume with a degree of Severity which I think had better been spared. I detest Mr Hume's Philosophy as destructive of every principle interesting to Mankind & I think the general spirit that breathes in his History unfavourable to Religion and Liberty, tho in other respects one of the most animated, entertaining & instructive Historys I have ever read. But I love Mr Hume personally as a Worthy and agreeable Man in private Life, & as I believe he does not know & cannot feel the mischief his writings have done, it hurts me extremely to see him harshly used (HL MO 1078, 1770, p. 2; Mossner, 1980, p. 580).

On the face of it, this letter indicates that, while Gregory may have liked Hume and enjoyed his social company, Gregory rejects Hume the philosopher. As John Dunn puts it, Gregory "was certainly strongly opposed

to the philosophy of Hume ... " (Dunn, 1964, p. 128). Stewart takes a
more pointed view of the matter, noting that Beattie and Reid

> ... were engaged in combating the teaching of David Hume, which had
> become very fashionable, and Gregory, though much attached to David
> Hume as a man, feared him as a teacher, and dreaded the growth of
> skepticism which marked the time (Stewart, 1901, p. 119).

Stewart adds that "Gregory's mind was deeply religious, but it was of that
sort that lives more by meditation than church-going" (Stewart, 1901, p.
121). Let us now consider this claim about Gregory's views of Hume's
philosophy.

We have seen already that Gregory does not reject skepticism *simpliciter*, because skepticism properly understood provides an essential investigational and analytical tool in Baconian experimental method. Gregory
formed himself intellectually around Baconian science and method as a
medical student and never departed from its basic tenets. Thus, it must be
excessive skepticism, skepticism not properly disciplined that is the
problem, just as Reid had argued before the Aberdeen Philosophical
Society.

Other evidence from the Wise Club record – the Club continued to
meet until 1773 – supports a reading that Hume's irreligiosity was the
problem. Beattie writes to Dr. Blalock in 1770, referring to Reid and
Campbell:

> I know likewise that they are sincere, not only in the detestation they
> express for Mr. Hume's irreligious tenets, but also in the compliments
> they have paid to his talents; for they both look upon him as an extraordinary person; a point on which I cannot disagree with them
> (Forbes, 1824, Vol. I., p. 123).

We know, however, that Reid and Campbell accepted Hume's principle
of sympathy and skeptical method, albeit in Reid's case with some revisions. This may be why Beattie goes on in this letter to complain that
Reid and Campbell should have gone further in "their researches" and
"expressed themselves with a little more firmness and spirit" (Forbes,
1824, Vol. I, p. 123). They had no reason to do so, however. Beattie did,
and more than made up for what he took to be timidity on the part of Reid
and Campbell.

Gregory's concern turns out to be Hume's irreligiosity. Beattie writes
to Mrs. Montagu on June 25, 1779, about an exchange between Gregory

and Hume when the latter was near death.

> Yet Mr. Hume must have known, that, in the opinion of a great major-
> ity of his readers, his reasonings, in regard to God and Providence,
> were most pernicious, as well as most absurd. Nay, he himself seemed
> to think them dangerous. This appears from the following fact, which I
> had from Dr. Gregory. Mr. Hume was boasting to the Doctor, that,
> among his disciples in Edinburgh, he had the honour to reckon many
> of the fair sex. 'Now tell me,' said the Doctor, 'whether, if you had a
> wife or daughter, you would *wish* them to be your disciples? Think
> well before you answer me; for I assure you, that, whatever your an-
> swer is, I will not conceal it.' Mr. Hume, with a smile, and some hesi-
> tation, made this reply: 'No, I believe that skepticism may be too
> sturdy a virtue for a woman.' Miss [Dorothy] Gregory will certainly
> remember, that she had heard her father tell this story (Forbes, 1824,
> Vol. II, p. 35).

Hume's skepticism taken into matters religious is "too sturdy a virtue for
a woman." Skepticism in other matters may not be. As we shall see when
we examine Gregory's *Legacy*, he believes that religion, i.e., religious
practice and devotion, was important to his daughters, though not to him
in his own life.

Gregory writes to Beattie after he publishes his attack on Hume. He
opens with a rebuke of Beattie: "Much woe has your essay wrought me"
(Forbes, 1824, Vol. I, p. 164). Gregory reports that Hume was quite angry
at Gregory, who had read Beattie's manuscript: "As it was known that the
manuscript had been in my hands, I was taken to task for letting it go to
press as it stands" (Forbes, 1824, Vol. I, p. 164). Gregory then criticizes
Beattie, mainly for style, noting that, although Beattie writes with
"warmth" and a style "uncommon on such subjects," Beattie lacks the
diffidence, the *measured*, forceful voice, that should mark philosophical
work (Forbes, 1824, Vol. I, p. 164).

> I wished, at the same time, some particular expressions had been sof-
> tened; but denied there being any personal abuse. In one place, you
> say, '*What does the man mean?*' This, you know, is very contemptu-
> ous. In short, the spirit and warmth with which it is written has got it
> more friends and more enemies that if it had been written with that
> polite and humble deference to Mr. Hume's extraordinary abilities,
> which his friends think so justly his due. For my own part, I am so
> warm, not to say angry, about this subject, that I cannot entirely trust

my own judgment; but I really think that the tone of superiority assumed by the present race of infidels, and the contemptuous sneer with
which they regard every friend of religion, contrasted with the timid
behavior of such as should support its cause, acting only in the defen-
sive, seems to me to have unfavourable influence. It seems to imply a
consciousness of truth on the one side, and a secret conviction, or at
least diffidence, of the cause, on the other. What a difference from the
days of Addison, Arbuthnot, Swift, Pope & Co. who treated infidelity
with a scorn and indignation we are strangers to. I am now persuaded
the book will answer beyond your expectations. I have recommended it
strongly to my friends in England (Forbes, 1824, Vol. I, pp. 164-165)

Gregory opposes skepticism taken to the extreme of attacking religion,
because, as we saw above, he apparently thinks that there is observational
evidence for a creator God – despite Hume's argument to the contrary.
Moreover, religion will play a key role in his moral instructions to his
daughters in his *Legacy*. Taking away such a foundation for female virtue
may well constitute the "unfavourable influence" that Gregory fears.
Gregory also expresses disappointment that no one has mounted an
effective defense of religion, implying perhaps that Beattie, in his intem-
perate style, had failed to do so.

The controversy between Beattie and Hume continued to simmer, and
Gregory writes again to Beattie on November 26, 1771. Gregory suggests
to Beattie that he opposes Hume, not because the latter is a "bad meta-
physician, but because he has expressly applied his metaphysics to the
above unworthy cause," namely attacking religion (Forbes, 1824, Vol. I,
p. 180). Beattie, Gregory goes on, does Hume no injustice in Beattie's
treatment of Hume on religion. But this topic is not the whole of Hume's
philosophy, the rest of which is presumably left intact after Beattie's
attack. Instead, all of the documentary evidence that we have considered
so far supports this reading.

Gregory does *not* reject Hume's philosophical method or the principle
of sympathy; he does reject Hume's *extreme skepticism*. In this Gregory
seems very close to Reid. Indeed, in his *Lectures* Gregory uses language
regarding sympathy that is much closer to Reid's than to Hume's. But, we
have seen, Reid adopts Hume's account of the principle of sympathy and
so does Gregory. Perhaps Gregory's choice of words in his *Lectures* was
made to gain him some distance from Hume and thus permit his medical
ethics to stand on its own and not be caught up in the anti-Humean views
of the day, a view which some of his medical students probably shared.[14]

Moreover, Gregory defends religion in *Comparative View* and religion plays an important role in his *Legacy*. Gregory thus has intellectual and personal commitments regarding religion that he wants to retain, even if he does not practice any religion himself. This position is entirely consistent with an Enlightenment commitment to relieve man's estate by improving religion itself.

E. Correspondence Between Gregory and Mrs. Montagu

While he lived in Edinburgh, Gregory frequently visited with Mrs. Montagu. He also sent his daughters to stay with her for periods of time, with Mrs. Montagu becoming a second mother of sorts to Dorothy, or "Dolly," as she was affectionately called. Gregory also saw to the education of his daughters through such visits, because Mrs. Montagu replicated Bulstrode in her own summer home. Dorothy Gregory lived with Mrs. Montagu after Gregory's death in 1773 (Myers, 1990, p. 249). Mrs. Montagu also visited Edinburgh and made a tour of Scotland with Gregory as her "knight errant and escort" (Blunt, 1923, Vol I, p. 143).

In a surviving fragment of a letter sent apparently after this visit Gregory writes to Mrs. Montagu the following:

> You have gained the Hearts of every body in Scotland. If you [corner of page missing] Gentleman I do not know how I should have bore it, but as you are a Lady, it gives me most sincere pleasure, because, as Lady Forbes has told you at great length, the Love & Esteem you procured in Scotland reflects some Honour on myself. This among many other reasons is an argument for Ladys chusing their friendships among the Men rather than with their own Sex. Our interests do not in the least interfere, & our own Vanity is concerned in supporting a Character which honours us with any mark of regard. I cannot account for the Love you have gained from your own Sex from any other principle than your bearing your facultys so meekly (HL MO 1083, after 1766?, p. 1).

Women in the company of self-assured men can display their "facultys," i.e., intellectual capacities, without fear of putting off an interlocutor. The company of men, as Mrs. Montagu herself demonstrated with her Bluestocking gatherings, played an important role in the education of women in an era when formal educational opportunities for them did not exist. We also note in this letter, as we shall in Gregory's *Legacy*, a reference to the concept of *asexual virtue* in women. This will become very important in Gregory's medical ethics, especially concerning sexual relationships between physicians and patients.

Gregory writes to Mrs. Montagu on October 2, 1766, thanking her for her "approbation of my little Work" (HL MO 1065, 1766, p. 1), namely *Comparative View*. He notes in this letter how the "finest Genius & most Elegant Taste" have been associated with "good hearts & good tempers" (HL MO 1065, 1766, p. 1), a theme by now familiar. He continues:

> But I must except many of them who have been engaged in the pursuits of fame & Ambition: Such pursuits have a tendency to corrupt the heart & sour the temper, by the clashing of Interest & frequent disappointments they subject us to (HL MO 1065, 1766, p. 1).

The pursuit of self-interest is antithetical to a good heart, because it closes us to others in a clash – not sympathy – of interests and therefore hinders the proper exercise of sympathy or humanity, a theme of prominence in *Comparative View*, as shall see shortly.

In a letter from October 11, 1766, Gregory makes it plain to Mrs. Montagu that she serves as an exemplar to him:

> There is a nobleness in your Mind which sets you above Ceremony & the Consciousness You must have of the Benignity of your own Heart should actually make it open & unreserved (HL MO 1066, 1766, p. 2).

Here Gregory sees in Mrs. Montagu the synergy of intellectual and moral virtues, the concept of which he had developed in Aberdeen. Nobility of mind, presumably a mind committed to learning and not ambition, leads to an open and unreserved heart, i.e., a heart open to the workings of sympathy. We shall see that this forms the template for Gregory's arguments in his *Lectures*, in which women of learning and virtue, such as Mrs. Montagu, also serve as exemplars.

In a letter from November 18, 1766, Gregory amplifies this theme of women as exemplars.

> For my own part I have always had a particular respect, rising to a sort of Veneration; for an Old Maid, who was so from choice & because she had not met with the Man whom her Heart could approve of; & whenever any Such play the fool, & depart from the Maxims I formerly thought them directed by, it gives me a very sensible pain. I think, I am sure I have always felt, that your Sex are the most interesting Subject in the World to ours, & therefore whatever lowers you in my Estima-

tion, hurts me because it hurts my pleasure & shakes those Sentiments of Esteem for your Sex, which however romantic, I have always enjoyed very Solid Satisfaction in indulging (HL MO 1068, 1766, p. 2).

Here Gregory displays the steadiness of his commitment about his views on women, which date, as we know, at least from his medical student days in Leiden. We also see here the negative exemplar that some women can be, namely, the Mrs. Modishes of this world who cause Gregory distress, out of sympathy for what they could be – a Mrs. Montagu.

On October 25, 1766, Gregory writes to Mrs. Montagu to thank her for the experiences and education that she provided to his daughters during their stays with her. Most of his surviving letters to Mrs. Montagu contain such expressions of gratitude, indicating that Mrs. Montagu functioned as a surrogate mother to his daughters. This letter addresses a particular experience Mrs. Montagu had given to Gregory's children.

I am greatly delighted with your allowing them to go sometimes among the Country people on their Joyous occasions. I have always thought that there was a Distance which most people of better fashion keept at from the lower rank of mankind, which was productive of very bad consequences. We can feel very little of the distresses of a set of people whome we never knew & whose pleasures & sorrows we are equally strangers to (HL MO 1067, 1766, p. 1).

Medical students at Edinburgh, of course, came from the ranks of "better fashion" and they may not have had their hearts open to the distresses of the working poor who became their patients in the Royal Infirmary, perhaps thinking of them as workers or servants from whose lives their own are greatly distanced. Here we see Gregory expressing a concern about the potentially negative effect of social class and stratification by class on the workings of sympathy. We also learn how increasing social stratification disconnected from the old paternalism can pose a serious threat to the workings of the social principle and thus to social cohesion and even national identity.

Gregory writes to Mrs. Montagu on January 3, 1767 about his daughter Dorothy, who had spent so much time with Mrs. Montagu.

Dolly I find has been so much affected at parting with you, that I have not chose to ask her many Questions on that Subject. That girl has a Heart too delicately sensible for her own future Quiet. I wish I could harden it sufficiently, without its appearing to be so. Sensibility &

Delicacy are very unworthily thrown away upon our Sex. Few Men know what they mean & therefore are perpetually shocking them, even when they do not mean to do so. The few men who have any Delicacy require more attention to keep their Hearts than any Mans Heart is worth. Delicacy is naturally Capricious, easily hurt, fond of Variety, so that it is a very dangerous Quality to have any connexion with, if not under the Controul of great Humanity, Good Sense & very steady Principles (HL MO 1071, 1767, pp. 1-2).

Women of learning have the capacity to exercise the "Controul of great Humanity" and "very steady Principles" over their sensibilities and so learning cultivates the virtue of self-discipline. This prevents sensibility and heart from running to excess and making one's life a misfortune. Mrs. Montagu herself provided an example for Gregory's daughters to follow, as he writes on February 10, 1767:

A thousand thanks for all your tenderness & attentions to my Girls. I Assure you they have not been thrown away. They are greatly improved in every respect & are constantly speaking of you with the most gratefull Affection. I observe with Pleasure that Dolly, in particular, has picked up a great number of your Sentiments, & imbibed from you many nameless Delicacy's, the full Extent and Foundation of which she is not yet capable of understanding. But I have always thought that much might be done to form the heart & even the Taste long before that Period when Reason exerts her fullest power. I have great confidence in early impressions of Virtue, & in early habits of Delicacy, of propriety, of Study, of Application to whatever is necessary, of Suffering & Self Denial. If these are not acquired when the heart is most sensible & the Mind most flexible, they will scarcely be attained when the Passions come to exert their Influence, as Reason and Philosophy are then too feeble of themselves to oppose them (HL MO 1072, 1767, pp. 2-3).

Suffering, he explains in a letter from August 14, 1768, "softens the heart & makes it sympathize more tenderly with the distresses of others" (HL MO 1075, 1768, p. 1).

When we are ill our sensibility is affected.

... but Sickness & Inactivity, at the same time that they increase Sensibility, they deprive us of the only means of tempering or resisting its bad effects (HL MO 1077, 1769, p. 1).

The ill can thus tax the patience and due attention of those charged with their care. Sympathy for the suffering and distressed serves to blunt the potentially irritating effect of a patient's bad behavior on a physician.

In this correspondence, then, we find further evidence that Gregory did indeed think that women of learning and virtue were moral exemplars, personified by Mrs. Montagu herself. Intellectual virtues of disciplined inquiry and thinking lead to disciplined passion and sensibility. As we discipline our passions we incrementally, progressively improve ourselves.

X. GREGORY'S WRITINGS:
COMPARATIVE VIEW, A FATHER'S LEGACY, AND PRACTICE OF PHYSIC

In addition to his writings on medical ethics, Gregory wrote three books: *A Comparative View of the State and Faculties of Man, with Those of the Animal World* (1765), *A Father's Legacy to His Daughters* (1774), and *Elements of the Practice of Physic, for the Use of Students* (1772a). The second of these appeared posthumously, but it would appear that Gregory had been working on it for some time before his death.

A. Comparative View

It seems apparent from the discussion in the previous section that the appearance of *Comparative View*, published first anonymously, with the help of Lyttleton, and then under Gregory's own name in later editions, was part of the trend of self-promotion described by the Porters (Porter and Porter, 1989). The publication of this book was meant to establish Gregory as an important intellectual figure, worthy of appointment to his country's best university. As we saw in the quotation from Stewart (1901) in the previous section, the publication of Gregory's first book had the desired effect.

Recall that Gregory, as a medical student, in his "Notes for a Medical Society" (AUL 2206/45, 1743), had already noted the promise of comparative studies. He also sets out as one of his four "Capitall Enquirys" the "Improvement of our Nature" (AUL 2206/45, 1743, p. 6). In *Comparative View* he takes up this youthful agenda (Gregory, 1772a). Gregory began work on *Comparative View* while he was in Aberdeen and presented his text in Discourses to the Aberdeen Philosophical Society (*vide* Section VII). Gregory takes the view that there are uniform, non-variant principles of human nature and its functions and that these principles can be discovered by means of experimental, Baconian method. He holds, also with the physiologists of his time, that body and mind "are so intimately connected, and such a mutual influence on one another, that the constitution of either, examined apart, can never be thoroughly understood" (Gregory, 1772a, p. 4). Thus, Gregory rejects the view that the body is merely a "machine," as Boerhaave had held, in favor of Stahl's view, who "united the philosophy of the Human Mind, with that of the body" (Gregory, 1772a, p. 6).

Gregory goes on to take the view that in some respects – notably the possession of instinct – human beings do not differ markedly from animals, while in others – notably the possession of reason and the moral sense – humans do differ from animals.

> But above all, they [humans] are distinguished by the Moral Sense, and the happiness flowing from religion, and from the various intercourses of social life (Gregory, 1772a, pp. 11-12).

In a way entirely typical of Scottish moral sense philosophers Gregory holds that there are two principles of mind, reason and instinct. Reason, "a weak principle" (Gregory, 1772a, p. 15), receives short shrift. On the other hand, instinct, in the form of the social principle, guides morality.

> The next distinguishing principle of mankind, which was mentioned, is that which unites them into societies, and attaches them to one another by sympathy and affection. This principle is the source of the most heart-felt pleasure which we can ever taste (Gregory, 1772a, p. 86).

The task of improving our nature, of improving man's estate, involves the proper development and exercise of the social principle and the other principles of instinct, with reason subordinate to instinct and serving as a corrective on it.[15]

> Reason, of itself, cannot, any more than riches, be reckoned an immediate blessing to Mankind. It is only the proper application of it to render them more happy that can entitle it to that Name. Nature has furnished us with a variety of internal Senses and Tastes, unknown to other Animals. All these are sources of pleasure if properly cultivated, but without culture, most of them are so faint and languid, that they convey no gratification to the mind. This culture is the peculiar prov-

ince of Reason. It belongs to Reason to analyze our Tastes and Pleasures, and after a proper arrangement of them according to their different degrees of excellency, to assign to each that degree of cultivation and indulgence which its rank deserves and no more. But if Reason, instead of thus doing justice to the various gifts of Providence, be unattentative to her charge, or bestow her whole attention on one, neglecting the rest, and if, in consequence of this, little happiness be enjoyed in life, in such a case Reason can with no great propriety be called a blessing (Gregory, 1772a, pp. 64-65).

We improve our nature, first, by recognizing that reason cannot be the foundation of morality; only instinct can serve this vital role. Instinct, however, requires proper development and regulation, else it lead to excess in the form of enthusiasm or insufficiency in the form of *dissipation*. Reason plays a vital role as an adjunct principle, regulative of instinct in the form of sympathy, the social principle, taste in the arts, and religion. The result will be properly cultivated instinct, i.e., virtues, that, in turn, serve as the basis of morality.

Gregory's confidence in nature is complete. Note that he uses the personal pronoun, "her," and not the impersonal pronoun in reference to nature.

Nature brings all her works to perfection by a gradual process. Man, the last and most perfect of her works below, arrives at his by a very slow process (Gregory, 1772a, pp. 55-56).

By contrast, cities, the man-made environment in which the majority of the population of Scotland was by then living, "are the graves of the human species" (Gregory, 1772a, p. 49). This is not just owing to crowded living conditions, which certainly existed for the less well-to-do, but also to the effects of commerce, which was becoming the principal activity and purpose of cities. He contrasts commerce with the "plain road of nature" guided by the social principle (Gregory, 1772a, p. x). Commerce promotes a "passion for riches, which corrupts every sentiment of Taste, Nature, and Virtue" (Gregory, 1772a, p. x), because virtue and religion yield to money (Gregory, 1772a, p. xii).

Comparative View became widely known and influential for its views on child rearing. Gregory notes with thunderous disapproval the staggering 60% mortality rate of children before their twentieth year, a phenomenon "unknown among wild animals" (Gregory, 1772a, p. 20). This mortality results from the diseases of infants and children, which in turn result from "the unnatural treatment they meet with, which is ill suited to the singular delicacy of their tender frames," a situation he characterizes as "evil" (Gregory, 1772a, pp. 20, 22). Children should be nursed as soon as they are born, rather than after a delay of several days. Nursing should continue until natural weaning occurs, usually after a year and a half or so. An infant, he says, should suck "as often as it pleases. It is then under the peculiar protection of Nature, who will not neglect her charge ... " (Gregory, 1772a, p. 32). Immediate and prolonged nursing promotes bonding with the newborn child and also increases the interval between pregnancies. Gregory was concerned that shortened periods of nursing led to pregnancies every year, which aged women prematurely, a result on which Gregory heaps his scorn, making him an advocate for women's health. Gregory apparently practiced what he preached; his wife's children were delivered at intervals no shorter than eighteen months (Lawrence, 1971, Vol. I, p. 167).

Children should not be constantly medicated, which leads to overmedication and weakens the child's constitution. Nor should children be kept wrapped in layers of clothing, which keeps them too warm. The reason is simple: "Nature never made any country [Scotland included] too cold for its own inhabitants" (Gregory, 1772a, p. 52). Gregory also attacks what he calls the "unnatural confinement" of children; they should be allowed more freedom to be outdoors. Crying of infants is neither unnatural nor to be discouraged, for "the cries of a child are the voice of Nature supplicating relief. It can express its wants by no other language" (Gregory, 1772a, p. 40). Cries, it also seems, naturally excite a sympathetic response of parents to the distress of their infant.

Children are not capable of abstract reasoning and so their early education should follow the "plan of Nature in bringing Man to the perfection of his kind ... to assist the successive openings of the human powers, to give them their proper exercise, but to take care that they never be over-charged," none of which the education of children then did, he thunders. Children should not be prematurely made into adults (Gregory, 1772a, p. 57). He expresses his educational philosophy in language that will appear at the core of his medical ethics. Education requires:

> such inflexible steadiness, – and at the same time, so much tenderness and affection, as can scarcely be expected, but from the heart of a parent (Gregory, 1772a, p. 61).

It is hard to resist the conclusion, given what he wrote to Mrs. Montagu upon his wife's death, that Gregory's exemplars here include his own wife and Mrs. Montagu. The goal of childhood education is to produce a morally well formed individual who will be able to resist "the gradual decay of the more humane and generous feelings of the heart," a decay that if unchecked will "leave us in a more helpless and wretched situation than that of any Animal whatever, ... surely the most humbling consideration to the pride of Man" (Gregory, 1772a, pp. 62-63).

The well-formed person, with properly cultivated faculties, exhibits this combination of steadiness and tenderness in taste and religion. The improvement of taste involves the "improvement of the powers of the imagination" (Gregory, 1772a, p. 104). Imagination involves the creative combination of ideas, pictorial forms, or musical notes. Reflecting perhaps his involvement in the Musical Society in Aberdeen, Gregory goes on at some length about music.

> Music is the Science of sounds so far as they affect the Mind. Nature, independent of custom, has connected certain sounds with certain feelings of the mind (Gregory, 1772a, p. 113).

> One end of Music is to communicate pleasure, but the far nobler and important is to command the passions and move the heart. [In the latter respect, music is] one of the most useful arts in life (Gregory, 1772a, pp. 114-115).

So, for example, fugues are not a preferable musical form, because they are more works of "great perfection" mathematically than they are works of the passions or feelings of the mind – this said by the scion of Scotland's most famous family of mathematicians (Gregory, 1772a, p. 140)!

He also opposes a fixed and systematic approach to literary criticism, favoring instead:

> ... assistance to the operations of Taste; as giving proper openings for the discernment of beauty, by collecting and arranging the feelings of Nature, they promote the improvement of the fine arts (Gregory, 1772a, p. 167).

Perhaps he had in mind Mrs. Montagu's literary criticism of Shakespeare.

In sum, the cultivation of taste and imagination in the literary, performing, visual, and musical arts should not be based on the demands of reason but, instead, heart, which is associated with "every softer and more delicate sentiment of humanity" (Gregory, 1772a, p. 164). Here he clearly associates "heart" with female virtues. On this basis he asserts that "a good Taste and a good Heart commonly go together," reflecting "the secret sympathy and connection between the feelings of Natural and Moral Beauty" (Gregory, 1772a, pp. 192, 193). The study of nature leads to and supports the study and appreciation of beauty, because the principles of both are the same and discoverable: the proper cultivation of instinct or the passions. That proper cultivation should result in a well-developed heart, i.e., "every softer and more delicate sentiment of humanity," of which women of learning and virtue provide the exemplar.

He brings this same analysis to religion, "that principle of Human Nature which seems in a peculiar manner the characteristic of the species" (Gregory, 1772a, p. 194). There is "something soothing and comfortable in a firm belief" in a provident God and the immortality of the soul (Gregory, 1772a, pp. 194-195). He rejects as genuine religion both superstition and the use of religion "to deprive Mankind of their most valuable privileges, and to subject them to the most despotic tyrany" (Gregory, 1772a, p. 197). Here Gregory shows his antipathy for the tight connection between the church and political power. And so it is no surprise that Gregory did not worship in any church and raised his daughters in the Episcopal Church, not the Church of Scotland.

He articulates his philosophy of religion in the following terms:

Religion may be considered in three different views.

1. As containing doctrines relating to the being and perfections of God, his moral administration of the world, a future state of existence, and particular communications to Mankind by an immediate supernatural revelation.

2. As a rule of life and manners.

3. As the source of certain peculiar affections of the Mind, which either give pleasure or pain, according to the particular genius and spirit of the Religion that inspires them (Gregory, 1772a, pp. 210-211).

Natural religion is "principally concerned" with the first and this is the foundation of the other two (Gregory, 1772a, p. 211). He opposes systematic views of religion, because a flaw in any one respect can bring the entire system down, because of the "gloom" of temper that accompanies attempts to follow the necessarily "perplexed and thorny path" of systematic reasoning, and because systematic theological views promote intolerance.

The second sense of religion "is very extensive and beneficial" and deals with matters of the heart (Gregory, 1772a, p. 219). The third aspect of religion "comprehends the devotional and sentimental part of it" and this depends on the "liveliness of Imagination and sensibility of Heart" (Gregory, 1772a, p. 230).

> The devotional spirit united to good sense and a chearful temper, gives that steadiness to virtue, which it always wants when produced and supported by good disposition only. It corrects and humanizes those constitutional vices which it is not able entirely to subdue; and though it may not be able to render men perfectly virtuous, it preserves them from being utterly abandoned. It has the most favourable influence on all the passive virtues; it gives a softness and sensibility to the Heart, and a mildness and gentleness to the Manners; but above all, it produces an universal charity and love to Mankind, however different in Station, Country, or Religion (Gregory, 1772a, p. 233).

"Good dispositions" thus do not alone constitute virtue; they must be steadied by intellectual discipline. Religion provides such a discipline, but not in the form of the threat of punishment as Kames had argued before the Aberdeen Philosophical Society. Devotional exercises train one's heart in the softer, tender virtues exemplified by women of learning and virtue. It thus comes as no surprise that Gregory thinks religion is very important for women. Recall, again, the letter he wrote to Mrs. Montagu upon his wife's death and his remarks there on his wife's devotional life and how important this was to him. In his medical ethics he retains the concept that there are "good dispositions," namely, of sympathy, and that these must be trained and disciplined by steadiness in the care of the sick.

Gregory's argument against religious infidelity plays an important role in his philosophy of religion.

> Absolute Infidelity, or settled Scepticism in Religion is not proof of a bad Understanding of a vicious disposition, but is certainly a very strong presumption of the want of Imagination and sensibility of Heart. Many philosophers have been Infidels, few Men of taste and sentiment (Gregory, 1772a, p. 201).

Infidelity would be destructive of virtue, especially for women who lived their lives for the most part in their homes and have no social outlet other than religion, for the most part, to sustain their moral lives. The professional man, living a public life, presumably could be a man of taste and sentiment with proper moral training and so didn't need the second and third aspects of religion, only the first.

Gregory's first sense of religion can be established, it seems, on Baconian grounds. With Bacon he accepts the argument from design for concluding that God exists. God originates both nature and the order that we observe in nature. God's works are good – they flow from his perfect nature – and so the order observed in nature, including the principles of human nature, is ordered to the good. This much can be established on secular, Baconian, philosophical grounds and provides, it seems, the source of Gregory's enormous confidence in the order of nature. This confidence rings through, for example, in his views on child-rearing, the "hospitable" climate of Scotland, and nature's capacities to correct conditions of disease in the direction of health. The last two elements of the first "view" of religion, and the second and third "views" are more theistic in character.

I can now state more precisely the role of religion for men and women, according to Gregory. Men have access to forms of moral discipline outside the home and they are required by secular, Baconian philosophical argument to be deists. If indeed men do find a sustaining moral discipline, they do not require the Kamesian correction of possible loss of salvation in a "future state of existence." Women, because they have no social outlets – there was no Bluestocking Circle in Edinburgh – yet are social creatures by nature and require religion – properly understood in its three "views" – as a source of moral discipline, a "rule of life and manners" sustained by devotion, as in the case of Mrs. Gregory (HL MO 1064, n.d.). Baconian, secular philosophy teaches by observation that enthusiastic religions harm the "affections of mind" and that the moral discipline of moderate religions prevents dissipation and false "affections of mind."

As we shall see in the next chapter, Gregory understands medicine in analogy with religion. When rightly grounded – in properly regulated sympathy expressed in tenderness and steadiness and in the intellectual virtues of diffidence and openness to conviction – medicine provides "rules of life and manners" that regulate "certain peculiar affections of the Mind," namely sympathetic responses to the pain and suffering of the sick.

In *Comparative View* Gregory further develops his views on women. We have noted his concern for women's reproductive health and the deleterious health effects of pregnancies spaced too closely together. He explicitly expresses the view that women are moral exemplars, because they exhibit the very virtues of heart that all of us should cultivate in the moral life. When he addresses the differences between the sexes (what we now call genders) he says of women: "They possess, in a degree greatly beyond us [i.e., men], sensibility of heart, sweetness of temper, and gentleness of manners" and the "sentiments of propriety" (Gregory, 1772a, pp. 93, 98). Reflecting, perhaps, on Mrs. Montagu's *converzatione* in London he writes the following:

> The fair sex should naturally expect to gain from our [men's] conversation, knowledge, wisdom and sedateness; and they should give us in exchange humanity, politeness, chearfulness, taste, and sentiment (Gregory, 1772a, p. 99).

Men need these virtues of heart from women because men live public lives, daily exposed to the corrupting influence of the pursuit of self-interest in commerce and city life and to constant human suffering in the practice of medicine.

Gregory also touches on medicine in *Comparative View*, treating medicine as "the art of preserving health, and restoring it when lost" but plagued with "useless theories, and voluminous explanations and commentaries on those theories" (Gregory, 1772a, p. 71), themes that echo his youthful "Medical Notes" (AUL 2206/24, 1743). Just as in religion, systems and abstract theories constitute a bane on the improvement of medicine. He also laments the "awful distance philosophers have kept from enquiries of genuine utility to Mankind," medicine included (Gregory, 1772a, p. 72). He commends the work of Sydenham and Boerhaave. The experimental method, he notes with approval, is beginning to influence medicine:

> The same liberal and manly spirit of enquiry which has shewn itself in other branches of knowledge, begins to find its way into medicine. Greater attention is now given to experiment and observation; the insufficiency of any idle theory is more quickly detected, and the pedantry of the profession meets with deserved ridicule (Gregory, 1772a, p. 76).

The physician must not be crippled by "gloomy and forlorn Scepticism," but, if he is to be useful to himself and to the public, the physician "must be decisive in his resolutions, steady and fearless in putting them into execution" (Gregory, 1772a, pp. 83, 84). Steadiness thus becomes a cardinal virtue of the physician.

Finally, as we saw in section VII, Gregory adopts whole Hume's account of sympathy: "it is a fact well established that such a thing exists, and that there is such a principle in nature as an healthy sympathy, as well as a morbid infection" (Gregory, 1772a, p. 102). Moreover, women of learning and virtue – those who have gained knowledge and wisdom – provide the exemplars of properly cultivated sympathy in the form of the virtues of tenderness and steadiness.

Women, and men too, also learned steadiness from the frequent and staggering experience of loss of parents, siblings, and children to early death. Gregory never mentions this, as far as I can tell, in the texts he published, but it seems an unavoidable influence on his thought, as witnessed in his letter to Mrs. Montagu upon the death of his wife (HL MO 1064, n.d.). The threat of loss of loved ones was constant. Gregory writes to Mrs. Montagu on May 3, 1771, about the illness of his daughter Elizabeth, then about fourteen.

> My own poor Betty has been in so declining a way for some weeks past that I have not had the spirits to write to you nor to do any thing I wish & ought to do. Her Disorder has turned out a Consumption & I have not the least hopes of her recovery. I will not hurt the Sensibility of your Heart by saying more on the Subject & I am incapable on writing on any other. In all Situations I ever think of you with every Sentiment of Esteem, Affection & Gratitude & I shall do to my latest breath (HL MO 1085, 1771, p. 1).

Physicians need the steadiness exhibited in this letter – his daughter died shortly thereafter – because they too will experience the loss of patients repeatedly, even more than their non-physician peers. To be sure, Gregory's clinical diaries indicate that most of his Infirmary patients were discharged "D^r cured," but not all of them. Gregory himself experienced the early deaths of his father, half-brother, wife, and one daughter. He struggles, he writes to Mrs. Montagu, not to be "unmanned" by his wife's death, i.e., to find a way to express his tender sentiments in regulated grief that would steady him. Tenderness and steadiness thus become central virtues in a life marked by common loss of loved ones, in life generally, or patients, in the moral life in medicine.

B. A Father's Legacy

Gregory began writing *A Father's Legacy to His Daughters* in anticipation of his premature death from hereditary gout (about which, more below, on his clinical lectures). We saw in the previous section some of Gregory's views on women expressed in his letters to Mrs. Montagu. We also saw that Gregory's interest in women and their social status began early in his intellectual career. *A Father's Legacy* represents in their mature form the culmination of his views on women and his feminism.

Gregory writes at a time when the man or woman of false manners and sentiment was recognized to be a problem, when the development of cities led to multiple class stratification – not the more simple laird (lord)-peasant stratification, and when the relationships between the genders was undergoing significant change. He did not want his daughters to be without guidance in their moral development and conduct. Gregory opens *A Father's Legacy* in the following way:

> You will see in a little treatise of mine just published [*Comparative View*], in what an honourable point of view I have considered your sex; not as domestic drudges, or the slaves of our pleasures, but as our companions and equals; as designed to soften our hearts, and polish our manners; and as Thomson finely says,
> > To raise the virtues, animate the bliss,
> > and sweeten all the toils of human life.

> I shall not repeat here what I have there said on this subject, and shall only observe, that from the view I have given of your natural character and place in society, there arises a certain propriety of conduct peculiar to your sex. It is this peculiar propriety of female manners of which I intend to give you my sentiments, without touching on those general rules of conduct by which men and women are equally bound. While I explain to you that system of conduct which I think will tend most to your honour and happiness, I shall, at the same time, endeavor to point out those virtues and accomplishments which render you most respectable and most amiable in the eyes of my own sex (Gregory, 1779, pp. 5-6).

He goes on to address manners and conduct in the areas of religion, amusements, and friendship, love, and marriage.

Men are at risk for "hardness of our heart," "uncontrouled license," and "more dissolute" manners that together "make us less susceptible of the finer feelings of the heart" (Gregory, 1779, p. 7). Because women do not live public but largely domestic, private lives, they find themselves in different moral circumstances, in which, as he pointed out in *Comparative View*, the devotional life is important.

> Your superior delicacy, your modesty, and the usual severity of your education, preserve you, in a great measure, from any temptation to those vices to which we [men] are most subjected. The natural softness and sensibility of your dispositions particularly fit you for the practice of those duties where the heart is chiefly concerned. And this, along with the natural warmth of your imaginations, renders you peculiarly susceptible to the feelings of devotion (Gregory, 1779, pp. 7-8).

Rather than hardness of heart, women are at risk for *dissipation*, as Hester Chapone had cautioned her niece. Gregory's view of dissipation is clinical and based on plain observation of its consequences: it "impairs your health, weakens all the superior faculties of your minds, and often your reputations" (Gregory, 1779, p. 9). Living a life of undisciplined feeling without improving one's faculties is morally repugnant, given everything that we have seen about Gregory so far. Religion provides the antidote to dissipation. It may be that, given the confinement of women largely to the home – and thus no opportunity for education and work outside the home – in Scotland, Gregory thought that women had no other antidote available to them. There were also no salons like Mrs. Montagu's. As Alexander Murdoch and Richard Sher (1988) noted (see Section VI, above), the intellectual life of the Scottish Enlightenment was closed off to women in very large measure.

> Religion, by checking this dissipation, and rage for pleasure, enables you to draw more happiness, even from those very sources of amusement, which, when frequently applied to, are often productive of satiety and disgust (Gregory, 1779, pp. 9-10).

Religious devotion encourages proper moral formation. So does religious tolerance, he argues (Gregory, 1779, p. 15). Religion also provides an antidote to false manners and feigned sensibility to the sufferings of others as well as to indifference, a set of concerns that Gregory apparently shares with Mrs. Chapone.

> There is a false and unnatural refinement in sensibility, which makes some people shun the sight of every object in distress. Never indulge this, especially where your friends or acquaintances are concerned. Let

the days of their misfortunes, when the world forgets or avoids them, be the season for you to exercise your humanity and friendship. The sight of human misery softens the heart, and makes it better; it checks the pride of health and propriety; and the distress it occasions is amply compensated by the consciousness of your doing your duty, and by the secret endearment which Nature has annexed to all our sympathetic sorrows (Gregory, 1779, pp. 16-17).

This is not Kames' theological appeal to the fear of punishment and promise of reward as an adjunct to virtue based on sympathy. Rather, Gregory takes the view that regularly exercised sympathy in response to human misery leads to self-improvement, that we can experience this self-improvement in the form of humility regarding our own health and prosperity and consequent preparation for their eventual loss, and that we will experience true happiness in the form of natural satisfaction from fulfilling the duties of humanity – a kind of incremental self-completion that potentially has no end point. Virtue is its own rich and complex reward, a claim that may seem trite in our time but to which Gregory gives considerable substance, meaning, and measurable content. If you doubt him, we can easily imagine him saying, go out and try both sympathetic and hard-hearted responses to human suffering and simply observe the differences in yourself.

Gregory returns to the theme of tenderness united with steadiness when he takes up amusements and generalizes his instruction to the whole of the moral life.

In this, as well as in all important points of conduct, shew a determined resolution and steadiness. This is not the least inconsistent with that softness and gentleness so amiable in your sex. On the contrary, it gives that spirit to a mild and sweet disposition, without which it is apt to degenerate into insipidity. It makes you respectable in your own eyes, and dignifies you in ours (Gregory, 1779, p. 48).

Insipidity is a base moral state; resolute behavior grounded in softness of heart and tenderness – the proper expressions of sympathetic response to others, especially to their misfortunes – represents exemplary moral development.

Gregory next turns to women's relationships with men: friendship, love, and marriage. Women should avoid men of "unmeaning gallantry," i.e., men of false manners who pursue self-interest in their relationships with women not the fulfillment of duty (Gregory, 1779, p. 60). Women should therefore not shy away from forming friendships with men of good moral character and intentions. The influence of Mrs. Montagu's salons and her example run deep in the following passage:

> There is a different species of men whom you may like as agreeable companions, men of worth, taste, and genius, whose conversation, in some respects, may be superior to what you generally meet with among your own sex. It will be foolish in you to deprive yourselves of an useful and agreeable acquaintance, merely because idle people say he is your lover. Such a man may like your company, without having any design on your person (Gregory, 1779, pp. 60-61).

Here Gregory seems to be calling on his experience with the Bluestocking Circle to instruct his daughters. Women can advance their education in conversation with men and can do so by means of asexual relationships with men of merit. There may be risk to one's reputation in doing so, but the risk is worth taking, apparently, because the views of "idle" – dissipated, debased – people shouldn't matter to a woman of good judgment and resolute self-confidence.

Men will benefit from such concourse with women. In a passage that echoes his remarks in *Comparative View*, he writes:

> His [a "man of delicacy"] heart and character will be improved in every respect by his attachment. His manners will become more gentle, and his conversation more agreeable ... (Gregory, 1779, p. 67).

Women as moral exemplars of virtue properly cultivated in resolute tenderness will have the desired effect on the moral formation of men of "delicacy," i.e., those whose hearts are not yet hardened beyond such influence. Proper moral development will prevent a woman from deceiving a man and from being deceived by a man regarding her own or his intentions (Gregory, 1779, p. 80). Virtue works to our true advantage, a demonstrable truth.

Gregory then addresses marriage in quite remarkable terms, even radical for his time. Consider Millar, for example, who notes that, as society has progressed from "ancient barbarous practices," "women become, neither the slaves, nor the idols of the other sex, but their friends and companions" (Millar, 1990, p. 89), views close to Gregory's own. Millar describes the role of women in domestic life in the following terms:

> As their attention is principally bestowed upon the members of their

own family, they are led, in a particular manner, to improve those feelings of the heart which are excited by these tender connections, and they are trained up in the practice of all the domestic virtues (Millar, 1990, pp. 90-91).

Women confined to the domestic setting are at risk of the "effects of idleness and dissipation" (Millar, 1990, p. 99). When women enter "public life,"

... they are led to cultivate those talents which are adapted to the intercourse of the world, and to distinguish themselves by polite accomplishments that tend to heighten their personal attractions, and to excite those peculiar sentiments and passions of which they are the natural objects (Millar, 1990, p. 100).

In public life women are at risk for "licentious and dissolute manners" (Millar, 1990, p. 101).

Keeping Millar's remarks in mind, consider Gregory on marriage. Women who did not marry and also did not possess considerable financial means were at risk of a life of poverty. Waiting to marry for love, a concept then gaining social currency, thus involved a high-risk choice and marrying for love could mean a life of reduced socio-economic standing or even penury – as it did for his daughter, Dorothy, who would seem to have followed her father's advice.

I know nothing that renders a woman more despicable, than her thinking it essential to happiness to be married. Besides the gross indelicacy of the sentiment, it is a false one, as thousands of women have experienced (Gregory, 1779, p. 81).

False sentiment is anathema and must be avoided, lest one simply chuck integrity over the side. But such an option is simply not available for the morally well-formed person. He goes on:

You must not think from this that I do not wish you to marry. On the contrary, I am of the opinion, that you may attain a superior degree of happiness in a married state, to what you can possibly find in any other (Gregory, 1779, p. 82).

This seems clearly to reflect his own experience and his wife's report on the happiness she knew, reported in his letter to Mrs. Montagu upon his wife's death. But marry for love, he admonishes, for the alternative is dire.

Instead of meeting with taste, delicacy, tenderness, a lover, a friend, an
equal companion in a husband, you may be tired with insipidity and
dullness; shocked with indelicacy, or mortified by indifference
(Gregory, 1779, p. 92).

Gregory, ever the practical Scotsman, provided for his daughters the
material means to carry out his recommendations.

As these have always been my sentiments, I shall do you but justice,
when I leave you in such independent circumstances as you may lay
under no temptation to do from necessity what you would never do
from choice. – This will likewise save you from that cruel mortification
to a woman of spirit, the suspicion that a gentleman thinks he does you
an honour, or a favour, when he asks you for his wife (Gregory, 1779,
pp. 85-86).

Gregory invokes his concept of sympathy when he takes up the matter
of how he should respond if his daughters were not to take his advice and
instruction to heart when they reached adult years and began to make
their own choices.

If you do not chuse to follow my advice, I should not on that account
cease to love you as my children, though my right to your obedience
was expired, yet I should think nothing would release me from the ties
of nature and humanity (Gregory, 1779, p. 87).

The obligations, i.e., ties, of sympathy and humanity are beyond choice;
they shape us at our core. We live according to them and find true happiness or defy them and plunge into the misery of mere self-interest.

Gregory's confidence in his moral sense theory leads him to the following summing up of his views.

I have no view by these advices to lead you to your tastes; I only want
to persuade you of the necessity of knowing your own minds, which,
though seemingly very easy, is what your sex seldom attain on many
important occasions in life, but particularly on this of which [i.e., mar-
riage] I am speaking. There is not a quality I more anxiously wish you
to possess, than that collected decisive spirit which rests on itself,
which enables you to see where your true happiness lies, and to pursue
it with the most determined resolution. In matters of business, follow
the advice of those who know better than yourselves, and in whose
integrity you confide; but in matters of taste that depend on your own

feelings, consult no friend whatever, but your own hearts (Gregory, 1779, pp. 93-94).

A properly formed heart reflects Nature improved and Gregory's confidence in Nature, especially Nature improved – i.e., functioning well. As we saw above in reference to his views on child rearing, a properly formed heart expresses itself in "that collected decisive spirit" of steadiness and tenderness, a spirit that "rests on itself" not on some external foundation such as fear of punishment or promise of reward. In this respect and to this extent Gregory's theory of virtue as developed in *A Father's Legacy* is *secular* in character.

So, too, is his remarkable and, for its time, robust, even radical feminism. Women are the equals of men intellectually and their superiors morally, in matters of the heart. Women of learning and virtue – educated women of resolute spirit and constant willingness to respond out of regulated sympathy to the misery of others – provide the moral exemplars for men. Indeed, men – their hearts hardened by city life and commerce, placing the social principle at risk of being deadened beyond revival – require women of learning and virtue if men are to develop and live in a morally proper fashion.

C. Medical and Clinical Lectures

For the use of his students, Gregory published his *Elements of the Practice of Physic* in 1772 (Gregory, 1772b). His lectures on the practice of medicine provide a nosology of diseases, although the printed version of them covers only fevers. In a way typical of his times, Gregory includes many clinical entities under the label, 'fever', everything from various febrile diseases though nephritis, gastritis, and gout to debility, dreams, and hydrocephalus. He also addresses the diseases of infants and women. For each, Gregory follows the pattern set down by Boerhaave and his followers, of defining each disease by its symptoms. Reflecting his longstanding antipathy to medical systems, Gregory eschews "artificial arrangement" or nosology (Gregory, 1772, p. 3r, Gregory's interleaved gloss). His complete nosology includes fevers, as well as "Hemorrhages," "tropical diseases," "spasmodic affections," "Diseases of the mind and its faculties" reflecting the influence of Whytt's work, "cachexis," "local diseases," and "Diseases of women and children" (EUL DC.6.125, 1772, pp. 6-8).

In a manuscript version of Gregory's "Lectures on Medicine," which may be Gregory's copy from which he gave the lectures, Gregory adopts the physiologic definitions of 'disease' and 'health' then prevailing in Baconian philosophy of medicine.

> A disease arises from a particular structure or change produced in the functions of the external and internal organs – When the structure is natural & the several organs perform their functions regularly & in a proper manner the body is said to be sound & in health. A disease is an unnatural structure, or disordered functions of the bodily organs (RCPE Gregory, John 1, 1766, p. 5).

In his "Lectures on Pathology" Gregory puts it this way:

> Many definitions have been given of the word *Disease*. The definition that I shall give is the most simple & easy, viz a deviation of the human constitution from its natural state. There is a particular structure of the body, and certain connection between it and the mind; and when the operations of the Body & mind go on according to the Laws of Nature it is then called a *sound state*. But when there is any deviation from or irregularity in the operations of nature, it is called *disease* (RCSE D 27, 1770, pp. 3-4).

He acknowledges his debt to his colleague, Cullen, whose views reflect "clearness and preciseness" (RCSE, 1770, p. 5). This account of disease and health seems to be taken in large measure from Boerhaave, who defined disease and health in the following terms:

> Every condition of the human body that wounds the vital, natural, or animal actions I call disease (Boerhaave, 1728, p. 1).[16]

> ... health is the faculty of the body adapted to exercising perfectly every action ... (Boerhaave, 1747, p. 362).[17]

In his lectures on the practice of physic, Gregory provides his students with guidance on how to work up and care for patients. Patients should first be urged to *tell their own stories*.

> It is best to let the Patient first tell his own story as there are often circumstances of consequence which might not occur to a Physician to ask yet which the patient from his own feelings will naturally complain of, & after this the Physician may ask what questions he thinks proper (AUL 129, 1769 & 1770, p. 2).

Human beings are not "machines" that simply break down but instead operate "in consequence of certain Laws latent in the constitution," laws that work toward the restoration of health (EUL DC.6.125, 1772, p. 14).

> It is the business of our Art then to assist these efforts, and sometimes necessary to restrain them, and sometimes even to ignore them, but these different methods must be followed according to the dictates of Experience ... (EUL DC6.125, 1772, p. 15).

Physicians should not confuse the corrective processes of nature with the disease process. Gregory's next prescriptions reflect the then prevailing theories of disease and health described by Risse (1986). In addition to noting symptoms, the physician should note age, sex, environmental factors, and the patient's bad habits. He urges that treatment be kept simple, a theme that he repeats again and again.

Gregory also lectured on the "institutions of medicine." These concerned normal anatomy and physiology. In these he takes up at some length the nervous system, and sets out views that he takes, with acknowledgement, from Stahl and that also seem to be influenced by Whytt and Hume. He addresses the influence of the mind and body on each other, in which he gives "an account of the state of the mind, as it depends upon the sensible changes made upon the corporeal system; But not treat it [mind] independent of the body" (EUL 2106 D, 1773, p. 77). These sensible changes involve sensation.

> Sensation is used in different senses. In common conversation it means any feeling: But here we mean, a feeling in consequence of a corporeal impression. There are some parts of the body which when an impression is made, they convey a sensation to the mind, or in other words are possessed of sensibility. There are other parts again insensible, or that convey no Idea when an impression is made on them; or are inconscious of the impression (EUL 2106 D, 1773, pp. 77-78).

The language here is strikingly Humean; or, rather, Hume's language is strikingly physiologic, and physical. The "nerves are the sources of sensibility" (EUL 2106 D, 1773, p. 78), a view that leads Gregory to develop a *physiologic account of sympathy.*

> It has been a question whether or not impressions are conveyed to any other part besides the Brain. They may be communicated from one Nerve to another which is called Sympathy. This term is used in different senses. When any action, any motion or any feeling is uniformly connected with any other action, motion, or feeling It is said to be

Sympathetic. This particular association is a fact, and it is certainly our business to determine the cause of it though we must allow in many cases it is impossible to give a cause for such association (EUL 2106 D, 1773, pp. 89-90).

Sympathy can arise from "irritation" of the nerves, from habit, and from contiguity. Hume adopts some of these categories, underscoring the strongly physiologic character of his account of sympathy. When Gregory comes to operations of the mind regarding sympathy, he declines to enter into the subject and commends to his students the work of Locke, Hume, Reid, and Beattie, who "have totally exhausted the subject" (EUL 2106 D, 1773, p. 102). Here we see, once again, Gregory embracing Hume's account of sympathy, just as he had in Aberdeen.

Throughout his lectures Gregory emphasizes the Baconian method in medicine.

The science of Medicine is principally founded upon facts and observations, which compiled together serve to furnish principalls and theories; and from these again principals of a more general kind are deduced (WIHM 2617, 1771-1772, p. 1)

Medicine is not a

Speculative science, to be acquired by reading: it is an active and practical art, the exercises of w^c [which] can be acquired by long and extensive practice only. This indeed is the case in every other practical art in life, and the education of them all is conducted accordingly (WIHM 2617, 1771-1772, p. 4).

In short, Gregory brings to his teaching at the University of Edinburgh and in the Royal Infirmary's teaching ward the Baconian philosophy of medicine that he had been developing since his medical student days.

Gregory's lectures sometimes become personal. For example, when he covers gout he first starts with an account of its (his) symptoms. He then turns to its causes, in the course of which he makes brief reference to himself.

The gout is a disease that is remarkably hereditary and often appears very early in life. Hereditary gout often attacks about twenty or thirty and *this was the case with myself*, but otherwise the gout is a disease that does not come on 'tell the middle period of life between forty and fifty (EUL Dc.7.116, 1769-1770. pp. 254-255, emphasis added).

Hereditary gout also causes premature mortality, of which his students were well aware. Gregory thus exposes himself to some extent, perhaps presenting himself as a role model of steady, resolute response to the threat or premature death. Gregory, it would seem, taught by example as well as by precept.

XI. GREGORY'S DEATH

Gregory wrote *A Father's Legacy to His Daughters* (1774) in anticipation of his premature death. He remarked often to his friends that he expected to die early, from gout, the disease that killed his mother in 1770 (Tytler, 1788). As we just saw, he also made revealing remarks on this subject in his lectures on the practice of medicine, which continues, "It seldom appears sooner unless brought on by an excess of eating and drinking or in vinery" (EUL Dc.7.116., 1769-1770, p. 255). He goes on to note that gout, while it often causes severe pain, in the foot particularly (EUL Dc.7.116, 1769-1770, p. 246), can also cause inflammation in the head and lungs (EUL Dc.7.116, 1769-1770, p. 344), in which symptoms gout can manifest as a life-threatening disease. Thus, response to symptoms of gout in the feet should be aggressive:

> When the gout is either improperly repelled or for want of natural strength it does not fix in the extremeties, we ought to do all in our power to bring it down as soon as possible (EUL Dc.7.116, 1769-1770, p. 303).

No one was available to "do all in our power" to assist Dr. Gregory on February 9, 1773, when gout took him in his sleep. William Smellie's account is terse:

> From his 18th year, he had been occasionally afflicted with the gout, which he inherited from his mother, who, in the year 1770, died suddenly, when sitting at table. Dr Gregory went to bed on the 9th of February, 1773, apparently in good health; but he was found dead the next morning (Smellie, 1800, p. 116).

Tytler provides greater detail of the end of Gregory's life (and was, no doubt, the source of Smellie's account):

> These letters [*Legacy*] to his daughters were evidently written under the impression of an early death, which Dr Gregory had reason to ap-

prehend from a constitution subject to the gout, which had begun to shew itself at irregular intervals, even from the 18th year of his age. His mother, from whom he inherited that disease, died suddenly in 1770, while sitting at table. Dr Gregory had prognosticated for himself a similar death; an event which, among his friends, he often talked, but had no apprehension of the nearness of its approach. In the beginning of the year 1773, in conversation with his son, the present Dr James Gregory [known for his directness, to the point of rudeness], the latter remarking, that having, for the three preceding years, had no return of a fit, he might make his account with a pretty severe attack at that season, he received the observation with some degree of anger, as he felt himself then in his usual state of health. The prediction, however, was too true; for, having gone to bed on the 9th of February 1773, he was found dead in the morning. His death had been instantaneous, and probably in his sleep; for there was not the smallest discomposure of limb or of feature, – a perfect *Euthanasia* (Tytler, 1788, pp. 78-79).[18]

Recall that Gregory was responsible for the clinical teaching of the medical students in the teaching ward at the Royal Infirmary. His last lecture survives in one of the student's notes (CPP 10a 44, 1771).

I have now resigned the Care of the Patients in the Clinical Ward to my ingenious and worthy colleague D^r. Cullen, who is to continue the Lectures upon the Patients Cases, – I have never seen the ward better supplied w^t [with] matter for observation, and I have marked w^t pleasure the part^r [particular] Attention given by most of you to these Cases, I have formerly observed, that med[icine] is ['not' seems to be missing here] to be considered merely as a speculative science, that is not to be acquired by reading and attending to Lectures & Professors, but that it is a practical art that is to be learned by accurate observ^n [observation] and intensive experience no man can become a tolerable Physician or Surgeon w^{ot} [without] this experience, either by close attention to the practice of others or f^m [from] those unhappy persons who are distin'd to be the fir't [first] victims of his Ignorance of his rashness, or timidity, some Gentlemen seem to think that a lib^l [liberal] Education supercedes the necessity of attending to practice. But this is a mistake, for not acuteness of understg [understanding] or Solidity of judgment can exempt them from this necessity, on the contrary a liberal Education, and the studying Physic upon a Systematic plan shows the disadvantages under which the Science labours, from the neglect of practice and leads us to distinguish between the real and pretended [page edge broken; probably 'physician'] such an Education enlightens and directs the conduct of observation, which w^{t}out [without] the assistance of science is commonly defective, erroneous, blind, and inconclusive, so that instead of being at variance w^t one another and independ^t. [independent] they are intimately and inseparably connected. What I mean by theory is the gen^l. [general] prin^{els}[principles] of science founded upon accurate Indu^n [Induction] from undoubted facts and not crude conjectures substituted in the place of facts, and Speculations which can have no more influence upon the practice of Physic, than upon Ship building. I hope none of you will look upon the simplicity of my Practice as less useful, than if I had varied my plan of cure, and Prescriptions. – The only method of improving the practice of Physic is to give a fair trial to each Remedy. Otherwise our Inferences from there effects are inconclusive, and if we persist when they do not succeed, it argues Stupidity, but if we desert them when they do succeed, we wantonly sport w^t the lives of our fellow Creatures. so far Doct[or] Gregory (CPP 10a 44, 1771, pp. 322-323).

This final lecture is numbered by an unnamed student who recorded it as being the twenty-second lecture in this series and dated, "Tuesday 10th February 1773" (CPP 10a 44, 1771, p. 305). Gregory died during the previous night and so we can safely surmise that this lecture was read to his students by his clerk, from Gregory's own lecture notes, which were usually written out fully.[19] The Baconian method and the intellectual virtues it requires suffuse this little lecture, as does the enlightened spirit of a liberal education, grounded in science and experience. Gregory's moral concerns are evident in his closing sentences, where he uses a metaphor that comes to play a prominent role in his medical ethics:

> The only method of improving the practice of Physic is to give a fair trial to each Remedy. Otherwise our Inferences from there effects are inconclusive, and if we persist when they do not succeed, it argues Stupidity, but if we desert them when they do succeed, we wantonly sport w^t the lives of our fellow Creatures (CPP 10a 44, 1771, pp. 322-323).

In a letter upon the occasion of Gregory's death, Mrs. Montagu provides a fitting epitaph:

> The hours I passed in his company were amongst the most delightful in

my life. He was instructive and amusing, but he was much more; one loved Dr. Gregory for the sake of virtue and virtue (one might almost say) for the sake of Dr. Gregory (Blunt, 1923, Vol. 1, p. 270).

CHAPTER THREE

JOHN GREGORY'S MEDICAL ETHICS

Notes taken from Dr Gregory

Nov^r. [November] 6th 1769 I am this day to mention to you some of the Moral Qualities of a Physician & ce. [et cetera] and the first I shall mention is *Humanity* or that compassion we ought to have for our patient. Some have alleged that our Profession are very hard hearted and void of feelings. but I hope the charge is groundless (with respect to the generality of Physicians[)] Real Sympathy is of the greatest advantage both to the Pat. [Patient] & the Physician: real Sympathy always trys to conceal itself – Another qualification I shall mention, is a certain ease & gentleness to be shewn towards the Pat. & not to be too rigid & stiff with regard to particulars which are sometimes not of the greatest Consequence (RCSE D27, 1769, pp. 1-2).

At the University of Edinburgh Gregory began giving lectures on medical ethics as early as 1767, as preliminary to his lectures on the practice of physick or medicine (RCPSG 1/9/5, 1767, WIHM 2618, 1767). He gives these lectures, he says, "agreeable to custom" (Gregory, 1772c, p. 2). In doing do, he makes reference to a practice that goes back at least to Boerhaave and Hoffmann. In this, as in so many ways, the influence of Leiden at Edinburgh remained strong.

In this chapter I provide an account of Gregory's lectures on medical ethics and philosophy of medicine. I begin with a consideration of his precedents in Boerhaave and Hoffmann and in his colleagues' teaching. I then examine briefly the publication history of *Observations* and *Lectures*. The stage is further set by considering Gregory's intellectual resources and the problem list in response to which he gave his lectures. I then provide a detailed account of the content of Gregory's lectures, including his definition of medicine, his defense of the "utility and dignity of the medical art," the qualifications – characterological moral prerequisites – of a physician, his clinical ethical topics, and his philosophy of medicine. I close the chapter with an argument that Gregory invents professional medical ethics in the English language and, with this invention, invents the concept of medicine as a fiduciary profession involving service to patients as a way of life.

I. THE CUSTOM OF GIVING PRELIMINARY LECTURES: LEIDEN AND EDINBURGH

Boerhaave attached "prolegomena" to his lectures on the institutes of medicine and to his "aphorisms" (Boerhaave, 1747, 1728). In the *"Prolegomena"* to his *Institutiones Medicinae* Boerhaave provides a history of medicine, starting in ancient Assyria and Babylon and culminating with the work of Harvey on the circulation of blood. On the basis of this history Boerhaave emphasizes the importance of observation and empirical science. Medicine concerns: the "life, health, disease, and death of man; the causes of these, from which they originate; and their media, by which they can be directed" (Boerhaave, 1747, p. 8).[20] The physician should use medicine to "conserve life and restore the sick to their previous well being."[21] From these, Boerhaave concludes that "the necessity, utility, and nobility of medicine spontaneously follow" (Boerhaave, 1747, p. 8).[22] In his *Aphorisms* Boerhaave provided a philosophy of medicine, which we examined in Chapter Two.

Hoffmann, in his *Medicus Politicus* (1749), provides concrete "rules" for the practice of medicine, which he gathers under a series of "heads." He labels the first head, "Concerning prudence on the part of the person of the physician" (Hoffmann, 1749, p. 3),[23] i.e., the rational calculation of one's interests and how they might be put at risk as a consequence of one's and others' decisions and actions. The first rule under this first head reflects the strong moral- theological orientation and method that Hoffmann employs in this text on medical ethics.

> Rule 1. The physician should be a Christian. And so the Christian practices humanity, toward which medicine will present the best occasion, when we consider what man would be other than fallen and fragile, from which he originates, for man is born between excrement and urine. What therefore should we make of the life of man? Nothing other than shadow.

> Rule 2. The physician should be moderate and not dispute very much religion and faith from things seen. ...

> Rule 3. The physician should not be an atheist (Hoffmann, 1749, p. 3).[24]

In defense of this explicit theism and the moral theology that it will generate for medical ethics, Hoffman adopts a Cartesian argument for the existence of God: "God is the cause, through which things are conserved altogether" (Hoffmann, 1749, p. 3).[25] As Descartes argues in his *Meditations* (1979), finite entities contain within themselves no causal explanation for their existing from moment to moment. Some cause must be sought external to finite being; only infinite being has the power to conserve all of reality from moment to moment, throughout time.

Hoffmann thus understands the Renaissance concept of humanity or *humanus* in explicitly theological, Christian terms. He then goes on to state as the fifth, not the first rule, of the first head: "The physician should be a philosopher. Through philosophy we acquire wisdom" (Hoffmann, 1749, p. 4)[26] Philosophy concerns both moral and physical matters and the physician should be moral both in intellect and will.

By philosophical means we can discover the natural laws by which God has ordered and directed his creation. These natural laws are philosophical principles, of which Hoffman derives three. First, "God ought to be worshipped"[27] and we owe God fear, honor, and love. We know this natural law or principle by the light of nature (another explicit reference to Cartesian reasoning about such matters). Second, "society ought to be served."[28] Mutual loves holds society together and we should therefore avoid whatever tears down or divides society. Third, "the order of nature ought to be served."[29] Man should live moderately and so "acquire long life and the end planned for man by God" (Hoffmann, 1749, p. 4).[30]

There then follow rules enjoining the physician to be learned, to engage in clinical practice, and to know surgery. He then addresses medical studies under a separate head. The third head concerns "the peculiar virtues necessary to preserve public opinion."[31] Under this head Hoffmann enjoins the physician to be "humble, not arrogant," "silent" – i.e., protect confidential information about patients, and not negligent of his duties.[32] In this last respect the physician should not be overly complex in prescriptions and should moderate visits, making them only when necessary (Hoffmann, 1749, p. 7). The physician should also exhibit the moral virtues of being "merciful, modest, and human," avoiding "as one would the plague a dissolute life, drunkenness, illicit games, and everything else that would lose the trust of the sick" (Hoffmann, 1749, p. 8).[33]

This essay continues with admonitions about simplicity of prescription, the reform of pharmacy, and the physician's relationship with apothecaries and surgeons. Surgery should be left to surgeons, but the physician has a role to play in deciding whether the sick require surgical management.

Hoffmann addresses the obligations of midwives, using the term *obstetrix* for a female midwife. The female midwife should be "pius, chaste, sober, not frightened, silent [i.e., bound by confidentiality], and experienced" (Hoffmann, 1749, p. 13).[34]

Hoffmann takes up the physician's duties to the sick in the third part of his *Prolegomena*. The physician should engage in clinical practice, so as to see diseases in their variety and gain experience in their diagnosis and treatment. In the face of serious illness the physician should be neither timid nor fearful. Nor should the physician fear challenging the views of the patient or other physicians when the situation is dangerous. In cases of incurable illness Hoffmann has this to say: "It is preferable not to manage incurable diseases" (Hoffmann, 1749, p. 25).[35]

Hoffmann also addresses ethical considerations in the care of sick women under the rubric, "concerning the prudence of the physician regarding sick women" (Hoffmann, 1749, p. 25).[36]

Rule 1. The physician should be chaste. The physician should be chaste in word and deed, when he is obliged to visit sick women; for this ought not to be the occasion for exciting the lust of the physician, especially concerning unchaste women ... (Hoffmann, 1749, p. 25).[37]

Under the head of prudence regarding acute and chronic diseases the first rule states that the physician should protect himself from "contagion of malignant diseases" (Hoffmann, 1749, p. 29).[38] The physician "should abstain from visiting the dying" (Hoffmann, 1749, p. 29).[39] At the same time, the physician "should always act most cautiously in malignant diseases regarding heroic medication" (Hoffmann, 1749, p. 29).[40] Hoffmann's discussion of communicating with patients reflects this general caution and sense of the limitations of medicine.

Rule 5. The physician should form cautious prognosis in malignant and acute disease. Hippocrates said it best. Uncertain are all predictions concerning the hope of health in acute illness. The physician therefore should speak circumspectly and neither prostitute himself nor discern immediately and absolutely that it will be necessary to die, but he should not be evasive, but always restrictively and with conditions admit that danger is near ... meditation on death should not be neglected, ... [and the sick] should be prepared for dying (Hoffmann, 1749, p. 29).[41]

Thus, Hoffmann opens his *Prolegomena* in an explicitly moral-theological fashion, interpreting Renaissance humanism in Christian terms, with a strong admixture of prudential protection of self-interest, e.g., regarding exposure to contagion. He then articulates natural laws that the physician can discover philosophically, the first of which is explicitly *theological*. There then follow many rules under various heads, explicating natural law-governed *prudentia* in the sense of practical judgment concerning probable futures and how to manage them well prospectively, as they might affect one's self-interest.

Hoffmann also connects prudence to natural law. His second and third natural laws emphasize the importance of maintaining mutual love or connectedness between people and avoiding anything that sunders human relationships. In maintaining the bonds of mutual love, we emulate God's relation to man and of man to God. Thus, prudence is also informed by an obligation to preserve functioning human relationships and to avoid harming them. Just as God in the person of Jesus is defined by the relationship of loving service and sacrifice to bind all men to God, so too the person of the physician is made human by service to patients, tempered by an admixture of self-interest.

The mutual love required by the natural law ordained by God for man's good nicely parallels Hoffmann's concept of the animal spirit that originates purposeful movement in living things. Mutual love originates purposeful activity on the part of the physician, namely, prudent management of the patient. Christian *humanus* in the form of law-governed prudential judgments controls the physician's activities and directs them to the good that God intended for medicine.

Johanna Geyer-Kordesch claims that Hoffmann, in his *Medicus Policitus* (1749), sees "medicine secularized in no uncertain terms" (Geyer-Kordesch, 1993a, p. 131). Moreover, she reads Hoffmann to be providing "the middle ground on self-assertive decorum. It is a mode of conducting oneself (effectively) with no recourse to a systematized ethics" (Geyer-Kordesch, 1993a, p. 131). Roger French provides a similar reading of Hoffmann, arguing that the concept of prudence at work in Hoffmann is "more than enlightened self-interest" (French, 1993b, p. 155). In light of the text itself (and my reading of it above), taking Hoffmann as providing a secular account of medicine and its duties cannot be sustained. Thus, French's reading of Hoffmann as a figure of the Enlightenment becomes difficult to sustain. Moreover, if I am correct about the connection between prudence and natural law, then Hoffmann's ethics do indeed count as systematized in the sense in which Gregory and other

Scottish thinkers of the eighteenth century used the term: Hoffman derives his admonitions to the physician deductively from the three natural laws applied to the clinical management of patients. His understanding of the philosophical authority of natural law as derivable by the light of nature depends inescapably upon Cartesian philosophy (i.e., metaphysics), which epitomized systematic philosophy for eighteenth-century Scottish thinkers. Gregory, as we shall see, takes up some of Hoffmann's topics, follows him at times, e.g., regarding improving relationships with surgeons and the simplicity of medication, and, more importantly, departs from him, sometimes sharply, e.g., in his views regarding the care of the dying and the misuse of the label "incurable."

Edinburgh professors also practiced the custom of giving preliminary lectures on philosophy of medicine and topics in medical ethics. As early as 1750, Rutherford – Gregory's mentor and sponsor for his position at Edinburgh – gave short preliminary lectures before his clinical lectures in the Royal Infirmary (Comrie, 1927, p. 199; RCPSG 1/9/3, 1750). Rutherford calls these preliminary lectures "Prolegomena," an apparent reference to Hoffmann and Boerhaave. As we shall see, he differs in important ways from Hoffmann both in method and content.

In these preliminary lectures Rutherford defines medicine, as well as the concepts of health and disease, sets out his clinical method and approach to patients, and discusses the non-naturals. Rutherford's distinction between a physician and a quack leads him to provide an account of how to recognize the true physician:

> A Quack always boasts of his Cures and Remedies, and is therefore more apt to deceive Credulous People, but the true Physician is acquainted with ye [the] fundamental Parts or Principles of his Art, who understands ye animal œconomy & not only knows w^t [what] Health is but can trace out the Causes of Disease & ye Rise of ye Symptoms; He likewise knows when & where Nature makes an effort, supports her w^{hn}: [when] weak, & cooperates with her in all her actions and operations as far as he can; He may be said to be her Minister; He follows her as his Guide varying his Practise as ye Indications change, in short he hits with reason in every thing ... (RCPSG 1/9/3, 1750, p. 2).

Rutherford understands 'art' here as did the ancient Greeks: the disciplined application through the hands – through human activity, in more general terms – of scientific knowledge. 'Art' does not mean some undisciplined, creative act, but something highly disciplined, in this case, by

Baconian science applied clinically to aid nature in its work: "... Regular Physicians admit experience to y^r [their] chief Guide and Disapprovae of all Theory y^t [that] is not founded on facts" (RCPSG 1/9/3. 1750, p. 4).

He lays out for his students the method that he will use, and then turns to topics. One concerns physicians without degrees:

I would not be thought to mean by y^s [this][42] y^t [that] you should all take out Degrees, for I am far from thinking that a Diploma furnishes a Man with Medicinal knowledge, His Improvement in ye [the] Art depends on his own study and industry, so y^t a good Physician may easily be made without y^t Price of Ceremony, but no one with y^t alone (RCPSG 1/9/3, 1750, p. 8).

The true physicians shows himself to be such in his scientific, experienced-based approach to medicine. A medical degree – recall that they could be purchased – therefore means little, compared to this more reliable criterion (Johnston, 1987) .

The physician should inquire into the circumstances of the patient. Rutherford uses 'patient' where Hoffmann uses '*aegrotus*', i.e., 'the sick'. We see in this change in usage the emergence of the social role of being a patient, which role is a function of the physician-patient relationship. This relationship, we shall see soon, Gregory understands to be founded on both intellectual and moral virtues of the physician.

Rutherford addresses the management of incurable diseases in patients in the following terms:

If you think it [disease] incurable you may use Palliatives to render the Disease tolerable to the Patient. If it's curable you are to use proper medicines, from the Diasgnostic of the Disease youll form proper Indications, after which it will not be difficult to find the Methodus Medendi, or Medicines to suit those Indications (RCPSG 1/9/3, 1750, p. 15).

Rutherford differs from Hoffmann, who, we saw above, recommends that the physician forego the care of patients with incurable diseases. Rutherford clearly means to imply that the physician should continue care of such patients, with the main therapeutic concern being palliation.[43]

Rutherford provides no clear indication of the intellectual authority for his remarks, as did Hoffmann, by explicitly adopting a moral theology and philosophy of natural law. Instead, Rutherford speaks from experience, just as Baconian method requires him to do. However, he provides no evidence for his remarks; his students were, it seems, to presume that experience teaches these things. As we shall see, Gregory is also a Baconian, but with more explicit philosophical foundations for his medical ethics.

Robert Whytt also adopted this pedagogical practice, with a preliminary, albeit brief lecture covering the definitions of 'medicine', 'health', and 'disease' and rules for the clinical management of patients (RCPSG 1/9/4, n.d.). With Rutherford, Whytt shares great confidence in nature's pointing the way to diagnosis and treatment:

> In every disease, we are to suspect something wrong in the Machine & it shows always that there is a right performance of the functions (RCPSG 1/9/4, n.d., p. 2).

The Baconian concept of an *observed normal* in human physiology obviously shapes Whytt's philosophy of medicine. In acute diseases, he says, either nature effects a cure itself or the patient dies. Medicine aims in acute diseases to "curb" their "violence" (RCPSG, 1/9/4, n.d., p. 3). In the case of chronic diseases, by contrast,

> ... nature can be of no service but the cure must either be attained by Medicine or by Art, and the Physician is obliged to be a bystander for several days, till he sees what course nature takes, then he is to assist her (RCPSG, 1/9/4, n.d., p. 5).

In his summary rules he includes an admonition: ask the patient what the patient believes to be the cause of his disease. We see here the effect of the competition between lay and medical concepts of disease, including their etiology. After all, Baconian method is open to anyone, not just physicians, and the observant patient may provide a reliable account of the cause of his or her disease, which may then suggest a cure.

Whytt's remarks are very brief, and he provides no methodological basis for them, although their Baconian spirit could readily be detected by his students. The core of this Baconian spirit shows through in Whytt's – and Rutherford's – great confidence in nature in the clinical setting, a confidence, we shall soon see, that Gregory fully shares.

Cullen also gave a terse preliminary lecture in which he simply mentions the purpose of his clinical lectures, his teaching method, and the temperament required of a physician (RCPSG 1/9/7, 1768). The temperament required of a physician, Cullen says, consists in "circumstances ... common to all men" (RCPSG 1/9/7, 1768, p. 3). But he does not elaborate on this concept. He seems to imply that the physician requires no special temperament, a view with which Gregory disagrees. Cullen's terseness obscures completely any philosophical basis he may have had for this remark. He closes with a consideration of the challenges of obtaining a history from the patient.

Finally, Thomas Young, who lectured on midwifery while Gregory was at Edinburgh (Comrie, 1927), concludes his clinical lectures with a brief reflection on the "Qualifications of a man midwife," some thoughts on clinical practice, and a prohibition against sexual abuse of patients (RCPSG 1/9/9, 1768). Young presents these lectures to physicians who hope to be man-midwives and so he does not concern himself directly, as did Hoffmann, with ethical topics regarding female midwives. Young also provides a preliminary lecture, which concerns a history of the "Science of Midwifery" (RCPSG 1/9/9, 1768, p. 7) culminating in the publications of Smellie's system of midwifery in 1751 and of Astruc in 1761.

Young's concluding lecture on the "qualifications" of man-midwives is brief enough to include here in its entirety:

> It is common to give a chapter on the Qualifications of a man midwife, but this is unnecessary as they are no other but a good operator, Patience & Composure of mind is necessary & in Midwifery we have not our own Eyes to direct us. Modesty is necessary especially in a place where the practice is only coming into the hands of men, therefore never propose any Rude Questions to your patient, especially where you learn it from the Nurse or the maid etc. Always write every case thay may occur to you, a Physician will get more by writing the Cases in 5 years practice than another will in 10 or 15 years. When you are called to Deliver never be too officious but stay by the woman, & let her rather Desire or wish for your assistance before you give it. Have as few assistants as possible in the Room more particularly when you are going to examine her, & let the Room be darkened, & the Curtains Drawn, these cautions are more requisite when it is the first child a Degree of Secrecy is also required. Never discover any oddities of particular women to others, I don't mean [by] this that you are not to tell a woman that you have met w^t [with] a Similar Case, because nothing will give her so much Satisfaction, but especially never reveal what the nature of your Profession has made you acquainted with to the other sex, nor never violate the trust that is placed in you, by harbouring any Immodest or unlawful Design. These who do act contrary to these

Principles, Ruin and Perdition will be their fate, but he who acts w^t
Decency & Decorum, Consistent with Honour, & the Character of a
Gentleman Success & Renown will crown his undertakings. That this
may be the happy lot of all the Gentlemen here present is the sincere
wish of Y [Young?] (RCPSG, 1/9/9, 1768, pp. 218-220).

The reference to it being common to address the "qualifications" may
allude to Gregory, who had been giving his own lectures for at least a
year. Basically, Young appeals to what his students already were or
aspired to be, gentleman. He also appeals to "your Profession." He assumes, apparently, that his students would know what this meant. This
must count as a risky assumption in the case of 'gentleman', given the
problems with this social role, especially the false man of manners and
the worry by patients of whom to trust. Young addresses the question of
trust, which should never be violated. But the reason for this – and for
everything else that Young says, including confidentiality, attendance at
examinations, etc. – may have as much to do with promoting the self-
interest of the man-midwife as with moral obligations to the patient. In
other words, Young's lecture on the "qualifications" of a physician lacks
an explicitly moral content, an omission that Gregory's lectures on medical ethics emphatically include. Whereas Young construes 'profession' as
lacking moral content, Gregory provides moral content of a philosophically substantively kind. "Qualifications" for Gregory became the moral
prerequisites for taking on the mantle of being a physician.

In summary, there was a tradition of offering *preliminary lectures* – on
philosophy of medicine, history of medicine, and the "qualifications" for
being a physician – going back to Leiden and Europe and continuing at
Edinburgh. Professors would address one or more of these general subjects. When they addressed more than one, the connections among them
are unclear. By contrast, Gregory clearly conjoined his philosophy of
medicine and medical ethics in a synergistic way, as we shall soon see.
Hoffmann has the most well-developed method, from which Gregory's
method differs significantly. Hoffmann also has a most well-developed
account of clinical topics in medical ethics, though the Edinburgh professors are less clear about their philosophical method, and usually provide
no argument for their particular views. Gregory, too, has a well developed
account of the topics that he addresses, and he advances what are clearly
to be recognized as philosophical arguments. His arguments display no
less rigor than those of Hume, Reid, or Beattie, for example. Finally,
Gregory provides a powerful philosophical content for such key moral

terms as 'gentleman' and 'profession'. His lectures thus emerge from an intellectual tradition but they do so in a unique way. In this respect, as we shall see, he contributes to the invention of professional medical ethics and the profession of medicine.

II. THE PUBLICATION OF GREGORY'S LECTURES

Gregory's medical ethics lectures first appeared in print anonymously, presented by a student editor with the following introduction:

> The following sheets contain two preliminary lectures, read not long ago, in one of the universities of a neighbouring kingdom, by a medical professor. Many copies, from a general satisfaction they afforded his audience, were taken down short-hand. Of these the reader is here presented with the most correct; and the editor flatters himself, that from the free and liberal spirit of enquiry which animates the whole of them, they will provide a most acceptable present to the public; and, of course, do not discredit the ingenious author (Gregory, 1770, p. iv).

Given the crucial role that Lyttleton played in seeing two of Gregory's other works into print – *Comparative View* and *A Father's Legacy* – it seems reasonable to suppose that Lyttleton played a similar role in seeing into print *Observations on the Duties and Offices of a Physician and on the Method of Prosecuting Enquiries in Philosophy*, which appeared anonynously (Gregory, 1770). Lyttleton may also have played a role in seeing the into the press *Lectures on the Duties and Qualifications of a Physician*, which appeared under Gregory's name (Gregory, 1772c). The evidence for this comes from a letter dated January 20, 1770; Gregory writes to Mrs. Monatgu:

> One of my students has published some of my lectures. I allowed any of my students to see and transcribe these Lectures that pleased, but should never have thought of publishing them for very evident ... Considerations. I had made several alterations, additions & Corrections in the Language this Winter which the printed copy wants. In the careless Dress they are in, I fancy there are some things that would please you. The Title is, Observations on the Dutys and Office of a Physician etc. If you think they will not discredit me, I beg you will shew them to Lord Lyttleton (NLS 3648, 1770, pp. 114v-115r).

We do not know the name of the student editor, but it may have been John Gregory's son, James, who was then a medical student at Edinburgh. I base this admitted speculation on a comparison of the style of the 1770 and the 1772 editions. The 1770 edition contains a hard-hitting "advertisement" at the beginning, which is completely excised in the 1772 edition. The following passage provides a sample:

> In the first lecture, the author has treated fully on the duties and offices of a physician: a path almost untrod til now. The noble and generous sentiments which are here displayed, will ever be a source of pleasure to minds unbiased by prejudice, self-interest, or the unwothy art of a Corporation. Whatever opposition this part of the work may meet with from those, who find their own foibles, or rather vices, censured with a just severity, the ingenious part of mankind, however, will not fail in bestowing that degree of applause so justly due its merit (Gregory, 1770, pp. iii-iv).

Moreover, the editorial changes in the 1772 edition display a consistent trend of correcting the often prolix and sometimes harsh writing in the 1770 edition. James Gregory became famous – infamous, if you will – for this prolix and bitter attacks on people and institutions in his published work, particularly his *Memorial to the Managers of the Royal Infirmary* (Gregory, J[ames], 1800). The bumptious style of the 1770 edition seems more from James' hand, while the measured and forceful style of the 1772 edition more closely matches other texts – published and unpublished alike – from John Gregory's hand.

This is not to say, however, that Gregory himself did not reign in some of his own writing. Consider the following three passages, near the beginning of the lectures, where Gregory notes the wit and satire to which dramatists had subjected physicians; the first from a student note set of 1767, the second from the 1770 edition, and the third from the 1772 edition.

> Physicians are much raged against, thus in a dramatic performance, they are always treated as solemn coxcombs & fools, but this is level'd at Particular persons Professors of the art, & not at Physicians in general (RCPSG 1/9/5, 1767, p. II).

> These arts, however well they might succeed with the rest of mankind, could not escape the censure of the more judicious, nor elude the ridicule of men of wit and humour. Accordingly, it has been pointed out

against them with as much keennesss, that we never meet with a physician in a dramatick presentation, but he is treated as a solemn coxcomb and a fool. But it is evident, that all this satire is levelled against the particular manners of individuals, and not against the profession of medicine itself (Gregory, 1770, p. 4).

These arts, however well they might succeed with the rest of mankind, could not escape the censure of the more judicious, nor elude the ridicule of men of wit and humour. The stage, in particular, has used freedom with the professors of the salutary art; but it is evident, that most of the satire is levelled against the particular notions, or manners of individuals, and not against the science itself (Gregory, 1772c, p. 5).

The 1770 version indicates that the student note set from 1767 records Gregory's words and sentiments. Note, however, that in the 1772 edition the third sentence is eliminated, in favor of a more mild account and "all this satire" becomes "most of the satire."

The *Observations* contain two lectures, which were modified – by subtractions as well as additions – into six lectures in the *Lectures*. In both versions Gregory addresses medical ethics *and* philosophy of medicine, although the title of the *Lectures* drops reference to the latter. Thus, Gregory continues the custom of providing preliminary lectures like Boerhaave, Hoffmann, Rutherford, Whytt, and Cullen, by addressing the philosophy of medicine, as it was then understood (King, 1978), and, with Young, medical ethics, too. He does so at greater length and with greater philosophical sophistication, however, than any of his predecessors or contemporaries. He also addresses both medical ethics and philosophy of medicine, because the two are so tightly related in his thinking. To see how this is so and to set the stage for the consideration of Gregory's medical ethics, I turn (in the next section) to a summary account, based on the previous chapter, of the intellectual resources that Gregory brings to his medical ethics and of the problem list that he confronted in then contemporary medicine.

Why did Gregory present and then publish these medical ethics and philosophy of medicine lectures? No definitive answer can be given, because there is no textual evidence to which one can appeal to provide an answer. Nonetheless, the following five-part explanation seems plausible.

First, we saw that Gregory took a strong interest, very early on, in improving medicine. The *Observations* and *Lectures* certainly contribute to this project by addressing the moral improvement of medicine, which, we saw in the previous chapter, stood in need of such improvement. Second, as we saw in the account of Gregory's appointment to his academic position, he may have been stung by the students' broadside and the attitudes they revealed in it about him. He may therefore have used these works in an attempt to differentiate himself from Cullen. In doing so, he relied on his strength, as a "thinker so just and original, and so universally acquainted with human nature" (AUL 404a, 1766, p. 5), as proven in *Comparative View* (Gregory, 1765). Third, as Robert Baker has suggested to me (in private communication), Gregory was teaching students, quite a few of whom could not read Latin and so lacked direct acquaintance with the original texts of Hoffmann and Boerhaave. Moreover, we know from the previous chapter that, while Gregory admired their work in medicine, he thought it in need of improvement. So, too, perhaps he thought their work in medical ethics and philosophy of medicine in need of improvement. Fourth, his students included young men who aspired to be gentlemen but who, in some cases, e.g., sons of merchants or colonials, were not born or bred to this social role. Moreover, this social role was destabilizing, as we noted in the previous chapter. Gregory's *Legacy* and his letters to Mrs. Montagu provide ample evidence of his interest in the moral formation of young people. What he did for his daughters in *Legacy*, he does for his medical students in *Observations* and *Lectures*. Finally, Gregory wants to improve medicine by "laying it open," with a particular view to breaking the hold of the medical corporations, the Royal Colleges, on medicine. The *Statua Moralia* of the Royal College of Physicians of London enunciate again and again the overriding concern for self-interest (Clark, 1964, Vol. I, pp. 393-417). Consider the following:

> No colleague will accuse by name another either of ignorance, malpractice, wickedness or an ignominious crime; not heap public abuse ... (Clark, 1964, Vol. I, p. 414).[44]

The penalty for such offenses was explusion from the College. Also the Royal College of Physicians of Edinburgh was pursuing its monopolistic interests at the time. In 1761 the College published a notice in the newspapers, that the College is

> ... Resolved to prosecute, as their Patent authorizes and directs them to do, all such who without their Licence ... shall ... assume the title of Doctor of Physick, and prescribe for the Internal diseases ... (Craig,

1976, p. 221).

Members of the College are not to consult with these upstarts. In the 1772 edition, even more so in the "Advertisement" to the 1770 edition, Gregory means to separate himself from the corporations – he was, we know, a Fellow of the Royal College of Physicians of Edinburgh – and provide a true account of what medicine ought to be. In this way, Gregory aims to improve medicine in a thoroughgoing way, by relieving its "estate" from the reactionary, anti-scientific and therefore harmful forces of the medical corporations.

III. SETTING THE STAGE: GREGORY'S INTELLECTUAL RESOURCES AND PROBLEM LIST

A. Gregory's Intellectual Resources

John Gregory brings to his medical ethics an unwavering twin commitment to Baconian method and the Baconian project of relieving man's estate. This twin commitment shapes him at the core as a physician, teacher, and philosopher. I therefore begin setting the stage for an account of Gregory's medical ethics with a review of Baconian method and, on this basis, consider Gregory's commitments regarding a Baconian science of man – and the moral philosophy, as well well as the philosophies of medicine and of religion, that the Baconian science of man generates.

The Baconian method involves *rigorous, disciplined observation*. It is also concerned with various degrees of reliability, where reliability is a function of the clarity of observations, the precision with which they can be stated, and their repetition over time. On the basis of observations, through which similarity is reliably detected, the natural philosopher makes rigorous and disciplined inductions to general accounts – again of varying degrees of reliability – of the causes of the observed phenomena – what Gregory calls 'facts'.

This method relies on a metaphysics of activity. That nature operates according to a given, invariant set of laws or principles forms the foundation of this metaphysics. God is the author of this given, invariant set of principles, as can be established on the basis of philosophical reasoning from effects to their causes, from a design in nature to its designer. That nature does not operate according to an invariant set of laws, or that these laws do not require a foundation, would strike Baconian thinkers, Gregory certainly among them, as absurd claims on their face – a widespread view in eighteenth-century Scotland. Thus, Gregory as a Baconian was a deist: God created the world and ordered it according to principles, or causal laws, and ordered it to the good of entities in it. All of these propositions can be known on philosophical grounds, independent of revelation.

> We shall consider what is meant by *nature* – It is sometimes meant by the word nature the general system of laws by which God governs the world at other times it is used to express the particular system of laws by which any part of the world is governed at other times to express that unknown principle in the animal œconomy by which the Body of itself may thro' off Diseases (RCSE D27, 1770, p. 2)

For Scottish thinkers in the eighteenth century, as this passage also indicates, Baconian method also involves a metaphysics of principles. Principles are real, constitutive, causal elements in things that produce their functions and so provide a real foundation for explanations of causality. These principles are really existing causes, observable in their effects, that constitute things *in re*. Scientific explanation for Gregory, then, involves reference to really existing causal principles: inherent, constitutive causes in things themselves. These principles have a normal function, expressed in a range of effects that can be observed with great reliability. The abnormal is defined as an insufficiency of causal efficacy, too little effect, or excess of causal efficacy, too much effect. The abnormal results from the effect of other principles on the operation of the principle in question. Both normal and abnormal structure and – far more important for medicine – function can be explained in this way. Moreover, living things possess the capacity to correct abnormal function; animals excel in this, while humans require assistance in the form of medicine.

Scientific observation of structure and function produce the following continuum of results. At one end of the continuum, Baconian method results in observations with little or no repetition of phenomena. Observations can sometimes exhibit great irregularity that does not exhibit patterns, or even emergent patterns. In response to this sort of observational finding, the Baconian scientist imposes on himself the discipline of continuing to gather observations, looking for similarities, and eschewing any precipitous move toward inductive generalizations. Baconian method at other times results in repeated, similar observations, suggesting the effects of a principle inherent in the entities observed, the middle of the continuum. In response to this condition the Baconian scientist imposes on himself the discipline of continuing careful observations, in order to carefully distinguish normal from abnormal function of the principle. Finally, at the far end of the continuum, Baconian method sometimes results in many repeated observations that exhibit a stable, highly consistent pattern, with variation in intensity from weak to strong. In response to this promising condition, the Baconian scientist follows rigorous rules for inductive generalization and proposes a principle as the causal explanation in the entities observed for the phenomena that have been observed. The Baconian scientist then goes on to test this hypothesized principle on the basis of further rigorous observation or carefully designed and executed experiments to establish the normal and abnormal functions of the principle with very high confidence, i.e., with negligible likelihood of being falsified. As a practical matter, such principles can be taken to be established with certainty in this pragmatic, clinical sense.

Diffidence defines the Baconian scientist. This capacity involves the disciplined habit of being informed by experience of the world, a source or origin of knowledge that is external to the knower, as opposed to requiring a match to preconceived or otherwise *a priori* ideas. The Baconian scientist lives by the discipline of having observations become corrected – even discarded – by future observations and of having one's causal explanations become corrected – even discarded – by future observation, hypothesis formation, and testing of the causal hypothesis. This is especially the case for principles established with certainty. The Baconian scientist should never act on self-interest – in fame, power, or money – and thus become dogmatic. Instead, he should always remain open to being shown that he is mistaken, what Gregory comes to call *being open to conviction*, the highest expression of the intellectual capacity of diffidence. The virtue of openness to conviction also immunizes the scientists against superstition, i.e., appeals to unfounded, untested, or – worst of all – untestable empirical claims.

In Gregory's hands Baconian method assumes a secular character in the following senses. I say 'sense*s*' because 'secular' is a term with multiple meanings and so must be used with great care. First, Baconian method does not appeal to revelation of any kind from any faith community. By definition, revelation originates outside the observable, natural order. Moreover, revelation does not meet the test for stable observations – revelation tends to happen to one or only a small group of individuals, as Hume points out. Second, Baconian method relies vitally on rigorous observation and induction, the criteria for which derive from the metaphysical realism of principles: There exist real causal principles *in things* and these principles express themselves functionally in a normal range of effects, which can be weak to strong depending on the influence of other principles. Third, this metaphysical realism is deist. Baconians such as Gregory do not take nature's order as surd, but as an explainable fact. God is the cause of the effect called the "order of nature," as is plain in the passage (above) from Gregory's clinical lectures.

To be sure, Hume departs from the Baconian "experimental method of reasoning" at this point. Hume rejects the argument from design in the *Treatise*, though not the metaphysical realism of the Baconians. Sympathy, for Hume, is a real, constitutive, causal, observable principle of human nature. In his rejection of deism Hume differs sharply from Gregory, which apparently did not constitute an obstacle to Gregory's adopting Hume's concept of sympathy, because the latter was the direct product of Baconian physiology, in which Hume and Gregory alike were steeped.

Thus, Baconian method, as Gregory understood and applied it, is secular in a meaning of the term that includes deism and a deist foundation for the origin of the principles discovered by his method. One could be secular and hold that God *originates* the principles of nature, including human nature. Hume's philosophy is also secular, but in a radical – and to Gregory unacceptable – way in that it excludes the deism of God as the creator of the world and the originator of the natural order.

Gregory also holds to a Baconian science of man and to the relief of man's estate. There are discoverable principles of human nature, just as there are of the "animal œconomy." There are two such principles: reason and instinct. Reason is a weak principle, though strong enough to play an important corrective function for instinct, the active principle. Instinct, the active, strong principle, provides the principle for morality, because morality concerns both judgment and action, not just judgment. Instinct is expressed in taste and religion and is paramount in the social principle.

Here I would like to offer an analogy to Baconian physics. Instead of a geometry of space in the old Aristotelian astronomy of interlocking spheres, rotating on fixed axes, Baconian method had produced a physics of space that explains regular change or motion in terms of an *active force*. Activity in inert matter requires an active principle; so too for human beings. The active force in dumb matter is gravity, a real functional something in things possessed of mass. Gravity "explains," because it is the cause of the natural *attraction* of matter that is otherwise inert. The social principle explains the natural attraction of humans to each other, an attraction that the weak principle of reason cannot explain. This natural attraction is especially strong among people in a contiguity shaped by common Humean "moral" causes. Baconian science of man thus neatly provides a solution to the crisis of Scottish national identity: The social principle naturally binds Scots together because it fosters interest in others, not self-interest. To what extent the crisis of Scottish national identity prompted this observational finding is a provocative question, although one beyond the scope of this book.

The social principle functions as sympathy, the first principle of human moral physiology. The physiology of sympathy is moral because it makes us other-regarding and therefore bound to each other. That is, we can observe a real physiologic principle of the human nervous system that causes both internal, intrapersonal phenomena, such as disease, and external, interpersonal, moral phenomena, such as fellow feeling. Sympathy thus provides an observable basis for morality, in the form of both judgment and behavior. Hume's account of the double relation of impressions and ideas explains the interpersonal sympathy widely observed by the physiologists but not explainable by them. The disciplined observer, Hume, discovers sympathy as an inherently normative principle: Sympathy originates morality. This may seem odd to twentieth-century thinkers but was a natural way of thinking for Baconian, eighteenth-century thinkers such as Gregory: The normative as a sub-species of human physiology – judgment and behavior count as functions in just the sense that heart rate and digestion do – was as open to observation as any other physiology. As Alasdair MacIntyre has pointed out, deriving "ought" from "is" statements does not involve the naturalistic fallacy when one performs the derivation on *functional* concepts (MacIntyre, 1981, pp. 55-56). Sympathy surely counts as such a concept.

Like all physiologic functions, sympathy exhibits an observable, normal range of function: the prompt, measured response to the distress and pain of others. Gregory understands pain in the following terms:

PAIN } This is a simple idea & therefore cannot be described – This sensation is known to every Body when one is pained he refers it to the part that ['is pained' is crossed out here] suffers violence. All impressions that are made upon the Body when they come to a certain degree give the sensation of pain. In order that pain be felt in any sensible part

of the Body, it is necessary that the nervous connection be not interrupted or obstructed – Pain is produced by whatever pre[ter]naturally
disturbs any of the sensible parts of the Body (RCSE D27, 1770, p.
323).

The causes of pain can be external, e.g., "puncture" (RCSE D27, 1770, p.
323) or internal, e.g., "such as calculi morbid ... " (RCSE D27, 1770, p.
323). Finally, for pain to occur, the brain must be in a "sound state, in the
feeling of Pain" (RCSE D27, 1770. p. 324). Pain is thus a self-report of
abnormality – a concept still current in human physiology – that is expressed in the natural language of pain, plainly discernible in adults and
children:

> The cries of a Child are the voice of Nature supplicating relief. It can
> express its wants by no other language (Gregory, 1772a, p. 40).

The natural language of pain and of pain behavior admit of ready and
reliable interpretation. Nature would no more mislead us in such matters
than create a climate too cold for us. Via properly working sympathy, one
automatically becomes aware of another individual's distress and pain.
Deism, as well as observation, teach one that pain and distress are abnormal and not good for that individual, whoever he or she may be. Thus,
sympathy puts one in a position to identify another's interests from a
reliable perspective external to that other individual, and sympathy moves
one to act to protect and promote another's interests by seeking to relieve
the cause of that individual's distress and pain. Success in this respect
returns that individual to – or at least puts him or her in the direction of –
a normal state, which is in his or her interest, again something known
from a reliable perspective external to his or her own. Because there is a
natural language of pain and a reliable translation, via sympathy, of that
natural language into compelling moral judgment and the behavior that it
therefore prompts, there are no moral strangers in the strong sense. It is
not the case, when it comes to pain, that no one can identify another's
interests and act with intellectual and moral authority to protect and
promote those interests.

Sympathy can function abnormally. An insufficiency of sympathy can
manifest in the form of dissipation, an inaction in response to the distress
and pain of others. An excess of sympathy can manifest in the form of
enthusiasm or of being "unmanned," losing self-discipline in response to
the distress and pain of others or one's own grief from death of a loved
one. Neither response advances the interests of the one who suffers and

causes degradation of one's inherent moral principle. Finally, sympathy can malfunction in the form of false sympathy, e.g., the suitor of false manners against whom Gregory warns his daughters in *Legacy*. Practicing false manners involves deep and dangerous corruptions of the moral principle because it leads one who suffers to place reliance in someone who will betray that trust – a moral calamity from the normative perspective of the social principle.

Sympathy can mislead us, when functioning properly, by someone who puts on, falsely, pain behavior or cries out in the natural language of pain. This, however, does not pose a problem for Gregory or any other physician in his clinical practice. Patients did not, it seems, feign pain and those alleged to do so – hypochondriacs and hysterics – experienced real distress, a form of psychic pain, and had diagnosable and treatable neurologic diseases or "nervous ailments." Thus, they are not, as it were, "faking it;" they exhibit symptoms of a disease.

Properly regulated sympathy finds expression in the virtues of tenderness and steadiness, creating as a matter of habit a well regulated regard for others. Properly regulated sympathy, as it were, makes one habitually open to the conviction that another individual may experience distress and therefore require relief. Tenderness expresses a form of compassion that opens one to, and engages one in, the suffering of others. After all, one will experience by the double relation of impressions and ideas their very impression of pain or distress. Tenderness provides the antidote to deficiencies of sympathy in the forms of dissipation and hardness of heart. Steadiness keeps that engagement regulated, well managed, so that one does not lose control in response to the threat or actual occurrence of great suffering and loss. Steadiness does not involved detachment, as in the sociological concept of detached concern (Lief and Fox, 1963). Rather, steadiness involves a kind of stoic self-mastery in direct openness and engagement in human suffering; in this respect Gregory seems to be borrowing from Adam Smith's theory of sympathy with its emphasis on self-command (Smith, 1976). Steadiness provides the antidote to excess of sympathy in the form of being unmanned. Together, tenderness and steadiness provide the antidote to false sympathy and to false manners. As Lyttleton (1760) puts it, tenderness should be seen as a regulated passion for the interests of others, which regulation steadiness provides. Tenderness and steadiness thus properly limit sympathy, producing true sympathy. True sympathy expressed habitually in judgment and behavior thus provides the basis for one's trust that another, i.e., the physician, will act in one's interest and not in the other's own – perhaps rapacious – interest.

Women of learning and virtue, as we saw in the previous chapter, provide the moral exemplars of the virtues requisite to the proper expression of the principle of sympathy in both tenderness and steadiness. Hester Chapone's admonitions to her niece fit this account exactly.

> Remember, my dear, that our feelings were not given us for our ornament, but to spur us to right actions. Compassion, for instance, was not impressed upon the human heart, only to adorn the fair face with tears, and to give an agreeable languor to the eyes; it was designed to excite our utmost efforts to relieve the sufferer (Chapone, 1806, p. 59).

In particular, in their social and intellectual intercourse with men, women provide the moral exemplars of how properly regulated sympathy results in asexual intimate human relationships, a theme that Gregory emphasizes in *Legacy* and that he experienced with Mrs. Montagu and the women of the Bluestocking Circle.

> They [women] possess, in a degree greatly beyond us [men], sensibility of heart, sweetness of temper, and gentleness of manners (Gregory, 1777, p. 93)

We do not choose sympathy; God the creator gives it to us as part of our nature. As an observable causal principle in the natural order of things, sympathy becomes a key element in Gregory and Hume's *physiologic essentialism of human nature*. Hume's language reflects his commitment to this physiologic essentialism:

> We may begin with considering a-new the nature and force of *sympathy*. The minds of all men are similar in their feelings and operations, nor can any one be actuated by any affection, of which all the others are not, in some degree susceptible (Hume, 1978, pp. 575-576).

> But *nature* may also be opposed to rare and unusual; and in this sense of the word, which is the common one, there may often arise disputes concerning what is natural or unnatural ... . We may only affirm on this head, that if ever there was any thing, which cou'd be call'd natural in this sense, the sentiments of morality certainly may; since there never was any nation of the world, nor any single person in any nation, who was utterly depriv'd of them, and who never, in any instance, shew'd the least approbation or dislike of manners. These sentiments are so rooted in our constitution and temper, that without entirely confound-

ing the human mind by disease or madness, 'tis impossible to extirpate and destroy them (Hume, 1978, p. 474).

Our task in the moral life is to sustain and deepen this pre-given moral principle, to live in and by its intellectual and moral discipline of our judgment, speech, and actions. We do this by training ourselves in the virtues of tenderness and steadiness.

There is, I think, an interesting link here to the constant experience of loss of loved ones to early death that marked Gregory's moral life and that of his contemporaries. Constancy of loss and of the threat of loss makes this moral training and self-discipline a necessity. Absent this moral discipline, dissipation in the form of hardness of heart or being unmanned in the form of profound melancholy become our fate – and degradation. Gregory's letter to Mrs. Montagu on the death of his wife, which we considered in the previous chapter, can be read as part of Gregory's work on moral discipline in response to his enormous loss. Gregory there and in his ethics lectures aims for the nobility of mind – a mind and heart open to the workings of sympathy – that he sees in Mrs. Montagu. Tenderness and steadiness become central virtues for human lives marked by regular, enormous loss, including the life of a physician.

A strong, moral-aristocratic sense of honor therefore works at the core of this science of man and the moral philosophy that it generates. I mean by this a sense of self-honor, or worth in our own eyes, of unsentimental and objective self-respect, that is sufficient to prohibit dissipation and false sympathy as repugnant and to discourage being unmanned as a deep and disabling embarrassment and form of shame.

To borrow a phrase from Stanley Hauerwas, we are involuntary constituted by sympathy and its moral demand that we cultivate tenderness and steadiness (Hauerwas, 1997). We do voluntarily train ourselves in these virtues, and that we would be intellectual and personally motivated to do so seems to be a function of honor, our sense of self-worth. In general, then, we are involuntarily constituted by reason and instinct; our job is to understand these principles, their roles and relationships to each other so that we can function well. In doing so, we relieve man's estate, the principal practical aim of the science of man.

Gregory's constant use of agricultural metaphors to explain what this improvement involves not only reflects his early experience in Aberdeen but something more substantive. In a national history before and during Gregory's life time, punctuated by periodic crop failure and resultant poverty, perfection in some strong sense of final completion of the human condition is not what we should aim for in relieving man's estate. Too much of the human condition, especially when it comes to health and disease – think here of Gregory's attack on the frightful pediatric mortality rate in Scotland in *Comparative View* – is simply not normal. We need to improve it, first by increasing the number of people experiencing normality, a major contribution to the relief of man's estate. Moreover, we will experience limits and setbacks in going after this ambitious goal. Gregory, as a medical student, began to emphasize the limits of medicine as it seeks to improve itself and never changes his mind on this subject. As a clinician, he reflects the understanding that we are not in control, especially the kind of control of the human condition required to perfect human beings by fully realizing their essence or nature as final cause. Gregory never uses such language. Control and its corollary, perfection, we can say for him, are illusions of pure reason, the product of systematic and speculative philosophy. Gregory has his feet squarely on the ground, in this world. He is a meliorist, to be sure, but not a utopian. For Gregory, relief of man's estate involves steady, incremental increase in the number of people experiencing normal health, lifespan, and social life so that they can train themselves in the social principle, which can then do its work in making the Scottish people a nation. The Scottish Enlightenment project of improvement is thus gradual "perfection," in the modest sense of making things whole as much and as often as possible, physically and morally, by habitual self-discipline of properly working sympathy.

Relationships formed by sympathy are asymmetrical downward. The one who acts from sympathy has obligations to the one in distress, the "downward" position vis-à-vis physical and mental normality. I pointed out in Chapter Two that Gregory models sympathy on the medieval, Highland, *moral-aristocratic obligation of paternalism*. The Highland social institution of the clan chief and, later, the social institution of the manor's laird are both constituted by asymmetrical moral relations downward: the chief owes more to his clansmen that they to him; so too for the laird and his tenant farmers. This becomes the template for the social principle in which each Scot has asymmetrical relations of obligation to other Scots.

The clan chief and the laird are born to their moral station and its obligations and so Scots, too, are born to sympathy as one of the constitutive principles as human beings. Paternalism obligates the clan chief and laird to cultivate and practice a regular regard and inclination to protect and promote the interests of those to whom he is bound by kinship and debt of service. And paternalism obligates the Scot to do the same toward those with shared national identity. Likewise, sympathy obligates one to protect and promote the interests of those close to one by virtue of moral causes or debt of service. When face-to-face with someone in distress – as the clan chief would be at a meeting of the clan – one responds from sympathy to those in distress, binding oneself to them across differences of class, gender, religion, and politics. Paternalism prevents exploitation – by avoiding harm and positively doing good (Reid, 1863, 1990) – of the inferior by the superior in hierarchical social and moral relationships such as that of clan chief and clansman. So too does sympathy function in the hierarchical social relationship of physician and patient.

Paternalistic sympathy becomes even more important in the increasingly socially and economically stratified Scotland of the eighteenth century. Physicians came more and more from the non-landed, upper classes, and patients came from many strata, including those higher than a physician might otherwise experience. Gregory's students would also encounter in the teaching ward at the Royal Infirmary, perhaps for the first time, patients from lower socioeconomic classes in roles other than servant or assistant to a merchant or worker in a colliery. Paternalistic sympathy provides the moral warrant for the growing power of the physician, especially in this institutional context of the Royal Infirmary. Increasing social stratification can drive people apart, weaken moral causes that join people naturally together, and promote self-interest. Gregory's reaction against city life and commerce in *Comparative View* is not simply reactionary; he has a corrective, moral agenda of considerable social and clinical urgency.

On Gregory's account of sympathy as a form of medical paternalism in its eighteenth-century Scottish sense there can be no moral strangers created by differences of class, gender, religion, politics, or any other factor. In this respect, sympathy goes beyond the model of clan chief and laird. Their duties of paternalism extend to those under their authority by virtue of kinship or land. Sympathy's duties of paternalism extend to any individual in one's purview who exhibits distress or faces danger. It is not the case that we are utterly without a reliable, morally authoritative perspective on the interests of others. Sympathy supplies a reliable perspective on the interests of others, because one experiences their interests just as they do. The consensus of experience is vivid, real, and binds us together morally. The "contagion" of an audience by actors, which Mrs.

Montagu describes, now characterizes all the world as a stage (Montagu, 1970, p. 30, pp. 31-32). Sympathy tells us that we should respond to the abnormal and repair it because it is not good for the individual who experiences it and we know the latter with moral authority from Baconian experience and its underlying deism. Others in distress are also not moral strangers, because one can know reliably enough about their interests to make reliable moral judgments about them and to take the action of offering to help. One treats all in distress as alike in need of succor, regardless of the social class of the one in distress. In sympathy thus understood Gregory has a powerful antidote to the man of false manners – an issue in private practice – and to the end of the age of manners when new social classes challenged accepted social hierarchical behavior – an issue in the Royal Infirmary. An ethic of moral strangers offers no help with these two problems; indeed, it only worsens them.

Here the difference between Smith's and Hume's accounts of sympathy and Gregory's reliance on Hume's becomes critical. Recall from Chapter Two that for Hume sympathy means that, when one sympathizes with another, one experiences the same impression – of pain – or emotion – of distress – as the other. For Smith, one *imagines* the other's pain or distress, at a remove from their immediacy in the experience of the other. Any move to abstraction incurs risk of erroneous conceptualization, as philosophers then and now know all too well. This is especially the case in Gregory's time, because philosophers regarded imagination – the faculty of combining the contents of different ideas into a new idea – as an inherently unstable human faculty, precisely because it involved exercise of reason unregulated, to a large degree, by experience. Gregory's account of sympathy does not have to deal with this problem. Sympathy, on Hume's account of it, gives the physician direct, vivid, highly reliable access to the patient's interests – in avoing the pain and distress of disease, injury, and early death – and generates obligations of paternalism – in the eighteenth-century Scottish sense of the term – to protect and promote the patient's interests, obligations that go far beyond the Hoffmannian concerns of prudence. These obligations, we shall see, can be defined in terms of the capacities of medicine.

Smith, obviously, had to have had confidence in the normal process of abstraction and combination involved in the workings of imagination. His account, once we strip it of his misplaced confidence in imagination – as we must – two centuries later, comes much closer to current accounts of *empathy*. Once we acknowledge the perils of unfounded or poorly founded imagination, the problem of patients as moral strangers does indeed emerge, as we shall see in the next chapter. But it was no problem for Gregory.

In making women of learning and virtue his moral exemplars Gregory becomes a feminine moral philosopher, because he molds sympathy through his Bluestocking feminist consciousness. He takes such women to be superior to men, who are at moral risk of being creatures of interest because of the pressures of commerce and city life. He also regards women as equal to men in having the prerogative to marry as they choose and not simply or largely for economic necessity.

In Gregory's feminine ethics women become the moral exemplars of a sympathy modeled on a medieval social institution, the Scottish clan, and a later, aristocratic institution, the agricultural estate. In contemporary feminism these tend to be seen as patriarchal, anti-egalitarian social institutions and so they almost certainly were, as we look back at them. It is therefore striking, indeed provocative – from our historical vantage point – that Gregory genders feminine the hierarchical social and moral relationships created by sympathy but retains the medieval, aristocratic structure that prevents their being moral strangers. This was not a problem for him; Mrs. Montagu was known, and held in considerable affection, for her paternalism toward those who worked in the collieries she inherited upon her husband's death (Myers, 1990). Gregory's sympathy is thus a non-patriarchal, medieval concept exemplified by then contemporary women of learning and virtue, who were advanced in their attitudes and behaviors and constituted a moral aristocracy.

Patriarchal social institutions – Gregory would, I think, say to us – tend to be rigid and hard-hearted. They will therefore not do as models for sympathy. Nonetheless, Gregory wants to retain the paternalism of Scottish patriarchy, because it *protects* those in an inferior or lower status by reason of gender, class, health, illness, or power. Private patients experienced inferior or lower status by reason of illness and, to some extent, loss of power, particularly female patients. The Royal Infirmary patients experienced these disadvantages, compounded by inferior social class and its diminished power. Gregory thus does not entertain or propose an egalitarian concept of the physician-patient relationship, because such a concept requires that in the relevant respects the parties to the relationship come to it as equals. Illness makes the patient vulnerable to pathology, to the physician – the physician of false manners, in particular – and, if admitted there, to the institutional strictures of the Royal Infirmary.

Gregory thus observed normative realities that exclude the twentieth-century concept of egalitarianism that has become so influential in contemporary bioethics. The aristocratic concept of paternalism, brought forward from its medieval roots, and the feminine virtues of tenderness and steadiness simultaneously shape Gregory's concept of sympathy (Haber, 1991).

From Baconian method and its capacity of diffidence Gregory derives the cardinal intellectual virtue of the physician as scientist. The physician qua scientist – which the physician must be if he is to contribute to the improvement of medicine – must be open to new ideas in the form of new observations, especially those that counter existing knowledge and in the form of newly proposed and supported principles. As Gregory puts it in his ethics lectures, the physician must be "open to conviction." From Baconian science of man Gregory derives a parallel moral virtue of the physician as sympathetic human being. Tenderness makes us open in compassion to the suffering of others and to the conviction that we should seek to relieve such suffering and thus contribute to the gradual relief of man's estate. Sympathy thus opens one to others, to a moral world in which there are no moral strangers, i.e., others whose interests must of necessity remain always opaque to one.

Gregory's philosophy of medicine builds on the Baconian method and science of man. Medicine, as a body of knowledge and clinical practice, springs from instinct, not reason.

> The art of medicine seems to be coeval with man, and at first to have taken its rise from the natural instinct implanted in men as well as other animals, from medicines first tried at random in order to give ease in pain or sickness and from observing the methods nature took to cure diseases (RCPSG 1/9/10, 1772, p. 1).

This view of the origins of medicine reflects Gregory's enormous confidence in the constancy of nature.

In Gregory's philosophy of medicine health is an observable norm of structure and function of human anatomy and physiology. Diseases constitute observable departures from this observable norm, a view that makes sense only on the assumption of the constancy of nature. Either the principle in question, e.g., that which regulates heart function, is weak in which case its effects are weak, e.g., fluttering or slowed pulse, and so the physician correctly diagnoses it as abnormal; or the principle in question is over-working in which its effects are too strong, e.g., palpitations or rapid pulse, and so the physician correctly diagnoses it as abnormal. Gregory dealt with infectious, acute, and in many cases life-threatening disease and injury and so required concepts of health and disease adequate to guide his clinical management of disease and injury, and clinical investigation into their causes and cures. Gregory also requires concepts to justify clinical response to disease, namely the attempt to restore normal structure and function. Medicine does so by assisting nature's self-corrective processes, which are built into us. When this fails, medicine should attempt to supplant nature's corrective processes. For both strategies the physician should be ever vigilant for the limits of medicine, i.e., for effects of clinical management that are unacceptably harmful.

These concepts also address effectively two of Gregory's "Four Capitall Enquirys" from his medical student days: "The Preservation of Health" and "The Cure of Diseases" (AUL 2206/45, 1743, p. 6). The first "Capitall Enquiry" involves regimens designed to prevent abnormal functions of the physiologic principles that together make up the human constitution. The second "Capitall Enquiry" directs the physician to restore normal function, always attentive to limits on medicine's improving capacities, i.e., to the iatrogenic risks that could either eliminate benefit gained by introducing another abnormal function or add risk by making the abnormal function worse. The physician understands 'normal' and 'abnormal' by reference to the observed normal function that defines health.

Gregory also committed himself as a medical student to the "Capitall" enquiry of "The Retardation of Oldage" (AUL 2206/45, 1743, p. 6). In Chapter Two we saw that Bacon's understanding of this capacity of medicine, the source for Gregory's "Capitall Enquirys," involved "the lengthening of the thread of life itself, and the postponement for a time of that death which gradually steals on by the natural dissolution and the decay of age ... " (Bacon, 1875c, p. 383). Dissolution and decay should be regarded as abnormal for advanced age; the careful observer notes the debility of aging as uncommon and a species of underfunctioning of a principle and so should label these debilities abnormal. Thus, *debility and decay associated with aging become diseases of aging*, the principles and cures of which should be sought by careful clinical observation and experimentation. For Gregory, aging itself appears not to be a disease.

These concepts of health and disease are not value-neutral; they are value-laden. Normal function reflects the God-ordained order which is good and so health can reliably be denominated a human good in Gregory's Baconian, deistic philosophy of medicine. Disease represents a departure from the divinely ordained order and so it should be disvalued. Nature, of course, is imperfect – observation teaches this above all – and so nature does not always serve human health.

> The word nature is variously used and explain'd; the most general sense in which it is applied is that of the system or body of laws by which the deity governs the world. The sense we take it in is that system of laws by which providence governs the animal machine. This machine tho' disordered, does not cease to move, like any factitious one, but nature of the system of laws by which it moves, makes some effort to rectify or regulate it's movements. These efforts however are sometimes very irregular and not at all uniform and sometimes so very violent as entirely to destroy the machine at other times insufficient to do any service from this weakness. Here art must be called in to regulate, to restrain and assist nature, according as she shall stand in need of either. When nature makes no efforts towards her own relief, physicians must be guided by experience and practice (NLM MS B7, 1768, pp. 61-63).

This provides ethical justification for medicine's efforts to correct nature's deficiencies or excesses, in order to restore the normal.

The moral authority for these concepts of health and illness and of the role of medicine is the following. Recall Gregory's condemnation in *Comparative View* of the staggering pediatric mortality rate in Scotland and his urgent call to correct it: Gregory uses concepts of health and disease that are designed primarily to address acute, life-threatening disease and these concepts do the job of justifying provision of "relief" of man's estate. These concepts provide sufficient warrant – the normal precludes early mortality – for clinical judgment and action to relieve patients of the distress of disease and injury. The sense of moral authority here is pragmatic: providing sufficient reason for clinical judgment and action that staves off the disaster of early mortality and crippling diseases and injuries. These concepts do not – and need not – claim the stronger moral authority of commanding obedience, at the price of engaging in contradiction (as in Kantian categorical imperatives). Nor do these concepts claim the weaker moral authority of concepts that only need to be taken into account in clinical judgment and action, but do not have the normative force to govern clinical judgment and action. Because these concepts are based on a powerful, discoverable normative concept of the normal, they possess adequate – and considerable – moral authority.

The goal of treatment in medicine is to cooperate with nature. Gregory, we saw in the previous chapter, has complete confidence in nature. Even Scotland, known still for its contrary weather, is not too cold for human habitation, Gregory claims in *Comparative View*. The physician uses treatment when it reliably appears that nature cannot do its self-correcting work. Nature should be assisted unless "her efforts are too violent" (RCPE Gregory, John 1, 1766, p. 11). The normative concept of *the normal* in Gregory's concept of disease provides the warrant for his philosophy of clinical management of patients. Thus, Gregory can, with moral authority, say of a patient who refuses relocation of a dislocated arm, that "No people would be stupid or obstinate as to leave a dislocated arm to nature" (AUL 37, 1762, p. 162r). Any patient's judgment to the contrary lacks moral authority, because it lacks intellectual authority: No one has observed dislocated arms spontaneously to self-correct. Nature lacks an adequate response to such a serious injury and so requires medicine's assistance.

Again and again, Gregory emphasizes the necessity of recognizing and respecting the limits of medicine and the limits of nature. Treatment carries its own risks and so Gregory does not subscribe to a view of medicine that warrants fighting death at all costs. *Respect for the limits of medicine's capacities is a cardinal virtue of clinical judgment and practice*:

> Method of cure. A chief part of the cure of diseases, consists in preventing any bad consequences after the disease itself is removed. Whatever symptoms in a disease point out a particular method of cure, we call the indications of cure, & require the most accurate discernment & solid judgment (RCPE Gregory, John 1, 1766, p. 11).

The physician should not substitute iatrogenic abnormality for the pathology of disease or injury, for doing so would violate the unavoidable limits of medicine imposed by the normative concept of normality that shapes medicine's concepts of health and disease.

Gregory's philosophy of religion turns on the distinction between those dimensions of religion that can be established on the basis of natural philosophy, Baconian method, and those that cannot and therefore derive from revelation, devotion, and prayer. As we saw in the previous chapter, Gregory does not denigrate the latter. Natural religion includes the propositions that God exists, that God designed and created the world, and that the order of the world is good – in Gregory's terms God provides "moral administration of the world" (Gregory, 1772a, p. 210). Gregory's philosophy of medicine depends on natural religion, for the former derives its normative concept of the normal from the latter. Not only is there no incompatibility between medicine and natural religion, the former leads to and requires the latter.

B. Gregory's Problem List

We saw in the previous chapter that Gregory lived and practiced at a time in which medicine was, to put it most diplomatically, in chaos. There was no stable pathway into medical or healing practice; anyone could offer services. There was no regulation in the form of uniform licensure of all physicians by the state, and the self-regulation and licensure that did exist were confined to the monopolistic activities of the corporations – which were to little or no effect. At the same time, to become a physician attendance at medical lectures was perceived to suffice.

In private practice concepts of health and disease and clinical philosophies competed in the marketplace, often with little or no scientific or other intellectual warrants by which potential patients might evaluate the intellectual competence of those purporting to help them. In the Royal Infirmary patients had to depend on the judgment of the trustees who appointed the physician staff. The trustees had some advantage in that they would have some idea of the type and level of medical education that physicians – whom the trustees considered for an appointment – had obtained. Nonetheless, what we now take for granted – that almost no one receives a medical degree without demonstrating mastery of relevant scientific and clinical knowledge – did not exist. There existed, as a consequence, rampant self-promotion, the unbridled pursuit of self-interest in the form of market share, income, and social prestige. The medical corporations attempted, with limited success, to monopolize treatment of the well-to-do sick; this only served to institutionalize the pursuit of self-interest.

Medical treatments failed as often, perhaps more often, than they succeeded. This was not an era – nor is ours – in which side-effects and other limits of medicine could be ignored (even if their causality was not understood). Moreover, many patients died or became worse despite medical interventions. Thus, the concept that the physician could be in control of natural processes and deliver promises and prognostications on such a basis did not have wide currency. Gregory's repeated emphasis on the limits of medicine reflects this therapeutic caution.[45]

Patients – both private and in the Infirmary – engaged in self-physicking. They diagnosed their own ills and treated themselves. Given the chaos of the medical marketplace and its wildly variant outcomes, the willingness to self-physick made a great deal of sense. Patients had rough and ready ideas of the nature of their problems and what would work to help them; they brought these ideas with them when they elected to seek out a physician. Moreover, as the Porters emphasize, private patients paid the piper and called the tune. The relationship with private patients, at least, was therefore a patient-physician relationship, as the Porters correctly argue (Porter and Porter, 1989). The patient's concept of severity of his or her problem prompted the call to the physician, and patients expected to play a role in the diagnosis and management of their problems. Thus, the patient-physician relationship involved subtle and unstable negotiations between oft-competing concepts of health and disease and between rival therapeutic philosophies and regimens. Dissatisfied patients simply discharged physicians – and spoke to their friends and acquaintances, no doubt. The market consequences for the physician of such badly managed relationships with patients could be and were severe.

In the Royal Infirmary there were perhaps the beginnings of a physician-patient relationship, at least on the teaching wards. But even in this setting physicians could not expect to be in complete control, either of the management of patients or of concepts of health and disease and therapeutic strategies based on those concepts. The trustees and lay managers had a great deal to say about the use of resources and the types of patients admitted. The trustees and managers, as Risse (1986) documents, practiced a bias against life-threatening diseases, with the result that the Royal Infirmary experienced an impressively low mortality rate – but only because it cleverly segmented the market to advance its own reputation and standing. This selection bias probably fed the confidence of Gregory and his colleagues in the existence of "cures" – in the sense of movement toward the normal – and in clinical research to develop and test more powerful cures – in the sense of restoring the normal most of the time with fewer unacceptable side-effects than those that existing remedies caused. This practice also involved an abuse of the label, 'incurable', an abuse that Gregory attacks in his lectures.

Scotland was, at this time, experiencing the end of the age of manners, as were England and Europe. Social relations and behaviors across the new economic and shifting class lines destabilized the Highland order of laird and cottar, clan chief and clansman or clanswoman. Moreover, Scotland also had the distinctive problem of the man of false sentiment and manners, a social pathology probably spawned by moral sense philosophy itself. As a consequence, patients lacked confidence in whom, among physicians, patients could count as morally trustworthy. This, combined with a lack of confidence in whom among physicians the patient could count as intellectually trustworthy, created a volatile context for transactions between patients and physicians – private and Infirmary alike. Moral character became of paramount importance. How could it be recognized? How could it be evaluated? Whom should a patient trust with his or her life and secrets?

The advent of man-midwives compounded these problems. The sexual abuse of female patients now became a live issue – both for them and for their husbands, as Roy Porter (1987) points out. Moreover, man-midwives appeared to be more willing to assist in dealing with illegitimate births – either through causing "stillbirth" or hiding non-paternity from the patient's husband.

Relationships between physicians and other health practitioners, apothecaries, and surgeons in particular, destablized toward the volatile, as well. The competition among these groups became severe and grew more – not less – intense. Patients often paid the price for this turbulence, as they sometimes still do. The more or less well understood system of *referral medicine* with which we have become familiar did not exist; no settled patterns of behavior existed to guide these practitioners in their dealings with each other.

Sometimes physicians abandoned the dying. Reflecting a long tradition (French, 1993c; Nutton, 1993), Hoffmann (1749) even thought it obligatory to do so. Sometimes, as in Gregory's case, they did not. In the latter circumstance, the physician-clergy relationships needed to be established and managed. Gregory's philosophy of religion gave him the intellectual and moral authority to criticize forms of religion that would cause individuals to become psychologically abnormal. In Gregory's judgment, "enthusiastic" religions did so, because they promoted an excess of passion, to the dilatation of both passion and reason, making reason too weak to do its modest but important work as the "weak" principle of human mental and moral constitution. On this intellectually and therefore morally authoritative basis, Gregory could and did reach the conclusion that clergy of such persuasion posed significant psychologic and spiritual danger for dying patients. In Gregory's mind the physical, mental, and spiritual all fell under the Baconian method, and the physician may well observe harmful effects of clergy on the patient's well being. The physician therefore should, with full ethical justification, interpose himself between the dying patient and such clergy, including, it would seem, clergy of the patient's choice. Friction became inevitable and there arose the ethical issue of *who should be in control* at the bedside of the dying patient.

In sum, in the late eighteenth century there was no profession of medicine in the sense of a fiduciary concept and practice governed by reliable intellectual and ethical standards. A fiduciary by habit blunts self-interest, focuses routinely and primarily on the interests of the one served, and acts to protect and promote that individual's interests. In Scotland, a fiduciary embodies the paternalism of the Highland chief and laird of the estate. For such an individual, much like Plato's philosopher-king, self-interest, or 'interest' as Gregory calls it, moves into a systematically secondary status. Scientific and ethical standards, as well as the intellectual and moral virtues their adherence requires, make this concept a reality in the moral life of the physician. A fiduciary in this strong sense earns and exhibits reliably the status of being trustworthy, both intellectually and morally.

Gregory repeatedly uses the word 'profession' in his medical ethics lectures. In private practice the available models of a profession were either the corporations or competition. Both enshrined the pursuit of self-interest with the difference being only corporate versus individual efforts to do so. In the Royal Infirmary the lay managers asserted and maintained control, dispensing charity to the worthy poor of the city of Edinburgh and its environs. Self-interest in this context took the form of a pursuit of reputation and prestige, often by experimenting on patients. Neither model would do as a solution to the problems Gregory confronted. Neither exhibits the intellectual rigor required by experimental method and neither exhibits moral rigor required by the new science of man and the concept of sympathy that this science produced in Hume's hands. In my judgment, Gregory responds to this *problem list* with the intellectual resources described above in an effort to conceptualize medicine as a profession in a meaningful ethical sense, as a fiduciary undertaking based on science and sympathy. In doing so, *Gregory invents professional medical ethics and therefore the profession of medicine in its intellectual and moral senses as a fiduciary social institution.*

Gregory also makes regular use of the term, 'gentleman'. By this he means normatively, we can now say, someone who is born or has formed himself according to the asymmetrical moral relationship of service to those in the lower social classes. In earlier times, one had to be born into the social role of gentleman, be the son of the laird of landed estate. One went on to the university to acquire a liberal education and manners appropriate to someone of this social standing and then one accepted a life of service and duty. With the advent of wealth independent of land, one could become a gentleman by accumulating wealth and social status, e.g., by becoming a university-educated physician successful in clinical practice. Now, one could do all of this and "purchase" the false manners of a gentleman. Gregory, I take it, rejects this option in favor of the concept of the gentleman as a man of true manners, true sentiment, rooted in properly cultivated and well functioning sympathy. To be sure, Gregory's makes an appeal to the social station of being a gentleman, to which his students aspired. And, I believe, he means to appeal to the moral-aristocratic, Scottish concept of a gentlemen who lives a life of service to others. His appeal, however, is *not* to patriarchy, because his moral exemplar for the virtues of being a gentleman is the woman of learning and virtue and the moral aristocracy they created.

IV. THE TEXTS

There exist the following texts as primary sources for Gregory's medical ethics: The 1772 *Lectures*, which appeared under his name, contains six lectures, each preceded by a summary list of topics. The first opens with a definition of medicine and some preliminary observations on the current state of medicine; it provides an introduction to Gregory's method, the moral sense theory of sympathy, and applies it to matters of virtue, including "openness to conviction" (Gregory, 1772c, p. 2). The second lecture addresses clinical ethical topics, such as communicating the gravity of the patient's situation, the "profits of his profession," manners, nostrums, and the "charge of infidelity against physicians" (Gregory, 1772c, p. 32). The third lecture discusses the relationship between the several sciences in medicine and its practice and the "ornamental qualifications" of a physician (Gregory, 1772c, p. 71). The fourth lecture addresses the philosophies of science and medicine. The fifth lecture continues the consideration of these topics, with particular attention to the "Causes that have retarded the advancement of the sciences" (Gregory, 1772c, p. 151). The final lecture addresses medical education and the "Advantages of laying the art open" (Gregory, 1772c, p. 195).

In the 1770 *Observations* there are two lectures, both lacking summaries. The first lecture of the 1770 version becomes the first three lectures of the 1772 version, while the second becomes the remaining three lectures of the 1772 version.

In addition to these two printed primary sources, students' versions of Gregory's ethics lectures survive, of which I have been able to find three (RCPSG 1/9/5, 1767; RCSE D 27, 1769; WIHM 2618, 1767). These texts have all been gathered together for the first time in a contemporary edition (McCullough, 1998). Gregory also addresses topics in medical ethics in his other lectures.

I will rely primarily on the *Lectures*, because these represent the authoritative text by John Gregory under his own name. I will make reference to and consider the *Observations* for topics not covered in or covered differently from the *Lectures*. Finally, I will supplement my consideration of the printed texts by reference to Gregory's lectures reflected in his students' notes.

A. Gregory's Definition of Medicine.

Gregory opens both versions of his medical ethics lectures with a definition of medicine:

> The design of the professorship which I have the honour to hold in this university, is to explain the *practice of medicine*, by which I understand, the art of preserving health, of prolonging life, and of curing diseases. This is an art of great extent and importance; and for this all your former studies were intended to qualify you (Gregory, 1772c, p. 2).

We see in this passage Gregory's affirmation of the Baconian definition of medicine that he had already embraced as a medical student more than two decades earlier. In Lecture IV Gregory offers an expanded definition:

> Medicine, or the art of preserving health, of prolonging life, of curing diseases, and of making death easy (Gregory, 1772c, p. 109).

"Making death easy" does not appear in the 1770 edition (p. 93), but reflects Bacon's concern with "outward euthanasia" (Bacon, 1875c, p.

387).

We saw in the previous chapter that in his "Medical Notes" (AUL 2206/45, 1743) Gregory accepts Bacon's three "offices" of medicine as the first three of four "Capitall Enquiries" (Bacon, 1875c, p. 383). Bacon treats these "offices" as obligatory capacities – rather than ends – of medicine as an intellectual and practical undertaking based on experimental method. That physicians should seek to improve these capacities as a matter of duty, particularly with respect to prolongation of life by increasing medicine's ability to increase life expectancy and postpone the delibilities that frequently accompany old age, shows that these are *capacities* and not ends. One cannot improve upon the end of an activity, because an Aristotelian *end* represents the perfection or improvement of that activity. One can, of course, improve a human capacity, as well as the capacity of a social practice like medicine.

Boerhaave's account follows Bacon's and emphasizes the capacities of medicine.

> ... the whole design of the Art is to keep off and remove Pain, Sickness and Death, and therefore, preserve present and restore lost health; so that every thing necessary to be known by a Physician, is reducible to one of these two Heads (Boerhaave, 1751a, Vol. I, p. 51).

The original text for this translation is the following:

> ...accipit haec Scientia, quae ut expellantur, considerandum omnen scopum Artis esse evitationem doloris, debilitatis, mortis; adeoque conservationem sanitatis praesentis, absentis restitutionem: ideoque quicquid sciendum & agendum Medico, uni tantum huic proposito inservire debet (Boerhaave, 1751b, Vol. I, p. 3a).

Medicine has an "object," the human body, toward and on which medicine exercises its capacities, according to Boerhaave.

> The *Object* therefore of Physic in the *Human Body*, is Life, Health, Disease, and *Death*, with the Causes from whence they arise, and the Means by which they are to be regulated, restored, or prevented (Boerhaave, 1751a, Vol. I, p. 53).

Medicine, according to Boerhaave, has a *scopum*, a target toward which it aims its capacities as an art, with 'art' understood to mean the disciplined application through the hands of scientific knowledge.

From these sources Gregory adopts the view that medicine possesses specific capacities, aimed at goals or targets that contribute to an end, human good, previously established as a given by the creator God. Gregory characterizes this end of medicine as "the convenience and happiness of life" (Gregory, 1772c, p. 151). His definition of medicine, then, assumes a given good for human beings and calls upon the capacities of medicine to contribute to the health-related aspects of that human good. Medicine can help to relieve man's estate by attempting to remove disease and prolong life, thus restoring health as a necessary condition for achieving human good. Medicine can also attempt to preserve health, thus contributing to one of the jointly sufficient and individually necessary conditions for achieving human good.

Note the content-minimal nature of this definition of the end of medicine, the goods that it has the capacities to seek for human beings. This comes as no surprise, given the urgency of the medical task of relieving man's estate by attempting to reduce the then very high mortality and morbidity rates in Scotland, and Gregory's deism and consequence confidence in nature. This will become important in the next chapter where I take up H. Tristram Engelhardt's (1986, 1996) challenge to "content-full" secular bioethics and whether his critique bears on Gregory's medical ethics.

B. The "Utility and Dignity of the Medical Art."

Gregory next takes up the "utility and dignity of the medical art" (1772c, p. 2). He points out that no one has seriously questioned medicine's utility, but he does note that individual physicians – more than medicine as a profession – have been the subject of "Much wit" (Gregory, 1772c, p. 3). His response to this underscores a central moral theme of the *Lectures* – the necessity of blunting interest.

> Physicians, considered as a body of men, who live by medicine as a profession, have an interest separate and distinct from the honour of the science. In pursuit of this interest, some have acted with candour, with honour, with the ingenuous and liberal manners of gentlemen. Conscious of their own worth, they disdained every artifice, and depended for success on their real merit. But such men are not the most numerous in any profession. Some impelled by necessity, some stimulated by vanity, and others anxious to conceal ignorance, have had recourse to various mean and unworthy arts, to raise their importance among the ignorant, who are always the most numerous part of man-

kind. Some of these arts have been an affectation of mystery in all their writings and conversations relating to their profession; an affectation of knowledge, inscrutable to all, except the adepts in the science; an air of perfect confidence in their own skill and abilities; and a demeanour solemn, contemptuous, and highly expressive of self-sufficiency. These arts, however well they might succeed with the rest of mankind, could not escape the censure of the more judicious, nor elude the ridicule of men of wit and humour. The stage, in particular, has used freedom with the professors of the salutary art; but it is evident, that most of the satire is levelled against the particular notions, or manners of individuals, and not against the science itself (Gregory, 1772c, pp. 4-5).

Some physicians pursue marketplace advantage by putting on intellectual airs, especially in their writings – a major form of self-promotion in the highly competitive market of medical ideas and practices. Physicians – and there were many of them – who pursue interest, i.e., self-interest, at the expense of patients "have an interest separate and distinct from the honour of the science" (Gregory, 1772c, p. 5). They commit a grave error: the physician as a true professional is to have no such separate interest. The physicians worthy of ridicule hide behind "mystery" and pretended "knowledge" and do not hold themselves accountable intellectually. Dramatists have been correct to ridicule such physicians, exposing their intellectual pretensions by mocking them.

Much Wit indeed, & criticism has at all times & all ages, been exerted against our profession, but that has always been chiefly aimed at certain ignorant Physicians Professors in the art, more than at y^e [the] Science of Physick itself. Some of its professors are Gentlemen of Honour, Modesty, candour, integrity & cet. [et cetera] but in this as in other Societys of men, these do no make the most numerous part, of it. Others again have professed this art to cloak ignorance, & cover their poverty (RCPSG 1/9/5, 1767, pp. I-II).

The dignity of the "profession" – Gregory always uses this term in a morally charged sense – turns on the learning required to practice it well and this involves the blunting of self-interest in favor of the interest of science and patients.

C. The Qualifications of a Physician.

Gregory then provides his account of the qualifications of a physician, the characterological moral prerequisites for the life of service to patients. He does so in the context of considering medicine "in two different views," as an opportunity (1) for the "exertion of genius" and (2) for the "exercise of humanity" (Gregory, 1772c, p. 8). He covers each topic in turn:

(1) Medicine is a field of endeavor without "established authority to which we can refer doubtful cases" (Gregory, 1772c, p. 14). This may, perhaps, be an oblique reference to Boerhaave's widely used publication of responses to letters about difficult cases sent to him by European physicians (Boerhaave, 1744, 1745). The exercise of *genius* in medicine should be subject to intellectual discipline. One uses genius – one's creative, imaginative intellectual powers – appropriately only when one uses them with strict intellectual discipline.

> Every physician must rest on his own judgment, which appeals for its rectitude to nature and experience alone. Among the infinite variety of facts and theories with which his memory has been filled in the course of a liberal education, it is his business to make a judicious separation between those founded in nature and experience, and those which owe their birth to ignorance, fraud, or the capricious systems of a heated and deluded imagination. He will likewise find it necessary to distinguish between important facts, and such as, though they may be founded in truth, are notwithstanding trivial, or utterly useless to the main ends of his profession. Supposing all these difficulties surmounted, he will find it no easy matter to apply his knowledge to practice (Gregory, 1772c, pp. 14-15).

Physicians also require genius in the form of understanding and dealing in a measured fashion with the "prejudices of his patient, of the relations ['of his own brethren" is added in the 1770 edition], and of the world in general" (Gregory, 1772c, p. 17).

Physicians also need to cultivate a "certain command of the temper and passions ... in order to give them their full advantage" (Gregory, 1772c, pp. 17-18).

> Sudden emergencies occur in practice, and diseases often take unexpected turns, which are apt to flutter the spirits of a man of lively parts and of a warm temper. Accidents of this kind may affect his judgment

in such a manner as to unfit him for discerning what is proper to be done, or, if he do perceive it, may, nevertheless, render him irresolute. Yet such occasions call for the quickest discernment and the steadiest and most resolute conduct; and the more, as the sick so readily take the alarm, when they discover any diffidence in their physician. The weaknesses too and bad behaviour of patients, and a number of little difficulties and contradictions which every physician must encounter in his practice, are apt to ruffle his temper, and consequently to cloud his judgment, and make him forget propriety and decency of behaviour. Hence appears the advantage of a physician's possessing presence of mind, composure, steadiness, and an appearance of resolution, even in cases where, in his own judgment, he is fully sensible of the difficulty (Gregory, 1772c, pp. 17-18).

The intellectual discipline of Baconian method applied to medicine, practiced over and over again, helps the physician develop the capacity to think quickly and reach reliable clinical judgments just as quickly – a crucial clinical skill in emergencies, then and now. This intellectual skill rests on moral virtues, including steadiness, developed in response to the sometimes trying behavior of patients who face emergent medical problems. As we shall see, sympathy itself requires the development of steadiness as one of the two cardinal virtues in which sympathy should be expressed in the physician's moral life. In emergencies steadiness requires at least the "appearance of resolution," a sort of steadiness in one's speech and behavior that can induce, reflexively, steadiness of clinical thought and judgment. One is reminded here of Descartes' account of one meaning of 'intuition,' namely, the ability to perform deductions quickly, from extensive practice (Descartes, 1979). Since Gregory's library included Descartes' works they may be the source of influence here.

(2) Gregory then turns to the "moral qualities required in the character of a physician" (Gregory, 1772c, p. 19) and in doing so presents his core views on sympathy or the exercise of humanity – recall that he uses the two terms interchangeably, as does Hume. The following passage is, in this respect, the most important in *Lectures*:

> I came now to mention the moral qualities peculiarly required in the character of a physician. The chief of these is humanity; that sensibility of heart which makes us feel for the distresses of our fellow creatures, and which, of consequence, incites us in the most powerful manner to

relieve them. Sympathy produces an anxious attention to a thousand little circumstances that may tend to relieve the patient; an attention which money can never purchase: hence the inexpressible comfort of having a friend for a physician. Sympathy naturally engages the affection and confidence of a patient, which, in many cases, is of the utmost consequence to his recovery. If the physician possesses gentleness of manners, and a compassionate heart, and what Shakespeare so emphatically calls "the milk of human kindness," the patient feels his approach like that of a guardian angel ministering to his relief: while every visit of a physician who is unfeeling, and rough in his manners, makes his heart sink within him, as at the presence of one, who comes to pronounce his doom. Men of the most compassionate tempers, by being daily conversant with scenes of distress, acquire in process of time that composure and firmness of mind so necessary in the practice of physick. They can feel whatever is amiable in pity, without suffering it to enervate or unman them. Such physicians as are callous to sentiments of humanity, treat this sympathy with ridicule, and represent it either as hypocrisy, or as the indication of a feeble mind. That sympathy is often affected, I am afraid is true. But this affectation may be easily seen through. Real sympathy is never ostentatious; on the contrary, it rather strives to conceal itself. But, what most effectually detects this hypocrisy, is a physician's different manner of behaving to people in high and people in low life; to those who reward him handsomely, and those who have not the means to do it. A generous and elevated mind is even more shy in expressing sympathy with those of high rank, than with those in humbler life; being jealous of the unworthy construction so usually annexed to it. - The insinuation that a compassionate and feeling heart is commonly accompanied with a weak understanding and a feeble mind, is malignant and false. Experience demonstrates, that a gentle and humane temper, so far from being inconsistent with vigour of mind, is its usual attendant; and that rough and blustering manners generally accompany a weak understanding and a mean soul, and are indeed frequently affected by men void of magnanimity and personal courage, in order to conceal their natural defects (Gregory, 1772c, pp. 19-20).

Humanity or sympathy directly engages us in the distress of others, and moves us to relieve that distress, just as Hume describes. Sympathy is the antidote to "rough manners" for which Scotsmen were then so well known.

Gregory describes the physician as a friend, someone is a close, voluntary relationship of caring for another through thick and thin. Risse takes
note of this, but underestimates its force in Gregory's thinking:

> John Gregory, for his part, [in response to the problems of "arrogance
> and condescension" among physicians] emphasized the importance of
> "sympathy," a human quality that allowed people to identify with each
> other's emotion. In his view patient and physician needs to become
> friends. The advantages accruing to both parties and to the therapeutic
> process were clearly recognized (Risse, 1986, p. 185).

Sympathy for a friend in distress is more demanding, in that friendship in
this context constitutes a true test of commitment to the well-being of
another. As we saw in the previous chapter, it is easy to take sympathetic
pleasure in the happiness of a friend, but far more draining and exhaust-
ing to suffer in sympathy with him or her. The paternalism of friendship
sometimes makes considerable demands of attention and self-sacrifice; so
too for the paternalism of sympathy.

Sympathy is expressed in "gentleness of manners, and a compassionate
heart" – the 1770 edition adds "softness" – all *female virtues*, as we saw in
Gregory's *Comparative View* and *Legacy*. He also appears to borrow from
Mrs. Montagu's work on Shakespeare, with his reference to the "milk of
human kindness" that should make the physician feel "the most violent
perturbation, and pungent remorse" (Montagu, 1970, pp. 164-165), should
he act against it requirements and pronounce the patient's doom.

Gregory seems also to have *Macbeth* in mind. In Act I Lady Macbeth
has to urge her husband to increase his ambition, in response to the
prediction of the weird women that Macbeth would be king:

> Glamis thou art, and Cawdor, and shalt be
> What thou art promised. Yet do I fear thy nature;
> It is too full o' th' milk of human kindness
> To catch the nearest way. Thou wouldst be great,
> Art not without ambition, but without
> The illness should attend it

> (Shakespeare, 1982, p. 50).

Here 'milk of human kindness' is a striking female metaphor, explained
by the editor as the "gentle quality of human nature." The editor explains
'illness' as "wickedness," or the relentless pursuit of self-interest, even to
the ruin of others and perhaps oneself. In other words, Macbeth lacks the

relentless ambition of unqualified self-interest, an ambition possessed in abundance by Lady Macbeth. She summons herself to act on ambition, precisely by abandoning female virtues that might restrain her:

> The raven himself is hoarse
> That croaks the fatal entrance of Duncan
> under my battlements. Come, you spirits
> That tend on mortal thoughts, unsex me here
> And fill me, from the crown to the toe, top-full
> Of direst cruelty! Make thick my blood,
> Stop up th' access and passage to remorse,
> That no compunctious visitings of nature
> Shake my fell purpose, nor keep peace between
> Th' effect and it. Come to my woman's breasts,
> And take my milk for gall, you murd'ring ministers

(Shakespeare, 1982, p. 51)

And so they do and so she infects Macbeth and they give themselves over to "[v]aulting ambition," i.e., self-interest untouched by regard for others, much less sympathy (Shakespeare, 1982, p. 55). Thus morally deformed, they propel themselves into destruction.

Physicians deal with the constant distress of their patients and loss of patients to death. Yet they can and do "acquire ... that composure and firmness of mind so necessary to the practice of physick," i.e., steadiness – a disciplined sympathetic engagement with patients. Such properly regulated sympathy neither enervates, i.e., causes dissipation, or unmans, i.e., causes loss of discipline. Physician who ridicule sympathy suffer themselves from hardness of heart, against which Gregory had cautioned near the beginning of this lecture.

A physician has numberless opportunities of giving that relief to distress, which is not to be purchased by the wealth of India. This, to a benevolent mind, must be one of the greatest pleasures. But, besides the good which a physician has it often in his power to do, in consequence of skill in his profession, there are many occasions that call for his assistance as a man, as a man who feels for the misfortunes of his fellow-creatures. In this respect he has many opportunities of displaying patience, good-nature, generosity, compassion, and all the gentler virtues that do honour to human nature. The faculty has often been reproached with hardness of heart, occasioned, as is supposed, but their being so much conversant with human misery. I hope and believe the charge is unjust; for habit may beget a command of temper, and a seeming composure, which is often mistaken for absolute insensibility. But, by the way, I must observe, that, when this insensibility is real, it is an misfortune to a physician, as it deprives him of one of the most natural and powerful incitements to exert himself for the relief of his patient. On the other hand, a physician of too much sensibility may be rendered incapable of doing his duty from anxiety, and excess of sympathy, which cloud his understanding, depress his spirit, and prevent him from acting with that steadiness and vigour, upon which perhaps the life of his patient in a great measure depends (Gregory, 1772c, pp. 8-9).

The physician who has correct sensibility of heart – a female virtue – does not suffer from "feeblenes of mind," an attempted insult that slurs women of learning and virtue and men who follow them as exemplars. Quite the opposite; the physician of such sensibility is following powerful, female virtues.

Gregory also cautions in this core text against false sympathy, the man of false manners who was becoming then so much of a problem – both in medicine and in society generally. This parallels Gregory's cautions in *Legacy* to his daughters against such men who might court them. One can use experimental, observational method to identify the physician of false sentiment and manners. False sympathy will show itself to be the opposite of real sympathy. Ostentatious behavior is diagnostic of false sympathy. True sympathy routinely engages the physician in response to distress and so "it rather strives to conceal itself" as habitual, routine behavior, nothing special. A physician who shows off or calls attention to his sympathetic response to the patient betrays false sympathy. The physician of true sympathy treats those of high and low social station alike. The physician of false sympathy defines himself as such by insinuating himself with false manners with the rich, in the pursuit of monetary interest and will "neglect those of Low rank" (RCPSG 1/9/5, 1767, p. VI).

But the real [sympathy] will never distinguish between those of High & Low rank, or if it does it will show even more assiduity in Low rank (RCPSG, 1/9/5, 1767, p. VI).

There is a story told about Gregory's views on women and the reception that these views received among the literati of Edinburgh, which was no better than that accorded his views by his fellow students in Leiden.

The story has it that Gregory presented a paper to the Poker Club – one of the most prestigious intellectual venues in the city – that advanced his feminine and feminist views. This paper provoked gales of laughter but Gregory left the meeting unbowed (Stewart, 1901, pp. 116-117). This negative reception of his feminist consciousness did not deter Gregory; he stuck very much to his intellectual guns. Perhaps Gregory's students knew of this incident – or others – and this may explain why Gregory provides such a strong defense of his feminine views of sympathy, including a counter-attack on his critics. Thus, at the end of this core passage Gregory turns the tables on critics of the female virtues of tenderness and steadiness. "Experience demonstrates" that unsympathetic physicians suffer from "weak understanding" – the antithesis of the genius that, he has already noted, is required of physicians – and are "void of magnanimity and personal courage;" they are mean-spirited, crabbed, ignoble cowards and exemplars only of a way of life that any person of honor should automatically and strenuously shun.

Medicine, then, is an emotionally and psychologically demanding and therefore risky calling, Gregory seems to be saying. Medical practice exposes the physician daily to the sufferings of others and many of one's patients will die from diseases that are truly incurable. Such experience with distress and with constant, almost unrelenting loss could harden one's heart, i.e., lead on to close oneself off from others as a measure of self-protection. To defend (unsuccessfully, of course) such dissipation, Gregory can be read as reasoning, the cowardly have to ridicule those who do risk engagement with patients and loss. If one comes to medicine, as did so many of Gregory's students, with the rough and ready manners of the day left uncorrected and undisciplined by sympathy, then one unnecessarily runs – even compounds – these risks.

Sympathy provides an antidote to hardening of the heart – for a moral sense theorist, a very serious characterological disorder – recall Hume's condemnation of ingratitude. Sympathy involves the disciplined engagement with distress and loss so that one can be warm hearted, a person from whom the "milk of human kindness" flows habitually. Properly developed sympathy involves steady engagement with distress and loss, so that one falls into neither dissipation nor unmannedness. One need not fear becoming a dissipated ingenue, lounging upon a couch with tears of false, immobile sympathy against which Hester Chapone cautioned. One's sensibilities can through sympathy remain open and responsive to the sufferings of one's patients.

D. The Duties and Offices of a Physician: Topics in Clinical Ethics

Gregory applies his moral philosophy of sympathy to clinical practice, in order to identify the clinical ethical obligations of physicians. He does so in response to the "problem list" discussed above and in Chapter Two. The first of these topics the extent to which physicians should seek to control a patient's regimen. Recall from the previous chapter that patients regularly self-physicked – whether at home or in the Royal Infirmary – and that private patients, at least, expected to have considerable say in the management of their problems. Recall, too, the prominent role played by the non-naturals in the management of patients' problems. The relationship with private patients was very much a patient-physician relationship. Against this background Gregory takes up the clinical ethical topic of the governance of the patient *by* the physician.

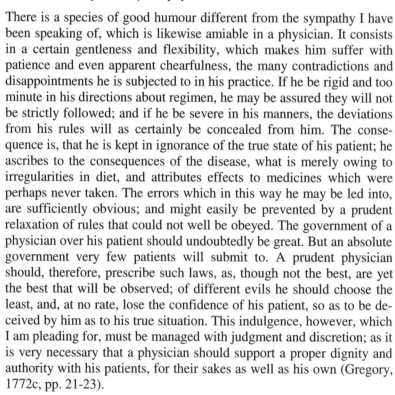

There is a species of good humour different from the sympathy I have been speaking of, which is likewise amiable in a physician. It consists in a certain gentleness and flexibility, which makes him suffer with patience and even apparent chearfulness, the many contradictions and disappointments he is subjected to in his practice. If he be rigid and too minute in his directions about regimen, he may be assured they will not be strictly followed; and if he be severe in his manners, the deviations from his rules will as certainly be concealed from him. The consequence is, that he is kept in ignorance of the true state of his patient; he ascribes to the consequences of the disease, what is merely owing to irregularities in diet, and attributes effects to medicines which were perhaps never taken. The errors which in this way he may be led into, are sufficiently obvious; and might easily be prevented by a prudent relaxation of rules that could not well be obeyed. The government of a physician over his patient should undoubtedly be great. But an absolute government very few patients will submit to. A prudent physician should, therefore, prescribe such laws, as, though not the best, are yet the best that will be observed; of different evils he should choose the least, and, at no rate, lose the confidence of his patient, so as to be deceived by him as to his true situation. This indulgence, however, which I am pleading for, must be managed with judgment and discretion; as it is very necessary that a physician should support a proper dignity and authority with his patients, for their sakes as well as his own (Gregory, 1772c, pp. 21-23).

Sympathy leads to the prudential, pragmatic judgment that patients who self-physick and who summon and dismiss physicians at will might harm themselves by resisting a regimen imposed by a tyrannical physician or by simply stopping it and switching to another physician and regimen that might be dangerous – either alone or in combination with the already initiated regimen.

> ... in this manner he will find his orders more regularly, & punctualy obeyed, if he is extremely rigid, & severe in his regimen, they then will be much more apt to disobey him, & deviate from his orders, & then perhaps he will be accounting for the surprising effects of Medicines w^h. [which] were every day thrown over the window (RCPSG 1/9/5, 1767, p. VI).

By adopting such a pragmatic strategy the physician maintains a sense of self-worth or dignity and also exercises appropriate – but limited – authority over patients. There is not a whisper here of the concept of patients' rights that have their origin in the autonomy of the patient as we understand these concepts in contemporary bioethics. Rather, the rights of the patient derive from the duties and offices of the physician. In other words, in forging an ethical concept of medicine as a profession, Gregory argues here and throughout that the physician's relationship with patients should be *a physician-patient relationship, originating in the moral response of the physician to the patient.*

We saw in Chapter Two that Gregory practiced and taught at a time when given the work of Whytt and others, nervous diseases became an important category in the nosology of disease. Gregory notes that patients with "nervous ailments" "put a physician's good-nature and patients to a severe trial" – as true then as now (Gregory, 1772c, p. 23). At the same time physicians have an interest in taking on such patients:

> It is said among the Tribe, that there is no disease so Lucrative as those of the nervous kind, as most people in high rank are troubled generally w^t [with] those complaints (RCPSG 1/9/5, 1767, p. VII).

Such diseases, one can delude oneself into thinking, clinically require many visits and expensive medications – altogether an attractive opportunity to maximize one's income. As an antidote to the powerful workings of self-interest – either to avoid such patients or to take full financial advantage of them – Gregory urges his students to treat "nervous ailments" as real, not fictions. The new science of neurology and diseases of the sympathetic nervous system teaches that these disorders are "as much seated in the constitution as rheumatism or a dropsy" (Gregory, 1772c, p. 23). He then invokes sympathy: "To treat patients with ridicule or neglect, from supposing them [nervous ailments] the effect of a crazy imagination, is equally cruel and absurd" (Gregory, 1772c, pp. 23-24). Sympathy also governs the clinical management of these diseases.

> Disorders of the imagination may be as properly the object of a physician's attention as those of the body; and surely they are, frequently, of all distresses the greatest [read 'most dreadful' in the 1770 edition], and demand the most tender sympathy; but it requires address and good sense in a physician to manage them properly. If he seems to treat them slightly, or with unseasonable mirth, the patient is hurt beyond measure; if he be too anxiously attentive to every little circumstance, he feeds the disease. For the patient's sake, therefore, as well as his own, he must endeavour to strike the medium between negligence and ridicule on the one hand, and too much solicitude about every trifling symptom on the other. He may sometimes divert the mind, without seeming to intend it, from its present sufferings, and from its melancholy prospects of the future, by insensibly introducing subjects that are amusing or interesting; and sometimes he may successfully employ a delicate and good-natured pleasantry (Gregory, 1772c, pp. 24-25).

Gregory next cautions his students against the perils of an exaggerated sense of self-worth and accomplishment that some physicians acquire later in their practice years, "when he is fully established in reputation and practice" (Gregory, 1772c, p. 25). His description of such physicians includes virtually all possible vices that sympathy properly nurtured should prevent.

> In the beginning he is affable, polite, humane, and assiduously attentive to his patients: but afterwards, when he has reaped the fruits of such a behaviour, and finds himself independent, he assumes a very different tone. He becomes haughty, rapacious, careless, and often somewhat brutal in his manners. Conscious of the ascendency he has acquired, he acts a despotic part, and takes a most ungenerous advantage of the confidence which people have in his abilities (Gregory, 1772c, p. 25).

Gregory also provides a detailed account of the physician's obligation of *confidentiality*. Physicians come to learn much about the "private characters and concerns" of private patients and their families (Gregory, 1772c, p. 26). The physician also sees patients in physical and mental conditions that others outside the home do not see. He also meets with people who are sometimes at their worst because of their illness. Sympathy here acts in the preventive, anticipatory way that Hume describes. "Hence appears how much the characters of individuals, and the credit of families may sometimes depend on the discretion, secrecy, and honour of a physician" (Gregory, 1772c, p. 26). By 'credit' here Gregory seems to mean not only an individual or family's good name and reputation, but quite literally their monetary credit. Landed and wealthy families borrowed regularly, e.g., against crops or to finance trade or business, and sickness of the male head of household was obviously of great interest to a creditor. The obligation of confidentiality means that the physician should keep these matters secret, an obligation that appears to admit of no exception.

Confidentiality has special application in the care of female patients.

Secrecy is particularly requisite where women are concerned. Independently of the peculiar tenderness with which a woman's character should be treated, there are certain circumstances of health, which, though in no respect connected with her reputation, every woman, from the natural delicacy of her sex, is anxious to conceal; and, in some cases, the concealment of these circumstances may be of consequence to her health, to her interest, and to her happiness (Gregory, 1772c, pp. 26-27).

Gregory does not provide any more detail of what he means here, but the discussions in chapter two help to fill in what his readers might well take from this passage; namely, confidentiality might extend to concealment of sexually transmitted diseases. Given the fact that Gregory's students could and did take Dr. Young's lectures on midwifery and add this skill to the marketing of their practice, confidentiality also might extend to the concealment of a pregnancy, not difficult to arrange when "fine ladies" often spent long periods in country homes away from their husbands in the city.

... There are also many Circumstances w^h. [which] women especially, from y^e. [the] Extrem Delicasy of their Nature w^d. [would] have conciled [concealed], & who depend intirely upon the Physician for this (RCPSG 1/9/5, 1767, p. VIII).

Confidentiality thus also might extend, as Roy Porter (1987) suggests, to concealment of non-paternity – either through the production of a still-birth by the man-midwife, through the concealment of pregnancy and birth followed by giving the child away, or simply by keeping non-paternity a secret.

In the 1770 edition, Gregory goes on to address sexual abuse of female patients, but the following passage is omitted from the 1772 edition, along with any other references to the topic.

> A physician, who is a man of gallantry, has many advantages in his endeavors to seduce his female patients; advantages but too obvious, but which it would be improper to recite. A physician who avails himself of these, is a mean and unworthy betrayer of his charge, or of that weakness which it was his duty, as a man of honour, to conceal and protect (Gregory, 1770, pp. 27-28).

Sexual abuse of patients violates *the asexual nature of the physician-patient relationship*. The physician should be steady, including when he is attracted to a female patient. Steadiness in this context includes concealing any attraction, much less acting on such personal feelings – whether positive or negative. The virtuous physician does not deny that he has such feelings; he subjects himself to the self-discipline of putting them aside, as a form of self-effacement, and not acting on them, as a form of self-sacrifice.

Gregory's remarks on this subject, as recorded in student notes, reveal his very strong views.

> Most profound secrecy where the women are concerned. Certain maladies, the concealment of which is of consequence. Has many opportunities of seducing; if he does mean and unworthy (WIHM 2618, n.d., p. 3).

> Again a Physician who is a man of Galantry injoys the best opp^{tys} [opportunities] of any person what ever for seducing his patients, but if he does such a thing he should be looked on as a base vile wretch, & not fitt to be trusted (RCPSG 1/9/5, 1767, p. VIII).

> Our profession have many opportunities to see the *private affairs* and actions of those they attend & particular among the *women*: A physician who takes the advantage of them, I mean one that is amorous is highly culpable (RCSE D27, 1769, p. 8).

The asexual relationship between virtuous women and men, of which women of learning and virtue provide the exemplar, should govern the physician-patient relationship. Just as the women of the Bluestocking Circle needed to be able to trust men whom they met at Mrs. Montagu's and other gatherings and just as his daughters needed to trust men, as Gregory points out in *Legacy*, so too women need to be able to trust their physicians to practice according to the self-control that self-effacement and self-sacrifice generate in response to the experience of sexual attraction. After all, while there may be such attraction, it has no part in sympathy for the *distress* of the female patient and therefore no part in the physician-patient relationship. Thus, the physician-patient relationship should be asexual, putting in place the basis for trust that was sorely missing so much of the time in eighteenth-century British practice of medicine.

Why is this passage from the 1770 edition omitted in the 1772 edition? This is an interesting question, given the evident force of Gregory's views as recorded in the student notes from this ethics lectures. As we saw in the previous chapter, William Smellie had come to prominence as a man-midwife and his techniques sparked controversy (Porter, 1987). The same William Smellie later wrote the second biography of Gregory and was among his friends. Perhaps Gregory did not want to cause offense or difficulty to his friend, one could speculate, and so this may be why Gregory omits this paragraph in the 1772 edition.

In the *Observations*, but not in the *Lectures* (where it is simply omitted), Gregory includes at the end of his discussion of changes in the physician's character as he ages (considered above) material that introduces confidentiality in the following terms:

> He not only takes a most ungenerous advantage of the confidence which people have in his abilities, but lives upon the effects of his former reputation, when all confidence in his abilities has ceased: because a physician who has once arrived at a very considerable practice, continues to be employed by many people or their friends, who think of him themselves with contempt; they employ him because it is fashionable to do so, and because they are afraid, if they nelgected it, their own characters may suffer in the world (Gregory, 1770, pp. 25-26).

Gregory's worry here is stark: Patients sometimes feel compelled to continue to employ a physician, whose intellectual and moral capacities they do not respect, for fear of the social pain that could follow if they did not, and the physician took revenge for his loss of income and prestige by violating their confidentiality. Gregory would reply that the physician who was a professional in the ethical sense would never violate the obligations of self-control and self-sacrifice, imposed by sympathy, for such crass self-interest – or any other form of self-interest for that matter.

Gregory argues that "[t]emperance and sobriety are virtues peculiarly required in a physician" (Gregory, 1772c, p. 27). We saw in the previous chapter that Gregory takes the view that drink in moderation can warm the hearts of those who live in the cold climate of Scotland. We also saw that the Aberdeen Philosophical Society had rules governing the taking of alcohol at meetings, so that intellectual discourse would be served, not hindered. So too, here, because in "the course of an extensive practice, difficult cases frequently occur, which demand the most vigorous exertion of memory and judgment" (Gregory, 1772c, p. 27), both of which are impaired by inebriation. It is no defense of insobriety that "some eminent physicians ... prescribed as justly when drunk as when sober" (Gregory, 1772c, p. 27) because they were then practicing "by rote" not by experience and its intellectual discipline (Gregory, 1772c, p. 27). Inebriation also affects the discernment that Gregory repeatedly underscores as a crucial feature of clinical diagnostic judgment, i.e., the ability to distinguish one disease from another and to distinguish the symptoms of a disease from signs of nature's response to that disease, and to distinguish the effects of regimen from the effects of disease. A failure in any of these respects could harm or even kill the patient. Finally, inebriation is inconsistent with the virtue of steadiness required by sympathy:

> Intoxication implies a defect in the memory and judgment; it implies confusion of ideas, perplexity and unsteadiness; and must therefore unfit a man for every business that requires the lively and vigorous use of his understanding (Gregory, 1772c, p. 28).

Intoxication can make one dissipated by deadening the senses and hardening the heart against suffering and loss; it can also unman one by overly warming one's heart, one's passions. Intoxication thus represents the antithesis of steadiness and therefore of sympathy.

Gregory ends his first lecture by underscoring the importance for the physician of "that candor, which makes him open to conviction, and ready to acknowledge and rectify his mistakes" (Gregory, 1772c, p. 28). As we saw earlier in this chapter, openness to conviction – to new knowledge based on experience, especially knowledge that requires one to change one's own knowledge and practice – constitutes the intellectual virtue of Baconian science that parallels sympathy and its moral virtues of tenderness and steadiness and that opens the physician in a disciplined way to the distress of patients and loss of them to death. Failure to be open to conviction makes one a deadly menace to patients.

An obstinate adherence to an unsuccessful method of treating a disease, must be owing to a high degree of self-conceit, and a belief of the infallibility of a system. This error is the more difficult to cure, as it generally proceeds from ignorance. True knowledge and clear discernment may lead one into the extreme of diffidence and humility; but are inconsistent with self-conceit. It sometimes happens too, that this obstinacy proceeds from a defect in the heart. Such physicians see that they are wrong; but are too proud to acknowledge their error, especially if it be pointed out to them by one of the profession. To this species of pride, a pride incompatible with true dignity and elevation of mind, have the lives of thousand been sacrificed (Gregory, 1772c, pp. 28-29).

Pride that follows from lack of openness to conviction links to vices of false sympathy and its attendant hypocrisy and of hard-heartedness and its attendant "weak understanding and mean soul" and lack of "magnanimity and courage" (Gregory, 1772c, p. 21). Intellectual vices go hand in hand with moral vices, because vices undermine and destroy the conjoint intellectual and moral virtues required of the physician of learning and virtue.

At the start of the second lecture Gregory takes up the "decorums and attentions peculiar to a physician" (Gregory, 1772c, p. 31) and does so on the basis of a distinction between decorums and attentions "founded in nature and common sense" and those generated by "caprice, fashion, and the customs of particular nations" (Gregory, 1772c, p. 32). The former ground "immutable" obligations while the latter grounds obligations that are "fluctuating and less binding" (Gregory, 1772c, p. 32).

On this basis Gregory addresses the question of the role that patients should play in their own care. We have already seen that in the first lecture Gregory argues that the physician should not seek absolute authority over the patient, on grounds partly of sympathy and partly of the prudential consideration that this strategy might not work and might also result in harm to the patient from failure to follow the physician's regimen or from self-physicking or from the interaction of the two. We also saw that Gregory argues that the physician should be open to conviction – from whatever worthy source, including the patient. Thus, Gregory has already put in place the foundations for the following:

> Sometimes a patient himself, sometimes one of his friends, will propose to the physician a remedy, which, they believe, may do him service. Their proposal may be a good one; it may even suggest to the ablest physician, what, perhaps, till then, might not have occurred to him. It is undoubtedly, therefore, his duty to adopt it. Yet there are some of the faculty, who, from a pretended regard to the dignity of the profession, but in reality from mean and selfish views, refuse to apply any remedy proposed in this manner, without regard to its merit. But this behaviour can never be vindicated. Every man has a right to speak where his life or his health is concerned, and every man may suggest what he thinks may tend to save the life of his friend. It becomes them to interpose with politeness, and a deference to the judgment of the physician; it becomes him to hear what they have to say with attention, and to examine it with candour. If he really approves, he should frankly own it, and act accordingly; If he disapprove, he should declare his disapprobation in such a manner, as shews it proceeds from conviction, and not from pique or obstinacy. If a patient is determined to try an improper or dangerous medicine, a physician should refuse his sanction, but he has no right to complain of his advice not being followed (Gregory, 1772c, pp. 32-34).

This version softens that which appears in *Observations*:

> If a patient is determined to try an improper or dangerous remedy, a physician should refuse his sanction, but he has no title to complain of his advice not being followed, as he has no right to hinder any man from going out of the world in his own way (Gregory, 1770, pp. 32-33).

One student note set indicates that this frankness is closer to Gregory's lectures as originally presented:

> ... but if he does not approve it let him say so, then if the patient or his friends persist in applying it, Let them do so, why not let a man die in his own way if he will (RCPSG 1/9/5, 1767, p. IX)

In the 1770 edition Gregory writes that every one has a "title" to speak when his health or life are at stake (Gregory, 1770, p. 32). Gregory says nothing about the origins of this right or title, but he asserts his claim in such a way that he means it to be an "immutable" obligation, not a function of "caprice, fashion, and the customs of particular nations." Common sense, i.e., the common sense or universal instinct of sympathy in all human beings, leads the physician to see that failing to let patients speak only adds to their distress and anxiety and may also block the avenue to effective treatment, a not unwise view in a time of routine self-physicking – Gregory's as well as ours.

Is this also an autonomy-based argument? Gregory's use of the terms 'right' and 'title' *prima facie* invites such an analysis. However, on closer view this analysis does not hold up. Gregory nowhere makes reference to conceptual sources that we would now use, e.g., respect for autonomy; nor does he make reference to a natural rights theory, such as Locke's, or a natural law theory, such as Hoffmann's, to ground this right or title to speak on the part of the patient. This right or title derives from the physician-patient relationship and functions within it: the right or title to speak has no origin independent of the physician-patient relationship. That this is Gregory's view seems more apparent when we consider what he has to say about his next topic – communicating with patients about their condition. Neither rights-talk nor title-talk play any role in Gregory's thinking on this important clinical ethical topic. As Ruth Faden and Tom Beauchamp have convincingly argued, autonomy-based accounts of the patient's rights regarding the physician's communications come well after the eighteenth century (Faden and Beauchamp, 1986). Again, the physician's moral commitment generates the physician-patient relationship.

Gregory does not hold, as an autonomy-based theory would require, that there is an exceptionless or nearly exceptionless obligation to communicate the *truth* to the patient about his or her condition and its prognosis. Instead, Gregory treats this obligation as a function of sympathy, i.e., limited by the serious harm that *direct* communication as a uniform practice could cause for some patients. One of his students recorded Gregory's remarks on this subject:

> A Physician is sometimes at a loss what to say to Patients concerning their diseases and y^e [the] state of their health now in this Case it would be very wrong to acquaint the patient that he was really on ye point of Death, as this would hasten his death so much the sooner, now this may be a very important time for to acquaint his friends, as some minutes longer in life, might be a dale [perhaps a variant on 'deal'.] of service to ye family (RCPSG 1/9/5, 1767, p. X).

Another student set of notes provides some insight into Gregory's concern about harming some patients by telling them they are dying:

> He who comes in to a patient & in a harsh and brutal manner tells him
> he is dying, is acting the part of one giving the sentence of death rather
> than a Physician Yet I would be loath to conceal from any person the
> state he was in, if any thing of great consequence depended on him, as
> the ruin of his family, the settling of his affairs, & ce. [et cetera] –
> (RCSE D 27, 1769, p. 7).

Harshly telling patients they are dying is never permitted, because it
violates the tenderness required by sympathy. Not telling patients "when
no provision" (WIHM 2618, 1767, p. 4), i.e., there is no will, will result
in calamity to the patient's family, which sympathy prohibits. Such
patients should be told of their condition, but not harshly. Thus, it is
"[n]ecessary not to tell the truth with regard to life sometimes" (WIHM
2618, 1767, p. 4), i.e., when the patient has made provision for his family
and when even a gentle communication is expected to hasten the patient's
death.

> There is one thing that particularly requires your attention, & that is
> telling a Pat. [patient] the true *state* & *hazard* he is in when labouring
> under any Disease & here a small *Deviation* from *truth* is sometimes
> requisite. He who comes to a Pat. & in a harsh & brutal manner tells
> him he is dying, is acting the part of one giving the sentence of Death
> rather than a Physician (RCSE D 27, 1769, pp. 6-7).

In another note set Gregory verges on endorsing lying to the dying patient:

> A Physician is sometimes at a loss what to say to Patients concerning
> their disease & y^e. [the] state of their health now in this case it would
> be very wrong to Acquaint the patient that he was really on the y^e.
> point of Death, as this would hasten his death so much the sooner
> (RCPSG 1/9/5, 1767, p. X).

Gregory's treatment of this topic in the *Lectures* is interesting:

> A physician is often at a loss in speaking to his patients of their real
> situation when it is dangerous. A deviation from truth is sometimes in
> this case both justifiable and necessary. It often happens that a person
> is extremely ill; but yet may recover, if he be not informed of his danger. It sometimes happens, on the other hand, that a man is seized with

a dangerous illness, who has made no settlement of his affairs: and yet perhaps the future happiness of his family may depend on his making such a settlement. In this and other similar cases, it may be proper for a physician, in the most prudent and gentle manner, to give a hint to the patient of his real danger, and even solicit him to set about his necessary duty. But, in every case, it behoves a physician never to conceal the real situation of the patient from the relations. Indeed justice demands this; as it gives them an opportunity of calling for further assistance, if they should think it necessary. To a man of a compassionate and feeling heart, this is one of the most disagreeable duties in the profession: but it is indispensible. The manner of doing it, requires equal prudence and humanity. What should reconcile him the more easily to this painful office, is the reflection that, if the patient should recover, it will prove a joyful disappointment to his friends; and, if he die, it makes the shock more gentle (Gregory, 1772c, pp. 34-35).

In the 1770 edition the third sentence of this passage reads: "It often happens that a person is dangerously ill, who, if he was to be told of his danger, would be hurried to his death" (Gregory, 1770, p. 33). We have here one of the first expressions of what has come to be known in American common law of informed consent as "therapeutic privilege" – a justification for an exception to the general obligation to be truthful with patients (Faden and Beauchamp, 1986).

Gregory notes that truthful communication with patients is unpleasant, but that it remains obligatory. After all, sympathy puts the physician in the place of the patient who is about to learn that he is dangerously ill and so the physician experiences the stress and anxiety that this might well cause. Hence the duty is "disagreeable," quite literally. The impression of distress and anxiety produced in the physician by the double relation of impressions and ideas is an impression no one would choose to have or find agreeable.

Gregory describes this as a "painful office" or duty. 'Office' derives from the Latin *officium* in the sense of a role-related obligation. Thus, a person in the moral role of physician in the physician-patient relationship has role-related obligations that originate in sympathy. The private patient may summon the doctor – indeed, as many physicians as the patient prefers and is willing to pay for – but this generates only the *patient-physician* relationship, which is contractual in nature. The patient initiates this contractual relationship because he thinks that it will advance his interests to do so. This is a commercial transaction, based on self-interest.

If the physician responds in kind, i.e., on the basis of his self-interest – in money, authority, power, market share, and prestige (so well documented by the Porters, 1989) – then there is no protection for the patient: no basis of trust, no assurance of an integrity-based commitment to scientific standards of care. Sympathy thus does not lead the physician to respond in kind out of self-interest, but to respond out of sympathy or interest in the patient; the physician's attention turns to and becomes rooted in the interests of the patient. Indeed, sympathy in the case of truthful communication with patients obligates the physician to do something disagreeable, i.e., *against* the physician's interest. For Gregory, the physician-patient relationship generated by the physician's sympathy-driven response to the patient's distress – not his or her money – becomes a moral relationship because it is essentially, necessarily other-directed.   Physicians can generate such a relationship regardless of whether the patient summons and pays the physician, and so Gregory's sympathy-based account applies both to private and Infirmary patients. This moral relationship is precisely the *opposite* of relationships based on self-interest – city life and commerce – that Gregory finds repugnant in *Comparative View*.

Gregory closes this passage with a pragmatic buttressing argument. Communicating the truth to patients amounts to a no-lose proposition in terms of its effects on the patient. Thus, as a rule, there is only clinical gain in honest communication when it is obligatory.

Recall from the previous chapter that physicians commonly withdrew form their patients when they were dying. Ludwig Edelstein's commentary on the Hippocratic injunction to respect the power of nature and the limits of medicine helps us to understand why physicians might take such a view: high mortality rates lead to a poor reputation, decreased market share, and compressed income (Temkin and Temkin, 1967). Physicians pursuing interest in contractual relationships with patients in a competitive market would be irrational to do anything other than withdraw from dying patients, especially since fees are hard to justify to a skeptical patient when the physician has nothing by way of an effective cure to offer. Hoffmann, recall, makes obligatory the leaving off of the care of the dying. This was also the prevailing custom at the time in Britain (Porter and Porter, 1989; French, 1993c; Nutton, 1993). Gregory's admonition to this students on this topic departs sharply from then current custom and tradition.

Let me here exhort you against the custom of some physicians, who leave their patients when their life is despaired of, and when it is no longer decent to put them to farther expense. It is as much the business of a physician to alleviate pain, and to smooth the avenues of death, when unavoidable, as to cure diseases. Even in cases where his skill as a physician can be of no further avail, his presence and assistance as a friend may be agreeable and useful, both to the patient and to his nearest relations (Gregory, 1772c, pp. 35-36).

In the 1770 edition Gregory views abandoning dying patients as a "barbarous custom" (p. 35). This language is appropriate because the custom in question violates the sympathy-based obligation to relieve the patient's distress. Dying patients experience distress – e.g., from fear of being alone or unattended – that the physician's "presence and assistance as a friend" can relieve. As one of his students records: "... even the presence of a Physician has great effect on easing the mind of the patient" (RCSE D27, 1769, p. 8), which is still true. Recall, too, that the test of sympathy comes when one's friends are distressed and therefore when one is even more deeply distressed in response. The obligation not to abandon the dying is entailed by Gregory's concept of sympathy. It also follows from Gregory's definition of medicine later in the *Lectures*, which includes "making death easy" (Gregory, 1772c, p. 109).

Moreover, the physician can often relieve the distress of the dying process itself. Reflecting perhaps the widespread use of analgesics, such as laudanum, and the ideal of a "perfect euthanasia" Gregory takes the following view:

There seldom or never occurs a case where we cannot give the patient some Remedies, at least to give relief if we cannot remove the disease: therefore in all cases that are incurable, I shall endeavor as much as possible to smooth the avenues of death (RCSE C36, 1771, p. 10).

Here Gregory appears to be adopting Bacon's concept of "outward euthanasia" (Bacon, 1875c, p. 387). Outward euthanasia concerns the "easy dying of the body," which Gregory interprets so as to create an obligation to relieve the pain and distress of the dying patient (Bacon, 1875c, p. 387).

Again, you ought never to leave your Pat. [patient] altho' you plainly see you can give him no relief but endeavour as far as is in your power to alleviate the Symptoms of his Disease; even the presence of a Phys^n [physician] has great effect in easing the mind of the Pat. (RCSE D27, 1969, pp. 7-8).

In another note set Gregory urges: "Do every thing that is not morally criminal for his patient" (WIHM 2618, 1767, p. 5), which appears in the published texts, perhaps opening the door to what we now call physician-assisted suicide.

We also saw in Chapter Two that Bacon identified a *second sense of euthanasia*, or a good death, namely "the preparation of the soul" (Bacon, 1875c, p. 387). Recall also that Gregory's philosophy of natural religion provides him with an intellectually and morally authoritative position from which to address what we may call "inward" or "spiritual euthanasia." This becomes an important consideration when we recall that there was considerable fluidity between "religious and naturalistic explanations" of disease and treatment (Wear, 1987, p. 241). Physicians thus have obligations to patients until their deaths.

> Neither is it proper that he should withdraw when a clergyman is called to assist the patient in his spiritual concerns. On the contrary, it is decent and fit that they should mutually understand one another and act together. The conversation of a clergyman, of cheerful piety and good sense, in whom a sick man confides, may sometimes be of much more consequence in composing the anguish of his mind, and the agitation of his spirits, than any medicine; but a gloomy and indiscreet enthusiast may do great hurt; may terrify the patient, and contribute to shorten a life that might otherwise be saved (Gregory, 1772c, p. 36).

Physicians, on the basis of a firmly established philosophy of natural religion and sympathy – both of which should also, of course, govern the obligations of clergy – can judge which clergy are good and bad for patient. A clergyman of "cheerful piety and good sense" is good for patients, but an "[e]nthusiast may injure a patient" (WIHM 2618, 1767, p. 5). Enthusiasts over-activate the passions and this will not be good for any seriously ill, dying patient – or anyone else for that matter. In addition, symptoms due to enthusiasm are difficult to distinguish from symptoms of nervous disease or reaction to one's imminent death, and so the enthusiastic clergy impedes the physician's ability to care for the dying patient – who might yet recover.

Recall that private patients summoned physicians at will and that physicians did not control the access of patients to physicians. There was therefore no settled process for consultation with one's potential or actual competitors. Nor were there any standard procedures for consultation in the Royal Infirmary. Patients can be harmed as a consequence, an outcome that sympathy forbids.

> There are often unhappy jealousies and animosities among those of the profession, by which their patients may suffer. A physician, however, who has any sense of justice or humanity, will never involve his patient in the consequences of private quarrels, in which he has no concern. Physicians in consultation, whatever may be their private resentments or opinions of one another, should divest themselves of all partialities, and think of nothing but what will most effectually contribute to the relief of those under their care. If a physician cannot lay his hand to his heart, and say that his mind is perfectly open to conviction, from whatever quarter it shall come, he should in honour decline the consultation. Many advantages arise from two physicians consulting together, who are men of condour, and have mutual confidence in each other's honour. A remedy may occur to one which did not to another; and a physician may want resolution, or sufficient confidence in his own opinion, to prescribe a powerful but precarious remedy, on which, however, the life of his patient may depend; in this case the concurring opinion of his brother may fix his own. But, if there is no mutual confidence; if opinions be regarded, not according to their intrinsic merit, but according to the person from whom they proceed; or, if there be reason to believe, that sentiments delivered with openness are to be whispered abroad, and misrepresented to the publick, without regard to the obligations of honour and secrecy; and if, in consequence of this, a physician is singly to be made responsible for the effects of his advice; in such cases, consultations of physicians tend rather to the detriment than to the advantage of the sick: and the usual and indeed most favourable conclusion of them is some very harmless but insignificant prescription (Gregory, 1772c, pp. 36-38).

Consultation should be guided, as we would now expect, by the cardinal intellectual virtue of a physician, namely, openness to conviction – which is essential to Baconian method, the advancement of science, and the improvement of medicine – and by sympathy directed to the "relief of those under their care" and not to the satisfaction of the physicians' interests.

Intellectual and moral virtues work in synergy to create the first ethics of consultation, where none yet existed. Consultation tends to be treated in recent scholarship as a matter of etiquette (Berlant, 1975; Waddington, 1984), reflecting the persistent influence of Chauncey Leake's distinction between ethics and etiquette (Leake, 1927). Leake wrote at a time when consultation had become a well-worked-out social practice in medicine, in sharp contrast to Gregory's time, when it was not even in embryonic form. Leake's distinction – when he made it and when it is used now – suffers from the crippling defect of presentism: reading the past as if it were not past but conforms to present practices and expectations. Intra-professional relationships may be matters of etiquette, but only when there is a profession in the ethical sense already in existence and only when the interests of patients are not at risk from such relationships. Neither was the case in Gregory's time. Moreover, the latter is not the case now, nor, one hopes, will it ever be (McCullough, 1983, 1984). Consultation-liaison psychiatry presents a compelling example of the ethically substantive issues involved in consultation (Engelhardt and McCullough, 1980).[46]

Gregory elaborates on what he sees those issues to be for patients.

> The quarrels of physicians, when they end in appeals to the public, generally hurt the contending parties; but, what is of more consequence, they discredit the profession, and expose the faculty itself to ridicule and contempt. – Nothing, in my opinion, but this cause, can justify any physician for refusing to consult with another, when he is required to do so. If he be conscious he cannot behave with temper, and that his passions are so ruffled as to impair his judgment, he may and ought to refuse it. But such circumstances, as the university where the person he is to consult with had his degree, or indeed whether he had a degree from any university or not, cannot justify his refusal. It is a physician's duty to do every thing in his power that is not criminal, to save the life of his patient; and to search for remedies from every source, and from every hand, however mean and contemptible. This, it may be said, is sacrificing the dignity and interests of the faculty. But, I am not here speaking of the private police of a corporation, or the little arts of craft. I am treating of the duties of a liberal profession, whose object is the life and health of the human species, a profession to be exercised by gentlemen of honour and ingenuous manners; the dignity of which can never be supported by means that are inconsistent with its ultimate object, and that tend only to increase the pride and fill the pockets of a few individuals (Gregory, 1772c, pp. 38-40).

Openness to conviction, which is the source of new knowledge, and sympathy, when that new knowledge is useful in the care of patients, make consultation obligatory. Thus, Gregory offers both an ethics *in* consultation and an ethics *of* consultation.

Gregory attacks the "private police of a corporation," i.e., the medical corporations and their sought-for monopoly, based on medical degrees from the "right" universities and on exclusive claims to medical knowledge. He also attacks those who would exclude from consultation those without degrees. The reader may recall that Gregory did not himself hold an earned medical degree; his was unearned and this was common and relatively easy to arrange (Johnston, 1987). Moreover, Gregory undoubtedly benefited from family connections: his brother was Professor of Medicine and his father-in-law Principal of King's College. Gregory, it seems, still smarted from the discrimination he must have experienced in London, being neither from the "right" university nor having an earned degree to boot – not to mention being a Scotsman from a provincial city. But Gregory's argument here is not personal; it is driven, as always, by sympathy and openness to conviction.

Gregory's earlier remarks in *Lectures* on the nature of medicine become relevant here. These concern his distinction between medicine as "an art the most beneficial to mankind" and as a "trade by which a considerable body of men gain their subsistance" (Gregory, 1772c, p. 10). He means his sympathy-based account to show how medicine "will most effectually maintain the true dignity and honour of the profession, and even promote the private interest of such of its members as are men of real capacity and merit" (Gregory, 1772c, pp. 10-11). True dignity and honor rest on the effacement and sacrifice of the physician's own interest as the primary concern, a theme that runs through the previous topics in clinical ethics. An other-directed practice of medicine becomes the ethical concept of a profession that Gregory forges in these texts. The physician's interests will be advanced as a positive side-effect of pursuing a true profession in its ethical sense.

Men of "real capacity and merit" live by the intellectual virtue of openness to conviction and the moral virtues of sympathy, namely, tenderness and steadiness. Gregory thus effectively addresses the central problem of trustworthiness of the physician in his time. He identifies a given foundation for moral trustworthiness, sympathy as a built-in principle of the human constitution. He provides an account of how that principle should be properly developed, trained, and expressed in virtues. On this basis he identifies both the intellectual and moral virtues of the professional physician in the true sense – what he calls a "liberal profession" – and does so in a way that patients can assess (Gregory, 1772c p. 39). Is the physician open to conviction? Does the physician consistently display tenderness and steadiness in asexual relationships with patients? Is the physician's attention and decorum to and with every patient the same, regardless of the socioeconomic status of the patient? Affirmative answers to these questions means that the physician is trustworthy. These questions need not be asked in *contractual* relationships – they are governed by *caveat emptor*, the guarded, prudential pursuit of self-interest – and so Gregory's ethical concept of medicine as a profession departs from contractual ethics and from Hoffmann's concept of prudence.[14] Gregory's concept also departs – decisively and radically – from the guild-oriented sense of profession promoted by the Royal Corporations, as evidenced in the passage from the 1770 "Advertisement" considered earlier.

Gregory continues his consideration of consultation by cautioning his students against the hubris of youth, often expressed in lack of respect for the clinical judgments of one's senior colleagues. Senior colleagues have seen the "revolutions ... of medical hypotheses" come and go and have gained perspective on novelty in science and clinical practice from this experience (Gregory, 1772c, p. 40). Moreover, when young physicians age they may have the tables turned on them, "when, perhaps, they are arrived at a time of life in which they have neither abilities nor temper to defend" the cherished beliefs of their youth (Gregory, 1772c, p. 41).

Gregory now takes up the relationship between physicians and surgeons. He takes a dim view of what he sees as a largely historical, traditional distinction not well grounded in reflection and experience. Indeed, when we look at matters with an open mind, he says, we will agree that the "separation of physick from surgery in modern times, has been productive of the worst consequences" for patients (Gregory, 1772c, p. 44). Experience teaches that diseases do not sort themselves into medical and surgical diseases. The lines between medical and surgical cases is difficult to discern. Thus, a sharp distinction between medicine and surgery lacks observational validity. This distinction is a function of caprice, not nature.

> Suppose a person to break his leg, and a fever and gangrene to ensure; the question occurs, whether the limb should be immediately amputated, or whether we should wait for some time till the effects of certain medicines, given with a view to stop the progress of the mortification, are known. It is evidently the business of a physician, in this case, to judge from the symptoms, from the habit of body, and from other

circumstances, whether the delay is prudent or not. – As to the performance of the operation itself, that is a different question. The genius and education requisite to make a good physician, are not necessary to make a good operator. – What is peculiarly necessary to make a good operator, is a resolute, collected mind, a good eye, and a steady hand. These talents may be united with those of an able physician; but they may also be separated from them (Gregory, 1772c, pp. 46-47).

Surgery will be improved when it is "confined to a set of men who were to be merely operators" and so too, *mutatis mutandis*, for pharmacy (Gregory, 1772c, p. 47). The problem with pharmacy is the lack of simple remedies, so that their effects can be properly evaluated – which compound remedies do not permit. Compound remedies, especially from secret formulae, may make an apothecary well off economically, but this represents shoddy clinical science. Physicians should relate with surgeons and apothecaries on the basis of "real merit," i.e., relate to surgeons and apothecaries who live the intellectual and moral virtues that Gregory has established:

> But a physician, of a candid and liberal spirit, will never take advantage of what a nominal distinction, and certain privileges, give him over men, who, in point of real merit, are his equals; and will feel no superiority, but what arises from superior learning, superior abilities, and more liberal manners. He will despise those distinctions founded in vanity, self-interest, or caprice; and will be careful, that the interests of science and of mankind shall never be hurt, on his part, by a punctilious adherence to formalities (Gregory, 1772c, pp. 50-51).

Such formalities are a function of "caprice" and other such factors, not "good sense." In effect, Gregory is proffering an intellectual and moral concept of the health professions that should apply to surgeons and apothecaries, not just physicians.

Gregory next takes up "peculiar decorums of a physician's character," including a "certain formality of dress, and a particular gravity in his bahaviour" (Gregory, 1772c, p. 51). These are non-trivial matters at a time when manners were for sale and when the man of false sensibility flourished, and physicians could and did insinuate themselves with patients by false manners, including dress and "gravity" of mien.

> Experience, indeed, has shewn, that all our external formalities have been often used as snares to impose on the weakness and credulity of

mankind; that in general they have been most scrupulously adhered to by the most ignorant and forward of the profession; that they frequently supplant real worth and genius; and that, far from supporting the dignity of the profession, they often expose it to ridicule (Gregory, 1772c, p. 52)

Which manners make a physician trustworthy? This has become for Gregory's time a compelling clinical ethical question for medicine.

Gregory appeals to his distinction between caprice and common sense to address this matter and dismisses dress and mien as "external formalities" not guided by common sense, with two exceptions:

This is an obligation, however, which common sense and prudence make it necessary he should regard. If the customs or prejudices of any country affix the idea of sense, knowledge, or dignity to any mode of dress, it is a physician's business, from motives of prudence, to equip himself accordingly. But, in a country, where a physician's capacity is not measured by such standards, and where he may dress like other people, without sinking in their estimation, I think he is at full liberty to avail himself of this indulgence, if he so choose, without being considered as deviating from the propriety and decency of his profession (Gregory, 1772c, pp. 52-53).

The second exception concerns children, who can be terrified by formal dress, as can some adults who suffer in their illness "feebleness and depression of spirits" (Gregory, 1772c, pp. 53-54). Physicians should, to protect such patients from unnecessary harm, refrain from formal dress and dress as a friend would.

There is, therefore, "a great impropriety in a physician having any distinguishing formality in his dress or manners" (Gregory, 1772c, p. 53). The physician should not cultivate manners that are in "any way singular" (Gregory, 1772c, p. 54). In particular the physician should eschew "any little peculiarities stealing into his manners, which can in any degree render him an object of ridicule" (Gregory, 1772c, p. 55). This would be inconsistent with the honor of the profession.

Recall that eccentricities and quirks of character might help the physician to stand out from the crowd and thus gain and hold market share and prestige (Lochhead, 1948). The ethics of a professional require the physician to forego such advantage. In the race of the tortoise and the hare, better to be the tortoise. Gregory continues, noting, that the physician should avoid "indulging himself in a certain delicacy, which makes him liable to be disgusted with some disagreeable circumstances he must unavoidably meet with in his practice" (Gregory, 1772c, p. 56). Gregory here echoes Mrs. Chapone's admonition to her niece to avoid the response of the dissipated to human misery. Moreover, this "certain delicacy" can blunt tenderness and lead to hardness of heart and the abandonment of patients – or even refusal to see them in the first place.

The physician should visit the patient "in proportion to the urgency and danger of his complaints," a matter of which the physician is the "best judge" (Gregory, 1772c, p. 57). "But some delicacy is often required, to prevent such frequent visits as may be necessary, from bringing an additional expense upon the patient" (Gregory, 1772c, p. 57). In other words, openness to conviction about the clinically-indicated need for patient visits means that the physician's monetary self-interest should be sacrificed for the patient's interest. The physician should also give the patient his "whole attention ... while he remains with him" (Gregory, 1772c, p. 57).

> That continual hurry which some of our profession seem to be in, is sometimes mere affectation; but it often proceeds from other causes. Some keep themselves constantly embarrassed by a want of œconomy of their time, and of a proper arrangement of their business; some, from a liveliness of imagination, and an unremitting activity of mind, involve themselves in such a multiplicity of pursuits as cannot be overtaken. But from whatever source it arises, it ought to be timely corrected, and not suffered to grow into a habit. It prevents a physician from doing his duty to the sick, and at the same time weakens their confidence in him (Gregory, 1772c, pp. 57-58).

These behaviors can promote an appearance of one's importance, but really reflect one's own self-absorption and self-aggrandizement – still a problem for us two centuries later!

Gregory next argues that the physician should not show "servility of manners toward people of rank and fortune" (Gregory, 1772c, p. 58). This behavior destablizes clinical judgment, because the "external magnificance and splendor attendant on high rank is apt sometimes to dazzle their understanding" (Gregory, 1772c, p. 59) and so undermine the physician's "independent spirit" (Gregory, 1770, p. 54). This is a compelling admonition, coming as it does from the King's Physician in Scotland and appealing as well to the tradition of *lehrenfreiheit* for which the University of Edinburgh was renowned. The theme of medicine's earned nobility, which surpasses inherited nobility, here becomes explicit.

So far Gregory has argued for openness to conviction, for clinical judgment based in the patient's interest and not the physician's, and for medicine as a true profession, i.e., against medicine as a trade. His attack on nostrums and secrets therefore come as no surprise.

> Great disputes have arisen in our profession, about the propriety of a physician's keeping secrets or nostrums. It has been said, with some plausibility, in vindication of this practice, that the bulk of mankind seldom attend, or pay much regard, to what is made level to their capacities; and that they are apt to undervalue what costs them nothing. Experience shews, that men are naturally attached to whatever has an air of mystery and concealment. A vender of a quack medicine does not tell more lies about its extraordinary virtues, than many people do who have no interest in the matter; even men of sense and probity. A passion for what is new and marvellous, operates more or less on the imagination; and, in proportion as that is heated, the understanding is perplexed. When a nostrum is once divulged, its wonderful qualities immediately vanish, and in a few months it is generally forgotten. If it be really valuable, the faculty perhaps adopt it: but it never recovers its former reputation. – It is likewise said, that this is the only way in which any good medicine can be introduced into practice; as the bulk of mankind will more readily follow the directions of a man who professes to cure them by mysterious means, than those of a regular physician, who prescribes plain and common remedies. It is further alleged, that some of the best medicines were originally introduced as secrets, though opposed by the regular physicians. But allowing this to be true, yet I am persuaded, that nostrums, on the whole, do more harm than good; that they hinder the advancement of our art, by making people neglect what is known and approved, in pursuit of what is unknown, and probably never to be divulged; and that, from their being generally dispensed by low and illiterate men, who prescribe them indiscriminately, they are become a public nuisance in these kingdoms. – In some places on the continent, however, physicians of honour and reputation keep nostrums. In such hands, the same abuses will not be committed, as we experience here; but still the practice has an interested and illiberal appearance (Gregory, 1772c, pp. 59-61).

In the following paragraph Gregory makes these remarks:

> Curiosity in a patient or his friends to know the nature of the medicines

prescribed for him is natural, and therefore not blameable; yet this is a curiosity which it is often very improper to gratify. There is a natural propriety in mankind to admire what is covered with the veil of obscurity, and to undervalue whatever is fully and clearly explained to them. A firm belief in the effects of medicine depends more on the imagination, than on a rational conviction impressed on the understanding; and the imagination is never warmed by any object which is distinctly perceived, nor by any truth obvious to common sense. Few people can be persuaded that a poultice of bread and milk is in many cases as efficacious as one compounded of half a dozen ingredients, to whose names they are strangers; or that a glass of wine is, in most cases where a cordial is wanted, one of the best that can be administered. This want of faith in the effects of simple, known remedies, must of necessity occasion a disregard to the prescription, as well as create a low opinion of the physician. Besides, where a patient is made acquainted with the nature of every medicine that is ordered for him, the physician is interrupted in his proceedings by many frivolous difficulties, not to be removed to the satisfaction of one ignorant of medicine. The consequence of this may be to embarrass the physician, and render him irresolute in his practice; particularly in the administration of the more powerful remedies. It should be further considered, that when a patient dies, or grows worse under the care of a physician, his friends often torment themselves, by tracing back all that has been done, if they have been made acquainted with it, and may thus be led, very unjustly, to charge the physician with what was the inevitable consequence of the disease. There are, indeed, cases where it may be proper to acquaint a patient with the nature of the remedies, as there sometimes peculiarities in a constitution in regard both to the quality and quantity of the medicine, which peculiarities a physician ought to be informed of before he prescribes it (Gregory, 1772c, pp. 61-63).

Sympathy directs the physician to relieve the patient, principally by doing what works clinically. Clinical experience teaches that gratifying the patient's natural curiosity about medicines leads to more harm than good. Compliance with regimen, Gregory points out, occurs more readily and regularly when regimen depends on the patient's imagination, not reason: an air of mystery is very effective in producing a "warm imagination," which, in turn, leads to a greater probability of compliance. Given its factual assumptions, this line of argument follows directly from a moral sense theory that treats reason as an inactive principle and instinct – to which imagination owes much – as an active principle. Twentieth-century concepts of informed consent, based on appeals to *reason*, not imagination and instinct, serve to distance us from Gregory's way of thinking.

Gregory closes his second lecture with his refutation of a "charge of a heinous nature, which has been often urged against our profession: I mean that of infidelity and contempt for religion" (Gregory, 1772c, pp. 62-63). This charge he regards as "ill-founded" (Gregory, 1772c, p. 64) and "absolutely false" (Gregory, 1770, p. 57). He first appeals to the examples of "real piety" in Harvey, Sydenham, Arbuthnot, Boerhaave, Stahl, and Hoffmann (Gregory, 1772c, p. 64). Hoffmann is explicit on this subject. The language change from the 1770 to the 1772 edition is interesting; in the former Gregory uses "regard for religion" (Gregory, 1770, p. 57). The root causes of this charge are found in the following:

> It is easy, however, to see whence this calumny has arisen. Men whose minds have been enlarged by knowledge, who have been accustomed to think, and to reason upon all subjects with a generous freedom, are not apt to become bigots to any particular sect or system. They can be steady to their own principles without thinking ill of those who differ from them; but they are impatient of the authority and controul of men, who would lord it over their consciences, and dictate to them what they are to believe. This freedom of spirit, this moderation and charity for those of different sentiments, have frequently been ascribed, by narrow-minded people, to secret infidelity, scepticism, or, at least, to lukewarmness in religion; while some who were sincere Christians, exasperated by such reproaches, have sometimes expressed themselves unguardedly, and thereby afforded their enemies a handle to calumniate them. This, I imagine, has been the real source of that charge of infidelity so often and so unjustly brought against physicians (Gregory, 1772c, pp. 64-65).

In addition, some suffer from "vanity" as their "ruling passion" and subscribe to a radical skepticism (Gregory, 1772c, p. 67).

Recall, however, that, in both *Comparative View* and *Legacy*, Gregory subscribes to the view that there is a natural religion that can be established on philosophical, secular grounds. He thus can distinguish natural religion and proper devotional forms of it, from superstition.

> But I will venture to say, that men of the most enlarged, clear, and solid understandings, who have acted with the greatest spirit, dignity, and propriety, and who have been regarded as the most useful and

amiable members of society, have never openly insulted, or insidiously attempted to ridicule the principles of religion; but, on the contrary, have been its best and warmest friends. – The study of medicine, of all others, should be the least suspected of leading to impiety. An intimate acquaintance with the works of Nature, raises the mind to the most sublime conceptions of the Supreme Being, and at the same time dilates the heart with the most pleasing views of Providence. The difficulties that necessarily attend all deep inquiries into a subject so disproportionate to the human faculties, should not be suspected to surprise a physician, who, in his practice, is often involved in perplexity, even in subjects exposed to the examination of his senses (Gregory, 1772c, pp. 67-68).

Moreover, physicians learn from the care of patients that religious practice, the devotional and prayerful aspect of revealed religion, can and does support patients in their distress. The physician should encourage patients of religious persuasion in their devotional response to disease and avoid the harms that can follow from disrespect for the religion of patients who suffer from serious, life-threatening conditions.

A physician, who has the misfortune to disbelieve in a future state, will, if he have common good-nature, conceal his sentiments from those under his charge, with as much care as he would preserve them from the infection of a mortal disease. With a mind unfeeling, or occupied in various pursuits, he may not be aware of his own unhappy situation; yet it is barbarous to deprive expiring nature of its last support, and to blast the only surviving comfort of those who have taken a last farewell of every sublunary pleasure. But, if motives of humanity, and a regard to the peace and happiness of society cannot restrain a physician from expressing sentiments destructive of religion or morals, it is vain to urge the decency of the profession. The most favourable construction we can put on such conduct, is to suppose that it proceeds from an ungovernable levity, or a criminal vanity, that forgets all the ties of morals, decency, and good manners (Gregory, 1772c, pp. 69-70).

The physician is "tied," i.e., bound, to the patient naturally and automatically by sympathy. As a natural philosopher the physician should admit to the validity of natural religion; intellectual discipline – Gregory holds, *contra* Hume, certainly – requires assent to natural religion. If the physician cannot assent, he should as a matter to tender regard for his patients

at least refrain from expressing his misplaced, unfounded agnosticism or
atheism.

> I shall make no apology for going a little out of my way in treating of
> so serious a subject. In an enquiry into the office and duties of a phy-
> sician, I deemed it necessary to attempt to wipe off a reflection so de-
> rogatory to our profession; and, at the same time, to caution you
> against that petulance and vanity in conversation, which may occasion
> imputations of bad principles, equally dangerous to society, and to
> your own interest and honour (Gregory, 1772c, p. 70).

Obviously, on this matter Gregory departs from Hume but does so on the
basis of the very "experimental method of reasoning" common to both.

Gregory places his discussion of human experimentation within the
larger context of laying medicine open. This concern links directly with
his earlier treatment of the intellectual virtue of *openness to conviction*.
Indeed, the first concern leads to and justifies the second: "I shall now
endeavour to shew, that the confinement of the study and practice of
physic, entirely to a class of men who live by it as a profession, is unfavourable to the progress of the art" (Gregory, 1772c, p. 213). Laying
medicine open would reduce the vulnerability of patients to a science that
"alone is kept so carefully concealed from the world" (Gregory, 1772c, p.
215) so that patients cannot evaluate the capacities of those who offer
their medical services.

The problems with medicine in this respect run very deep.

> A physician, when he sets out in the world, soon perceives that the
> knowledge most profitable for him, is not that merely of his profession.
> What he finds more essential, are the various arts of insinuation and
> ostentation. This leads to views very different from those of genius and
> science. To his real merit as a physician, he cannot easily find a patron;
> because none are judges of this but those of his own profession, whose
> interest it often is to have it concealed. By what I have said, I mean
> only to represent the disadvantages naturally consequent on leaving it
> to physicians to judge of the merits of their brethren. It is putting hu-
> man virtue to too severe a trial, and indeed it is a trespass against the
> most obvious maxims of prudence and humanity, to suffer people to be
> tried by judges whose interest it is to condemn them. Nor do I mean, in
> making an observation which is equally applicable to every class of
> men, to include all the individuals in our profession. There is a virtue
> found among many of them, which can stand the severest test, and

there is an elevation of mind, that generally accompanies genius, which renders those who possess it equally superior to the suggestions of envy or interest, and to all the low arts of dissimulation (Gregory, 1772c, pp. 216-217).

The lack of public, testable, intellectual foundations for medicine in science, which would allow patients to assess any physician's intellectual and clinical capacities more reliably, leads to problem of trustworthiness, because there is no check on "the low arts of dissimulation." Thus, scientific investigation becomes part of the intellectual response to the crisis of trustworthiness described by the Porters (Porter and Porter, 1989).

Moreover, Baconian experimental method itself requires the constant improvement of medicine and this, in turn, requires scientific and clinical investigation. This is especially the case for someone such as Gregory who harbors a deep skepticism about the use of the term, 'incurable', by physicians acting more often from interest than obligation. However, when physicians undertake research they are subject to the attacks of their brethren, attacks motivated by interest in the form of protecting market share, prestige, and power – all in violation of sympathy and therefore to the detriment of patients.

It were to be wished, that ingenious men would devote half the time to the study of nature, which they usually give to that of opinions. If a gentleman have a turn for observation, the natural history of his own species is surely a more interesting subject, and affords a larger scope for the display of genius, than any other branch of natural history. If such men were to claim their right of enquiry into a subject that so nearly concerns them, the good effects in regard to medicine would soon appear. They would have no interest separate from that of the art. They would detect and expose assuming ignorance, and would be the judges and patrons of modest merit. Cases often occur, where a physician sees his patient hastening to dissolution; he knows a remedy that affords some prospect of saving his life: but it is not agreeable to common practice, and is dangerous in its operation. Here is an unhappy dilemma. If he give the remedy, and the patient dies, he may be ruined; for his conduct will be watched with a malignant eye. But if the scheme of gentlemen of fortune applying to the study of physic should take place, the encouragement and assured protection of knowing and disinterested judges, would animate a physician in his practice. Such judges, not fettered by early prejudices, unawed by authority, and unbiased by interest, would canvass with freedom all the commonly received principles of medicine, and expose the uncertainty of many of those maxims of which a physician dares not seem to doubt. (Gregory, 1772c, pp. 218-219).

In his clinical lectures Gregory places human experimentation within the larger context of his practice philosophy. As we have seen, his main view is that nature should be assisted unless "her efforts are too violent" (RCPE Gregory, John 1, 1766, p. 11). Patients should be given the opportunity to self-physick, under the restraints of limited physician authority that he already has identified in his first ethics lecture. Medications should be simple – an admonition he repeats – and the physician should attend to the patient's "spirits" as well as his disease (RCPE, Gregory, John 1, 1766, p. 13). Only when these prescriptions fail can more aggressive measures be considered.

> Desparate remedies should be used in some cases, where every other method has been proved ineffectual. In such circumstances we should have recourse to medicines which under more favourable circumstances might be thought dangerous (RCPE, Gregory, John 1, 1766, p. 13).

For example, when bleeding and blistering have failed to effect relief of "Violent pain in the side" and "exceeding Violent Cough," and when the "ususal Remedies" have been tried and have failed to effect relief, then opium may be used, despite the "general prejudice" against it. The important goal is to achieve relief of the patient's suffering and "procure several hours of refreshing sleep" (MBO 7568, 1771, pp. 8-9). And, ideally, such intervention should take the form of organized clinical investigation, undertaken in a way that neutralizes bias based on interest and thus allows a fair test of the regimen.

Gregory also addresses human experimentation in the context of the teaching ward at the Royal Infirmary. We saw that Risse (1986) documents that clinical research was undertaken on this ward and that there were numerous criticisms of this practice, some quite pointed.

> In treating the Patients under my care I am only to give you my Common Practice, & only prescribe such Medicines as I have had experience of their good effects in similar cases. I know very well that it is a common opinion with many young Gentlemen, that the Physician who attends an Hospital should always try Experiments on the Patients; this

I think contrary both to Justice & Humanity: I shall therefore give you my common Practice & not sport with the lives of poor people: I would not give any Person a Medicine that I would scruple to take myself. I shall always have in my Eye that moral precept, "Do as you would be done by" (RCSE C36, 1771, p. 9).

We know when that would apply: when other remedies have proven ineffectual. Thus, experimentation is not the standard of care and, therefore, should not constitute the physician's initial response to patients' problems.

Gregory here uses some of his strongest language, the metaphor of "sporting" with patients. One sports in hunting wild animals by torturing and killing them for one's own pleasure, out of interest. One goes to a prostitute out of interest, not love. Sporting with animals or humans thus violates every aspect of sympathy. When the physician, from the upper social reaches, sports with the patients in the Royal Infirmary, who come from the lower social reaches, he acts against sympathy and the fair treatment of his social inferiors, in violation of the Highland ideal of paternalism and even justice.

Gregory goes on to consider inoculation, an area of considerable interest and importance.

Some I know have tried dangerous Experiments, such as the innoculating people with matter taken from one who has the very bad kind of confluent Small pox, or from patients labouring under other diseases at the same time, as Sues Niniria etc: Altho' I believe it is of no consequence from whom we take the Nariolous matter, yet I would ask that man if he would have done so to his own child: if he would not do it to his own Child, why should he indanger the lives of other people. When the Medicines that I commonly use fail, I shall try other substances, but the Common practice must always have a fair trial (RCSE C36, 1771, p. 10).

To do an experiment on one's child as the first clinical response violates sympathy because an experiment will almost certainly increase rather than relieve the patient's distress, when compared to the "common practice," which has become common because experience has taught physicians that it does often relieve the distress of patients. Thus, Gregory's argument here does not fall prey to a possible critique of this line of argument: a reckless investigator might well risk himself or his child and so, without some independent account of when clinical investigation counts as reckless, one cannot argue for limits of human experimentation based on the golden rule or the investigator's regard for his children. But Gregory's concept of sympathy provides him precisely with the requisite independent account.

Gregory touches only briefly on animal experimentation. Recall from chapter two that even as a medical student he had noted the value of comparative studies and his own *Comparative View* contributed to comparative psychology. He notes, for example, that we can observe sympathy of a sort in animals, of a sort we observe in humans (Gregory, 1765, 1772a). Comparative basic science, as we call it, also has value.

> In order to illustrate the human physiology, a knowledge of the comparative anatomy of some animals that most nearly resemble man, is requisite, Several important discoveries in the animal œconomy have been illustrated by experiments first made on brutes, many of which could not have been made on the human subject, e.g. the experiments relating to the circulation of the blood, respiration, muscular motion, sensibility and irritability of different parts of the body, and the effects of various medicines. The instincts of brutes have sometimes given the first hint of valuable remedies, and might throw light on the subject of regimen, and the cure of diseases, if they were properly attended to. At the same time it must be acknowledged, that the comparative anatomy has often led into great mistakes, by too hastily applying it to the human body (Gregory, 1772c, pp. 77-78).

Because animals do not possess the moral sense, nor reason as in humans, they do not enjoy the same moral status as humans, though sympathy does give us some moral regard for them, because they experience animal-type pain and distress. Thus, in principle sympathy does not prohibit animal experimentation. When such experiments are reliably expected to contribute to the relief of human suffering, sympathy supports, indeed, requires, them.

Gregory enjoyed, deservedly, the affections of his students and was justly known for his generosity to them, e.g., giving the profits of the 177 edition of the *Lectures* to a "poor and deserving pupil" (Lawrence, 1971, Vol. I, p. 162). It comes as no surprise that he would turn his moral philosophy on himself and his colleagues in medical education. He writes a normative philosophy of medical education.

> It is much in the power of a teacher of the art, to obviate the inconveniences commonly chargable upon systems. It is his duty, in treating of

any subject, to give a full detail of facts; to separate real from pretended ones; and to arrange them in such a manner as may lead to the discovery of causes and general principles. If these cannot be established by a just induction, he may, with propriety, suggest an hypotheses; but, while he gives his reasons for its probability, he should, with equal impartiality, state every objection against it. So far from throwing a veil over the numberless imperfections of his art, he should be solicitous to point them out, and at the same time direct to such observations and experiments as may tend to remove them. Sensible of the warm imagination and credulity of youth, of their proneness to admiration, and their eagerness to have every fact accounted for, he ought to guard against the errors into which these dispositions may lead them, and should endeavour to direct their ardour in the pursuit of knowledge to proper subjects; not to those that merely amuse the fancy, but to such as exercise the powers of useful observation and invention; to subjects of real and permanent importance (Gregory, 1772c, pp. 210-212).

In short, the professor of medicine should be an exemplar of the intellectual virtues of the Baconian scientist committed to the improvement of medicine in his own work and in educating students to the proper skeptical spirit of medical science. Sympathy for the "warm imagination" of youth and an experience-informed observation of its perils obligate the professor of medicine to school his students in intellectual and moral discipline.

In his lectures on the institutions of medicine Gregory touches briefly on the concept and clinical determination of death. Since we have no reliable such concept, our clinical criteria should be the most conservative, he argues. Gregory writes at a time when fear of premature burial concerned many people.

In short in what Death consists we canot say, because we do not know in what life consists. Haller says that Death takes place when the Heart ceases to be irritable: But we do not know in what the irritability of the heart consists. If a person has all the appearances of death where the Vis Vitae has been gradually extinguished in consequence of any disease & we have not doubt of burying him. But putrification is the only mark that should determine sudden Death, and this particular happens in consequence of nervous disorders where they have had all the marks of apparent death ... (EUL 2106 D, 1773, pp. 168-169).

In an era in which people lived in fear of being buried prematurely, still alive, such caution seems well placed.

Despite the brevity of this treatment, putrefaction still meets the criteria for an adequate determination of death, as set out by the President's Commission for the Study of Ethical Problems in Medicine and Biomedical and Behavioral Research (1981). Very few putrefying corpses will be the living bodies of individuals and there will be very few classifications of dead bodies as alive. The determination can be made without "unreasonable delay," especially of funeral rites, especially when people did not want burial to be premature. This clinical criterion applies to all clinical situations and is "explicit" and "accessible to verification" (President's Commission, 1981, p. 161).

E. Philosophy of Medicine

The reader should recall that in Gregory's time there was no profession of medicine and no established educational and training path into practice as we know them. Physicians could and did enter practice with or without a university education, with or without an earned degree, although these pathways were then regarded as sufficient for becoming a physician. No standard medical curriculum existed in universities, nor did there exist an entity such as the Liaison Committee on Medical Education that reviews and accredits every (allopathic) medical school in the Unites States. Indeed, there existed competing concepts of what the medical curriculum should include. Nor was there any board certification for speciality and subspecialty practice, much less the increasingly sophisticated quality controls now being applied to the managed practice of medicine. As the Porters (1989) and Wear (1987) note, there existed no stable, standardized body of scientific knowledge and patterns of clinical practice known as "Medicine."

In short, not only did medicine as a fiduciary profession not exist, the social institutions of education, training, certification, and licensure that support such a social practice did not exist. Gregory, therefore, could not address the "duties and qualifications" of a physician without also addressing their intellectual underpinnings: the experimental method of science and the content of the medical curriculum. He devotes his remaining four lectures to these topics. Four central themes form the core of Gregory's philosophies of science, medicine, and medical education.

First, Gregory takes the view that nature is a whole, that things are defined by their multiple relations and interconnections with each other. This theme reflects, in my judgment, the deep influence of Leibniz's metaphysics. Leibniz took the view that *relations* to all of the rest of reality constituted monads, the basic building blocks of reality (McCullough, 1996). To be sure, the Scottish experimentalists reject the concept that the basic building blocks of reality were non-material substances, but they embrace wholeheartedly the concept of interconnectedness as one of the fundamental features of nature, a theme, we saw above, that also interested Hoffmann.

> I PROCEED NOW to explain the connexions of the several branches of physic with the practical part of it, and inquire how far a previous knowledge of these is necessary, in order to practise with reputation and success. Here I must previously observe, that all the works of Nature are so intimately connected, that no one part of them can be well understood by considering and studying it separately. In order, therefore, to be qualified for the practice of physic, a variety of branches of knowledge, seemingly little connected, are nevertheless, necessary. As this is the case, it is proper that a student should be on his guard, not to waste his time and labour in pursuits which either have no tendency, or a remote one, to throw light on his profession. Life is too short for every study that may be deemed ornamental to a physician; it will not even allow time for every study that has some connection with physic. Every one of the sciences I am about to name, is of great extent; but it will be necessary for a physician to confine his application to such parts of them as are really subservient to practice. If a student's genius incline him more particularly to any of these preliminary sciences, he may, if he please, indulge himself in it, taking care not to impose on himself, and consider this as studying physic (Gregory, 1772c, pp. 71-72).

Gregory interprets the doctrine, that the works of nature are connected, in clinical terms. We have seen above that he bemoans the separation of physick and surgery because diseases come connected and not separated into medical versus surgical diseases. Gregory's emphasis on the interconnectedness of nature at once reflects and reinforces his commitment to the social principle in political philosophy and sympathy in medical ethics. By nature, humans come interconnected, not as isolated, complete social atoms or moral strangers to each other.

His clinical interest also leads him to be practical with his students.

Thus, his curriculum includes anatomy, physiology – especially of the nervous system – chemistry, the "laws of union between soul and body" (Gregory, 1772c, p. 71), comparative anatomy, pathology, materia medica, and some botany. He includes as "ornamental qualifications" knowledge of the history of medicine, of Latin and Greek, of "our native language," and "the style and composition proper for medical writings" (Gregory, 1772c, p. 71). His comments on the last two topics provide an interesting and useful insight into his overarching concerns for simplicity, concrete thinking, and his abhorrence of abstract, unfounded metaphysical systems.

> It may appear at first view superfluous to recommend an attention to your own language. But it is well known, that many physicians of real merit, have exposed themselves to ridicule by their ignorance of, or in attention to, composition. It might be expected, that every one who has had the education of a gentleman, should write his native language, with at least grammatical exactness; but, even in this respect, many of our writers are shamefully deficient. Elegance is difficult to attain; and, without great taste, dangerous to attempt. What we principally require in medical writings, is perspicuity, precision, simplicity and method. A flowery and highly-ornamented style in these subjects is entirely out of its place; and creates a just suspicion, that an author is rather writing from his imagination, than copying from Nature. We have many bulky volumes in medicine, which would be reduced to a very small compass, were they stripped of their useless prefaces, apologies, quotations, and other superfluties, and confined to the few facts they contain, and to close inductive reasoning. – What I would principally recommend to you in every species of medical writing, next to a simple and candid history of facts, is a strict attention to method. I am no admirer of that display of system and arrangement, so remarkable in some writers, who split their subject into endless divisions. This may strike a young reader, not accustomed to such kind of writing, with an high opinion of the author's ingenuity and accuracy; but in general it is a mere deceit. It is a mode of writing easily attained, and was in the highest perfection when the scholastic logic, which consisted rather of nominal than real distinctions, was held in admiration (Gregory, 1772c, pp. 97-99).

Gregory's second general theme, no surprise, concerns the governance of nature by discoverable principles or laws. These principles or laws originate in God the creator and so experimental method is "favourable to religion" (Gregory, 1772c, p. 104) as a matter of conceptual necessity. We come to learn the principles of nature by following the intellectual discipline of the experimental method.

I proceed now to lay down certain general principles, which require our attention in the investigation of Nature, and shall endeavour to apply them more particularly to the science of Medicine. When we look around us, we find objects connected together in a certain invariable order, and succeeding one another in a regular train. It is by observation and experience alone, that we come to discover this established order and regular succession in the works of Nature. We have all the evidence which the nature of the things admits of, to persuade us that nothing happens by chance: nay, we have every reason to believe, that all events happen in consequence of an established and invariable law; and that, in cases similar, the same events will uniformly take place (Gregory, 1772c, p. 112).

The metaphysical commitments of the Scottish philosophers include the reality of natural laws or principles and their effects. These philosophers reject reason as the basis for discovering principle and their effects. They also reject the reason-based explanation of causal judgments that effects are contained in their causes, a connection that can be discovered upon conceptual investigation without the aid of experience. Gregory also follows Hume and Reid (he cites the latter by name(Gregory, 1772c, p. 114)) in his *epistemological* account of how it is that we come to discover causal connections that are regulated by natural principles.

Here I must observe, that, antecedent to all reasoning and experience, there is an original principle implanted in the human mind, whereby it is led to a belief that the course of Nature is regular. In consequence of this principle, whenever a child sees any event succeeding another, he has an instinctive persuasion, that the same event will succeed it afterwards in the same circumstances. This persuasion does not flow from any connection he sees between the cause and the effect, nor from experience, nor from reasoning of any kind (Gregory, 1772c, pp. 112-113).

Gregory goes on to note that "instinctive persuasion," if left to its own devices, can become undisciplined, which will result in confusion and be of little utility. And so, much in the manner of Bacon and the Scottish philosophers he lays down rules for the proper prosecution of inquiries using the experimental method. His remarks follow largely from those of Bacon and the other figures we considered in the previous chapter.

Gregory's concern with the regularity of nature and its principles leads him to enunciate a third main theme, that the principles of nature are simple and few in number. Gregory is acutely aware that the demand for simplicity arises mainly as a requirement of reason and failure to be disciplined by experience can lead us to the mistaken conclusion the nature is simpler than disciplined experience teaches. Experimental method therefore requires great care in the level of simplicity with Nature's principles are expressed; oversimplification is a problem to be avoided.

> I OBSERVED before, that in our inquiries into human nature, an impatience to acquire a knowledge of her laws, and a natural love of simplicity, lead us to think them fewer and simpler than they really are. The more we know, the more we discover the uniformity and simplicity of the laws of Nature, when compared with the vast extent and variety of her works; but still we must not imagine that they are confined within the narrow limits of our knowledge, or even perhaps of our comprehension (Gregory, 1772c, p. 152).

Abstract reasoners mistake the simplicity required by reason for the simplicity of Nature; this involves a grave error of departure from the experience-based discipline of the experimental method. Abstract reasonings produce medical systems based on overly simple causal explanations – a cluster of effects are contained in one cause when often several causes are functioning simultaneously. This excessive simplicity makes these systems defective scientifically and clinically, where effects in the form of a cluster of symptoms can and often do have more than one cause.

Gregory's clinical experience seems to be informing this theme. There is a great deal that we do not know, he emphasizes. Positing overly simple formulations of Nature's principles of health and disease could deeply mislead the physician and result in great harm to the patient. Gregory's concern for intellectual and clinical rigor in scientific thinking and clinical judgment leads him to examine "causes that have retarded the advancement of the sciences," including "inattention to their end, viz., the convenience and happiness of life, ... useless subtlety, ... credulity, ... attachment to great names, ... blind admiration of antiquity, ... fondness of novelty, ... hasty reduction of the sciences into system, ... [and] too great attention to elegance of language, or an affected obscurity of style" (Gregory, 1772c, pp. 151-152). Note the last: affected sympathy will not do, nor will affected learning. The Scottish and clinical character rings through once again: be practical, be useful, and do so by being disciplined by what experience teaches about the world. No flights of fancy through systems of logic based only on reason or undisciplined analogy that detaches itself from experience; just the steady, unglamorous work of the natural philosopher. Indeed, Gregory urges his students in several places in *Observations* and *Lectures* to study and follow Bacon's example in their mastery of experimental method.

As a consequence, students require intense and regular exposure to the best scientific books and also to patients. Clinical experience provides the crucible in which the results of experimental method in medicine are put to Nature's test. Thus, Gregory's philosophy of medical education includes a strong emphasis on the necessity for students to have extensive clinical experience. It is in this context that he addresses the obligations of a professor of medicine.

The first three themes, taken together, provide the basis for Gregory's complete confidence in Nature. In medicine this confidence in Nature leads him to view health as an observable norm and to observe that nature includes corrective responses to departures from this observable norm. Medicine's task is to assist Nature when her responses are too weak and to temper them when they are too strong. As a colleague in pediatric critical care puts it, echoing – unwittingly – Gregory's confidence: his job is to keep patients alive while Nature heals them.

Anyone willing to do the work can master experimental method. The method and its results do not belong to any individual or group, but rather to anyone possessed of the intellectual virtues of the experimental methodologist. These core intellectual commitments lead Gregory to his fourth main theme, that there are significant "[a]dvantages of laying the art open, and of gentlemen of science and abilities, who are not of the profession, studying it as an interesting branch of philosophy" as a corrective to the "[i]nconveniences arising from the absolute confinement of the study and practice of physic to a class of men who live by it as a profession" (Gregory, 1772c, p. 195). We saw earlier that Gregory points to the interest of men who live by medicine as a trade. This influence of interest, for example, creates problems for the proper evaluation of medical experiments. In other words, Gregory wants to distinguish medicine "as an art the most beneficial and important to mankind" from medicine as a "trade by which a considerable body of men gain their subsistence" (Gregory, 1772c, p. 10). Laying medicine open blunts the deleterious effects of self-interest in favor of the intellectual virtue of diffidence.

> It were to be wished, that ingenious men would devote half the time to the study of nature, which they usually give to that of opinions. If a gentleman have a turn for observation, the natural history of his own species is surely a more interesting subject, and affords a larger scope for the display of genius, than any other branch of natural history. If such men were to claim their right of enquiry into a subject that so nearly concerns them, the good effects in regard to medicine would soon appear. They would have no interest separate from that of the art. They would detect and expose assuming ignorance, and would be the judges and patrons of modest merit (Gregory, 1772c, pp. 218-219).

On this basis, medicine could advance as a science based on diffidence and openness to conviction and contribute to the relief of man's estate.

In the 1772 edition this commitment to laying medicine open leads Gregory to attack the "selfish corporation-spirit" (Gregory, 1772c, p. 237) that leads to attempts at monopoly and control of the medical market by limiting membership to physicians from the right universities and keeping secrets and nostrums – both of which Gregory attacked vigorously early in his medical ethics lectures. The 1770 edition's "advertisement" at the front of the volume includes a more pointed attack.

> The noble and generous sentiments which are here displayed, will ever be a source of pleasure to minds unbiased by prejudice, self-interest, or the unworthy arts of a Corporation. Whatever opposition this part of the work may meet with from those, who find their own foibles, or rather vices, censured with a just severity, the ingenious part of mankind, however, will not fail in bestowing that degree of applause so justly due to its merit. At present there seems to be a general disposition in mankind to expose to their deserved contempt, those quackish, low, and illiteral artificies, which has too long disgraced the profession of medicine. It is therefore hoped, that the general spirit of this lecture will have a remarkable tendency to promote this laudable end; and that it will excite men of influence and abilities to exert themselves in crushing that arrogance, which hath frequently served to correct the ignorance of many practitioners of medicine, and by means of which alone, they acquire such a share of practice as they are by no means entitled to. In consequence of this, real merit, which is very often ac-

companied with great modesty, will meet with its due reward (Gregory, 1770 pp. iv-v).

I noted at the beginning of this chapter that the writing of the 1770 edition has a harsh, indeed, hard edge, uncharacteristic of every other publication of Gregory, as well as of the numerous manuscript materials. Thus, this harsh attack on Royal Corporation physicians as quacks, arrogant, and low may reflect the student editor's views – whom I believe to be his son – more than those of Gregory himself. That Gregory means to attack self-interest in the form of group self-interest, or "corporation spirit," we should not doubt, however.

This commitment to laying medicine open shapes Gregory's consideration of the concept of an incurable disease – a concern rooted in Bacon's work and held by Gregory from his medical student days. The commitment to experimental method and to diffidence – or being open to conviction – as the cardinal intellectual virtue on the clinical setting will help patients.

A physician who has been educated upon this plan, whose mind has never been enslaved by systems, because he has been a daily witness of their insufficiency, instead of being assuming and dogmatical, becomes modest and diffident. When his patient dies, he secretly laments his own ignorance of the proper means to have saved him, and is little apt to ascribe his death to his disease being incurable. There are indeed so few diseases which can be pronounced in their own nature desperate, that I should wish you to annex no other idea to the word, but that of a disease which you do not know how to cure. How many patients have been dismissed from hospitals as incurables, who have afterwards recovered, sometimes by the efforts of unassisted nature, sometimes by very simple remedies, and now and then by the random prescriptions of ignorant quacks? To pronounce diseases incurable,* ['*Bacon' appears at bottom p. 209] is to establish indolence and inattention, as it were, by law, and to skreen ignorance from reproach. This diffidence of our own knowledge, and just sense of the present imperfect state of our art, ought to incite us to improve it, not only from a love of the art itself, but from a principle of humanity. I acknowledge, however, that such a diffidence as I have described, if it be not united with fortitude of mind, may render a physician timid and unsteady in his practice; but, though true philosophy leads to diffidence and caution in forming principles, yet, when there is occasion to act, it shews how necessary it

is to have a quickness in perceiving where lies the greatest probability
of success; to be decisive in forming a resolution, and firm in carrying
it into execution (Gregory, 1772c, pp. 209-210).

Recapitulating Gregory's philosophies of science and medicine and his
medical ethics, we can say that diffidence unites with "fortitude of mind"
– the cardinal moral virtues of the physician, tenderness and steadiness as
the proper expression of disciplined sympathy – to produce the professional physician.

> I hope I have advanced no opinions in these Lectures that tend to
> lessen the dignity of a profession which has always been considered as
> most honourable and important. But, I apprehend this dignity is not to
> be supported by a narrow, selfish, corporation-spirit; by self-
> importance; by a formality in dress and manners, or by an affection of
> mystery. The true dignity of physic is to be maintained by the superior
> learning and abilities of those who profess it, by the liberal manners of
> gentlemen; and by that openness and candour, which disdain all arti-
> fice, which invite to a free enquiry, and thus boldly bid defiance to all
> that illiberal ridicule and abuse, to which medicine has been so much
> and so long exposed (Gregory, 1772c, pp. 237-238).

V. GREGORY'S INVENTION OF PROFESSIONAL MEDICAL ETHICS AND THE PROFESSION OF MEDICINE IN ITS INTELLECTUAL AND MORAL SENSES

As the eighteenth century dawned, Scotland found itself at risk of being
divided along geographic, social, religious, and class lines and of being
subordinate more than just politically to the English crown. Cities and
their commercial and professional practitioners began to have prominence
over the Highlands and their clans, and the great agricultural estates of
the Lowlands. Religious differences ran deep and people lived with the
legacy – still only tentatively in the past – of religious warfare and blood-
shed. Socio-economic classes multiplied and society stratified along class
lines. The rough and ready self-identity as a nation born of clans, as
Millar noted, could not sustain itself against these unyielding realities.
    As the crises of Scottish national identity deepened in the first third of
the eighteenth century, Scottish philosophers turned to the Baconian
experimental method that had begun to prove itself in natural philosophy,

particularly in what we now call physics. In physics the experimental method had led to the discovery of the laws or principles, real in things, that caused the regular, stable patterns of change. Scottish law also found its basis in identifiable principles of human conduct. The philosophers thus *looked to see* if they could observe *moral* principles that structured and therefore provided a normative basis for cognition and conduct; they quite self-consciously pursued the new science of man. Their core discovery in this new science they termed the "common sense" – observed, shared instinctual functions – far more powerful and reliable than unaided reason, among which senses they found the social principle. They described the normative reality that they observed: the shared natural affinity of human beings for each other in groups defined by shared moral, as distinct from physical, causes with the result that the group exhibited a national character.

> Different reasons are assigned for these *national characters*; while some account for them from *moral*, others from *physical* causes. By *moral* causes, I mean all circumstances, which are fitted to work on the mind as motives or reasons, and which render a peculiar set of manners habitual to us. ... By *physical* causes I mean those qualities of air and climate, which are supposed to work insensibly on the temper, by altering the tone and habit of the body, and giving a peculiar complexion, which, though reflection and reason may sometimes overcome it, will yet prevail among the generality of mankind, and have an influence on their manners. That the character of a nation will much depend on *moral* causes, must be evident to the most superficial observer; since a nation is nothing but a collection of individuals, and the manners of individuals are frequently determined by these causes (Hume, 1987a, p. 198).

These common moral causes draw people who share them together into a national identity. Scottish philosophers, just as Hume demonstrates in the above passage, discovered the observable normative principle, the social principle, something real and functional in the human constitution that causes the regular, stable patterns of change that encompass and define national identity. Scotland need not be in disarray or crisis; her citizens need only use their God-given faculties to discover the solution to the problem of national identity, the prevention of the sequelae of an unmanaged crisis of national identity.

We know that from his medical student days, Gregory found medicine to be in methodological disarray and confronted therefore with its own crisis of identity as a human enterprise. Gregory's "Proposall for a Medicall Society" proposes a series of prescriptions, drawn largely from Bacon and Gregory's own clinical reading and concerns, to remedy this crisis (AUL 2206/45, 1743). Gregory found his solution – starting in his medical student days, continuing in Aberdeen through its Philosophical Society, and culminating in the medical ethics and philosophy of medicine lectures that he gave at the University of Edinburgh and at the Royal Infirmary of Edinburgh – in Bacon's observational method and the openness to conviction that it required. By following Baconian method – adjusted to medicine by Gregory with his insistence on respect for the limits of medicine and risk of iatrogenic harm when enthusiasms substitute for scientific discipline in medical research – the physician can discover the observable, normative function that is health and observable normative departures from it that are disease and injury. Experimental method will lead to reliable accounts of the causes of the regular stable changes that constitute health and the regular, stable changes that will count as effective attempts to restore health, cure disease, prolong life into old age with minimum debility, and ease the dying process.

Gregory found the practice of medicine in disarray. Private patients summoned doctors and other practitioners at will; boundaries among physicians, surgeons, apothecaries, female midwives, irregulars, and clergy were unsettled at best and volatile at worst; patents self-diagnosed, self-treated, and shared prescriptions and remedies with no scientific information or discipline; physicians put on manners and tested the confidence of patients in their trustworthiness – female patients put themselves at particular and serious risk in these respects; the lay managers and funders of the Royal Infirmary exerted considerable control over physician access to resources and patients; and patients in the Infirmary risked abuse, particularly as research subjects – to name only the most troubling items on Gregory's problem list.

Gregory turned to the physiologists and philosophers of *sympathy* as his sources for his response to the crises in the practice of medicine. He takes from them – principally Hume's concept and account of sympathy in Reid's linguistic dress – the observable, normative principle of human constitution that, when properly functioning, naturally disposes one to be concerned with the fortunes and – crucially for medicine – misfortunes of others and moves one to respond to them; because one experiences just what the misfortunate experience – pain and distress – along with the normative, natural inclination to be rid or misfortune and so relieve man's estate. Properly functioning sympathy causes the regular, stable changes in judgment and conduct that bind the physician to the offices or duties of relieving and preventing human pain and suffering. Gregory also discovers, from his observations of the people around him, *a compelling moral exemplar of sympathy rightly and routinely functioning in the moral life: women of learning and virtue.* Mrs. Montagu, especially, and also her Bluestocking Circle friends, as well as Elisabeth Gregory, provided the compelling models of the diffidence-based intellectual virtue of openness to conviction and the twin sympathy-based virtues of tenderness and steadiness, especially in response to repeated, and often great loss.

Clinical openness to conviction obligates and inclines the physician to advance the capacities of medicine and thus relieve man's estate always with a view toward taking only ethically justified iatrogenic risk; Gregory is neither therapeutic imperialist, nor therapeutic nihilist. Clinical tenderness obligates and inclines the physician to remain focused on the patient's health-related interests and on the impact of the physician's conduct on other interests – such as good name and credit – that the patient risks in the clinical encounter. Tenderness blunts self-interest – in fame, power, market share, and money – in favor of the interests of the patient. Clinical tenderness obligates and inclines the physician to respond to the suffering of patients, day-in and day-out, with disciplined calm, clear-headedness, and the willingness to take desperate measures when desperate measures are *ethically* justified. Clinical steadiness disciplines the physician against dissipation, hard-heartedness, and being unmanned in the face of human pain, suffering, misery, and death.

Gregory then goes on to provide an account of the offices or duties of the physician in response to the problem list that defined the crisis he observed in the then-current practice of medicine. He bases this account on sympathy and its requirements. In doing so, he effectively addresses the problems of his time: he provides his students and his reader with concrete, clinically applicable moral instruction that issues with moral authority from the requirements of sympathy.

Gregory's medical ethics is professional, secular, philosophical, feminine, and clinical. He did what we in bioethics now do, two centuries before we thought of doing it.

In response to a world of medical practice marked, if not defined, by the pursuit of self-interest, he argues that the physician's conduct should be based on properly formed, i.e., systematically other-regarding, intellectual and moral character. His secular, philosophical method leads him to identify both a systematically other-regarding, self-effacing, self-sacrificing intellectual capacity of human nature, diffidence, and its intellectual virtue of openness to conviction *and* a systematically other-regarding, self-effacing, self-sacrificing moral capacity of human nature, sympathy, and its moral virtues of tenderness and steadiness. These twin capacities and their attendant virtues work in synergy to form the intellectual and moral character of the physician as fiduciary of the patient. The physician can claim to know with intellectual authority what is in the patient's interest. The physician's own self-interest is systematically blunted, so that he focuses on the patient's interests as his primary consideration. The physician is moved, as a matter of habit, to protect and promote the patient's interests, accepting self-sacrifice whenever required to do so. This ethical concept of the physician as professional has a substance and content lacking in sociological concepts (Berlant, 1975; Waddington, 1984).

In forging this ethical concept of the physician as a true *professional* – a word then in currency was thus given substantive philosophical content – Gregory breaks decisively from the Royal Colleges. He also breaks from the previous tradition in medical ethics (French, 1993c; Nutton, 1993), epitomized in Hoffmann, who was explicitly theological and whose admonitions include a strong dose of self-interest. Further, Gregory breaks from Boerhaave and his own teachers and colleagues at the University of Edinburgh by providing a thoroughgoing philosophical basis for the duties and qualifications of a physician. In doing so, he advances medical ethics far beyond the handbooks and admonitions that characterize the preceding English-language literature (Wear, 1993). Gregory therefore can be credited, in the history of medicine, with inventing secular, philosophical, professional medical ethics in the English-language literature. He writes the first English-language medical ethics for the physician-patient relationship, grounded in the intellectual and moral virtues of the physician.

Gregory also writes the first feminine medical ethics in the history of medicine as a whole. He genders sympathy and its virtues feminine and argues that the physician-patient relationship should be asexual and disinterested – though far from detached. The physician should, when it comes to matters of self-interest, be systematically self-effacing and self-sacrificing. Women of learning and virtue provide the moral exemplars for physicians, who should, in response, be gentlemen of gallantry and honor, just as men were expected to be at Bluestocking gatherings and just as they should be in courting his daughters.

Finally, Gregory's medical ethics is clinical throughout. His abhorrence of "systems" and their abstractions keeps him firmly rooted in clinical experience, a leitmotiv of his thought that takes root in his medical school writings and culminates in *Lectures*. For Gregory medical ethics was necessarily clinical, making ethical theory and its clinical application seamless. The term, "applied ethics," we can now say, is just an artifact of "systems" and should not be used to refer to Gregory's accomplishments: The distinction between "theoretical" and "applied" ethics is simply foreign to Gregory's medical ethics.

Gregory's original and formative contribution to the history of medical ethics and therefore to the history of medicine was to combine Baconian openness to conviction and Humean sympathy to provide medicine with an *intellectual and moral* identity – scientific discipline and a life of service to patients – that it had until then lacked, and to do so in a secular fashion that both religious and non-religious individuals would have to accept because openness to conviction and to the results of experimental method requires them to accept it. In other words, Gregory writes the first English-language, philosophical, secular ethics of the physician-patient relationship as a truly *professional* relationship. Before Gregory, this literature comprised advice to the "good and learned physician" and admonitions based on Christian ethics (Wear, 1993). Gregory thus provides for medicine the intellectual and moral identity that it lacked. As the Porters put it, there was only a *patient-physician relationship* in eighteenth century British medicine (Porter and Porter, 1989). Gregory brings all of his intellectual resources and experience to bear to create the *physician-patient relationship* as fiduciary.

The virtues of the physician as fiduciary – openness to conviction, tenderness, and steadiness – provide the moral causes that will gather physicians together into a group worthy of the name, 'profession', in its intellectual and moral senses. Hume defines the professionals precisely in terms of such moral causes, just as he defines national identity:

> The same principle of moral causes fixes the character of different professions, and alters even that disposition, which the particular members receive from the hand of nature. A *soldier* and a *priest* are different characters, in all nations, and all ages; and this difference is founded on circumstances, whose operation is eternal and unalterable (Hume, 1987a, p. 198).

Gregory's enormous and original intellectual accomplishment was to identify the "eternal and unalterable" functions of diffidence and sympathy in the care of patients and to show how the virtues of openness to conviction, tenderness, and steadiness form the basis for the existence and ethics of the physician-patient relationship, and therefore for the profession of medicine, and to do so on philosophical, secular grounds that everyone committed to openness to conviction – i.e., everyone of minimal honor – must accept. Gregory creates the ethical basis for the social role of the physician as an intellectual and clinical authority for the sick, those for whom the physician cares. Gregory also establishes the ethical basis for the social role of being a patient, someone whose problems the physician can diagnose and manage with intellectual and clinical authority, and whom the patient can trust to act for his or her interests as the physician's primary concern, adding *moral* authority to *intellectual* and *clinical* authority.

The neglect of Gregory's contribution to the history of medicine and the histories of medical ethics and philosophy of medicine should, I hope by now to have persuaded the reader, come to an end. I now turn to Gregory's subsequent influence, as well as his importance for contemporary bioethics.

CHAPTER FOUR

ASSESSING GREGORY'S MEDICAL ETHICS

I turn now to the task of assessing Gregory's medical ethics. I do so in three parts. First, I consider how Gregory's work was received and understood by his contemporaries, for which there is limited documentary evidence. Second, I examine briefly his influence – mainly on subsequent work on medical ethics in Britain and North America, and, very briefly, on European medical ethics – during the eighteenth and nineteenth centuries. Third, I will argue for the importance of Gregory's medical ethics for contemporary bioethics.

I. THEN-CONTEMPORARY VIEWS OF GREGORY'S MEDICAL ETHICS

Benjamin Rush was among the most prominent of Gregory's students from the British colonies in America. The wonderful drama on the Surrender Field at Yorktown lay in the future, an event that Rush and his cosignators of the Declaration of Independence from English tyranny helped in no small measure to precipitate. Rush went to Edinburgh to study medicine in 1766, and he kept a journal of his student days. That journal contains the following entry:

> The Practise of Physic is taught by *Dr. John Gregory*, a Gentleman of considerable Genius and Learning. The greatest part of his life has been spent in *Aberdeen* in the prosecution of studies that have no Connection wth [with] Physic; upon this Acc:^{t} [account] he is not much esteemed as a Professor, nor do his [lectures] contain anything new or interesting in them. The students were so little pleased w:th them, that they presented a Petition to Dr. Cullen to give a Course of Lectures upon y^e [the] Practise of Physic, w:^{h} [which] the Doctor very kindly complied w:th – had it not been for these invaluable Lectures of Dr. Cullen's I sh: [should] have returned but little wiser in the practical parts of Medicine y^n [than] I came here. *Dr. Gregory* is however much esteemed & beloved as a man. – he possesses a full belief in Christianity, & used occasionally to introduce into his lectures such Reflections upon Infidelity as had a good Tendency to convince young minds that

true Honour could only consist in a Regard to Religion. He has wrote a very ingenious little Essay called "A Comparative View of the Faculties of Man w:th Regard to the Animal World" – it contains many original tho'ts [thoughts] upon Education – Music – & misscelanious Subjects of that Nature (EUL mic. m. 28, 1766, pp. 71-73).

Gregory's "full belief in Christianity," as we saw in Chapter Three, had its basis in his Baconian deism and its argument from design for God's existence. As we saw in Chapter Two, Gregory himself did not attend church services and so his "Regard to Religion" should be read as a regard to *natural* religion.

Rush takes a somewhat more positive view of Gregory as a clinical teacher in a letter to John Morgan of January 20, 1768:

> Dr. Gregory's lectures abound with excellent practical observations, but they are by no means equal to the unrivaled Dr. Cullen's, whose merit is beyond all praise (Butterfield, 1951, p. 51).

By 1768 Gregory had begun to give his ethics lectures and this may have influenced Rush's kind words. Rush thought enough of Gregory's other work to send a copy of *Legacy* to John Adams for Abigail (Butterfield, 1951, p. 192, p. 218).

Rush's understanding of Gregory in his *Journal* may reflect more Rush's own views on religion than Gregory's. In 1782, near the end of the War of Independence from English tyranny Rush, writes the following advice to William Claypoole on July 29, 1782:

> The following short directions to Dr. Claypoole were given as the parting advice of his old friend and master. If properly attended to, they will ensure him business and happiness in North Carolina. ...
>
> 2. Go regularly to some place of worship. A physician cannot be a bigot. Worship with Mohamitans rather than stay at home on Sundays (Butterfield, 1951, p. 284).

Gregory would accept the second of the admonitions in item number 2; he reserved the first only for his daughters; he would, I believe, have nothing to do with the third.

Rush, on May 13, 1794, writes to James Kidd, professor of Oriental languages at Marischal College in Aberdeen:

> I have just committed to the press a history of our late epidemic; it will contain nearly four hundred octavo pages. Reverberate over and over

my love to Dr. Beattie. I cannot think of him without fancying that I see Mr. Hume prostrate at his feet. He was the David who slew that giant of infidelity (Butterfield, 1951, pp. 748-749).

We saw in chapter two that Gregory's reaction to the Beattie-Hume controversy was not quite so clear-cut or vehement. Perhaps Rush's view of Gregory in his *Journal* reflects more Rush's own religiosity than it does Gregory's.

This becomes plain in Rush's medical ethics, the title of which indicates his intellectual indebtedness to Gregory: *Observations on the Duties of a Physician, and the Methods of Improving Medicine. Accomodated to the Present State of Society and Manners in the United States* (Rush, 1805), which Rush first delivered "February 7, 1789, at the conclusion of a course of lectures upon chemistry and the practice of physic" (Rush, 1805, Vol. I, p. 385). Rush followed the example of Young, by giving his ethics lectures at the end of his clinical lectures. Rush divides his *Observations* into two sections, the first on "the most profitable means of establishing yourselves in business, and of becoming acceptable to your patients, and respectable in life," and the second on "a few thoughts which have occurred to me on the mode to be pursued, in the further prosecution of your studies, and for the improvement of medicine" (Rush, 1805, Vol. I, p. 387). These followed, roughly, Gregory's plan of two lectures in his own *Observations* (1770).

The two texts exhibit considerable similarity, with Rush admonishing his students to "[a]void singularities of every kind in your manners, dress, and general conduct" and to "[s]tudy simplicity in the preparation of your medicines" (Rush, 1805, Vol. I, p. 390, p. 404). Rush departs from Gregory's text when it comes to the topic of infidelity. Contrast, for example, the way in which the two introduce the topic:

> I shall conclude this subject with some observations on a charge of a heinous nature, which has been often urged against our profession; I mean that of infidelity, and contempt of religion (Gregory, 1772c, pp. 63-64).

> It has been objected to our profession, that many eminent physicians have been unfriendly to christianity (Rush, 1805, Vol. I, p. 391).

Gregory provides, in support of his case, physicians of "real piety" (Gregory, 1772c, p. 64) or "regard to religion" (Gregory, 1770, p. 57). Rush provides the same list, with this introduction:

But I cannot admit that infidelity is peculiar to our profession. On the contrary, I believe christianity places among its friends more men of extensive abilities and learning in medicine, than in any other secular employment (Rush, 1805, Vol. I, p. 392).

Already, Rush has given Gregory's secular medical ethics a *religious* emphasis, an emphasis that American medical ethics goes on to retain in its subsequent history in the nineteenth century, and even into our own. Indeed, this emphasis on religion, indeed on Christianity, as a (the?) source for medical ethics emerges as a distinctive American theme in the nineteenth century, becoming explicit in the American Medical Association Code of Ethics of 1847. (See below.) It is perhaps therefore no accident that the texts that marked the recrudescence of medical ethics in the mid-twentieth-century United States (Sperry, 1950; Fletcher, 1954; Ramsey, 1970) came from moral theologians, not philosophers (Jonsen and Jameton, 1995).

Tytler's *Life* provides another contemporaneous account of Gregory's medical ethics. Tytler summarizes the purpose of Gregory's *Lectures* under two "objects":

> *First*, To point out those accomplishments, and that temper and charac- ter, which qualify a Physician for the practical duties of his profession; and, *secondly*, to lay down those rules of inquiry, which, as he judged, were necessary to be observed in prosecuting the study of Medicine, considered as a branch of natural science (Tytler, 1788, p. 57).

Tytler claims that Gregory addressed the first of these two topics on the basis of both "the author's acquaintance with human nature, and that humanity of temper and elegance of mind, which distinguish all his moral writings" (Tytler, 1788, p. 58). The first reference is, of course, to Greg- ory's *Comparative View*.

Tytler provides an insight into Gregory's personal interest in the topic of medicine as a trade rather than an art (Gregory, 1772c, pp. 10ff.).

> Nor does his character appear to less advantage in his liberal and disin- terested remarks on physic, considered as a *lucrative trade*, which are expressed with the spirit and animation natural to one who felt for the real dignity of his profession, and was ashamed of the unworthy arti- fices, and the servile manners by which it has been too often degraded (Tytler, 1788, p. 58).

In light of the previous two chapters we can read Gregory as responding self-consciously and with deep personal conviction and commitment to the crisis of trustworthiness in physicians, brought on by the intense competition in the medical marketplace, and the man of false manners who insinuates his way into a patient's good graces, or who treats the well-to-do very differently from those of lower economic and social classes. Gregory, according to Tytler, took one of the central topics of the *Lectures* personally, intellectually, and professionally – just the way an engaged moral sense philosopher should, it would seem.

Of Gregory's philosophy of medicine, as Gregory presents it in the *Lectures*, Tytler makes the following, very interesting claim:

> The last three lectures related chiefly to Medicine, considered as a branch of natural knowledge; and they will probably be regarded by the more intelligent of his readers as the most valuable part of the volume. They display more fully than any of the author's other works, the extent of his philosophical views; and it is perhaps from them that we are best enabled to form a judgment of the loss which the science of Medicine sustained by his death (Tytler, 1788, p. 60).

Tytler here sees the priority that ought to be assigned to Gregory's philosophy of medicine: the central role that it plays in Gregory's thought, including his medical ethics. However, Gregory's philosophy of medicine is only incompletely developed. Thus, for those who came after him in the nineteenth century and, indeed, for us, however, Gregory's medical ethics remain "the most valuable part of the volume."

Smellie basically rewrites and repeats Tytler's view's of Gregory's *Lectures* (Smellie, 1800, pp. 93-94). Stewart considers the *Lectures* and *Elements of the Practice of Physic* together:

> He thought medicine required a more comprehensive mind than any other profession, and often brought much besides mere technical knowledge into his lectures. As a speaker, he was simple, natural and vigorous. He lectured only from notes, 'in a style happily attempered,' said one of his contemporaries, 'between the formality of studied composition, and the ease of conversation.' On one thing he insisted, that every student should appreciate the limitations of medicine, for only so could they learn to extend its borders (Stewart, 1901, pp. 118-119).

On the basis of these sources we can say that Gregory's contemporaries, Tytler most especially, appreciated the personal investment that Gregory made in his teaching, and in his teaching on medical ethics and philosophy of medicine in particular. Rush, putting Gregory through what appears to be a distinctive American filter, reads more theism into his work and life that was there. Finally, Tytler notes the importance of Gregory's *Lectures*, but not for medical ethics. Yet, Gregory's medical ethics had more influence on the subsequent history of Anglo-American medical ethics than did his philosophy of medicine.

II. GREGORY'S INFLUENCE

Evidence for the immediate influence of Gregory's medical ethics took the form of translations, into French, Italian, and German. One M. Verlac translates the *Lectures* into French in 1787 (Gregory, 1787). In his introduction Verlac summarizes Gregory lectures. In the course of doing so, he says that Gregory's first two lectures set out the "duties and imperceptible different nuances for whomever has not been endowed with very great sensibility, unknown to the grosser and mercenary spirits who do not see the moneyed nature of their profession ... " (Gregory, 1787, p. xxiii).[47] Here Verlac picks up Gregory's criticism of medicine as a trade, in perhaps stronger terms.

Verlac appreciates that the basis of medical ethics is not "amour propre" but the "pleasure of the practice of duty" (Gregory, 1787, p. xxiv).[48] This expresses medical ethics in language with which Gregory could readily agree. '*Amour propre*' means a kind of self-love or vanity and the prickly honor it generates. Gregory would say that *amour propre* involves too much "delicacy and sensibility," of the kind his daughter, Dorothy, exhibited when she was young. *Amour propre* lacks steadiness. Moreover, the honor of *amour propre* is a false honor, because, Gregory would say, it proceeds from self-interest, not duty to others. This theme apparently held considerable appeal for Verlac and for what he thought to be the potential audience for his translation.

To continue in this vein, for Verlac, the physician of false manners and sentiment represented as important an ethical challenge as it did for Gregory. Verlac points out that, on the basis of humanity, the public can discern the "artifices" of "flattering doctors," recognize "true merit," and cut off the "route to reputation" of these false physicians, to "unmask" such men and their "loincloth of modesty and immunity" (Gregory, 1787, pp. xxv-xxvi).[49] This calls to mind the spirited tone of the introduction to the 1770 edition.

Verlac summarizes the value of the *Lectures* in the following terms. "[P]ersonnes de la profession," practicing physicians, will find "new and interesting discussions" on various topics pertaining to the art of medicine (Gregory, 1787, p. xxix).[50] Medical students will gain from the opportunity to absorb "sentiments of humanity, nobility and generosity" (Gregory, 1787, p. xxix).[51] Here Verlac picks up the medieval concept of *paternalism* that lies at the heart of Gregory's medical ethics, couched in a nobility expressive of humanity and earned by generosity. The public, who are prey to diseases, will "learn what the duties of physicians are" (Gregory, 1787, p. xxx).[52] Medical ethics thus becomes a public matter. Patients need to know what the duties of physicians are because they need to be able to identify physicians who are worthy of their trust. Gregory, Verlac says, provides the answer: those who live by and practice the duties of humanity.

Finally, Verlac applies Gregory to the new institution of the hospital, indicating the advent of the ethics of medical institutions, on which Percival (1803) writes the first extended treatise. (See below.) Verlac states that the hospital is an "object of public beneficence" (Gregory, 1787, p. xxxiv);[53] as such, the hospital should be managed for a "class worthy of compassion," as a place to test medicine in the clinical setting, i.e., to conduct research, and for the training of students of medicine and surgery (Gregory, 1787, p. xxxiv).[54] Thus was started in Europe the ethics of the research-teaching hospital, already articulated in similar terms by Samuel Bard in America in 1769 (Bard, 1769). Gregory's influence includes providing the basis for the ethics of an institution that has become central to medical education. That this Gregorian account has become attenuated – through neglect of its requirements and vast changes in society that make medieval notions of paternalism difficult to sustain – represents a largely unexplored topic in contemporary bioethics.

The introduction to the Italian translation speaks in terms of the duties that each physician has in virtue of belonging to a profession. Again, there appears to have been a similar crisis of confidence in medicine motivating the translator, reflected in his emphasis on "the virtuous quality judged necessary in a physician ... to attain greater reliability of medicine" (Gregory, 1789, pp. vii-viii).[55] Echoing the French, the introduction urges: "the physician ought to abandon that amour propre ... " (Gregory, 1789, p. xx).[56] Here again we see the role that sensibility-based – the sensibility being the faculty of sympathy – virtue plays in replacing self-interest as the basis of the physician's behavior, making them duties in the true sense and creating an ethical foundation for a group of practitioners worthy of trust by patients, i.e., creating a profession.

The German translation of the *Lectures* also includes a preface that provides a brief biography of Gregory and a brief account of the publication of the *Observations* and *Lectures* (Gregory, 1778, pp. 3r-4v). This short introduction closes with the comment that the *Lectures* are a "text which one can recommend with justice to all young physicians" (Gregory, 1778, p. 4v).[57] The introduction provides no clues as to why the translator found the *Lectures* significant, other than as a complement to the 1777 translation of *Elements of the Practice of Physic*.

The first copy of *Observations* (1770) arrived in the American colonies of Britain in the year in which it was published (Truman, 1995). Rush also championed Gregory's work and Americanized it, as we saw above. In addition, Gregory's *Lectures* were published in an American edition in 1817 (Gregory, 1817). Translators had made Gregory's medical ethics available in three of the major languages of the Continent, with the French and Italian introductions picking up and emphasizing central themes of Gregory's medical ethics. In both France and America Gregory's work had become the basis for nascent work on the ethics of the new institution of the hospital. Tytler's evaluation of the importance of Gregory's *Lectures* quickly proved wrong. Events in the nineteenth century proved Tytler even more wrong, happily for the history of Anglo-American medical ethics. I cannot relate this story here, for its details would carry us far beyond the scope of the present book. I wish, instead, to highlight aspects of Gregory's nineteenth-century influence.

Gregory's subsequent influence appears in Thomas Gisborne's (1794, 1795) writings on the duties of physicians and particularly in Thomas Percival's *Medical Ethics* (1803). Percival writes the first extended treatise on the ethics of the new medical institution of the Royal Infirmary, the model on which the American hospital developed. Robert Baker reads Percival as a syncretic medical ethicist, bringing then contemporary moral theory – largely Gregory's and Thomas Gisborne's (1794, 1795) – to bear on the problems of this new institution (Baker, 1993a). I think that this reading is essentially correct, but would emend it to suggest that the main philosophical source for Percival's ethics is Richard Price (1948), the English moral realist and, like Percival, a dissenter, Unitarian, and sophisticated student of English moral realism. Percival learned his moral realism (as a college student at the Warrington Academy) from John Taylor (1760), a student of Price, who taught the required ethics course. Percival blends Gregory into this moral realism, and adds a powerful dose of practical wisdom and experience.

Percival opens his *Medical Ethics* with the following passage:

> ... Hospital physicians and surgeons should study, also, in their deportment, so to unite *tenderness* with *steadiness*, and *condescension* with *authority*, as to inspire the minds of their patients with gratitude, respect, and confidence (Percival, 1803, p. 9).

The first two virtues, of course, are the twin virtues of sympathy that Gregory sets out in his *Lectures*. The third, condescension, Baker correctly reads as the physician lowering himself from his lofty social station to that of the patient, who, in the Royal Infirmary, came from the new working class. Baker sees condescension and tenderness as *egalitarian* virtues, but I disagree. *Condescension*, for Percival, comes from the example of Christ; God condescended to become man. Authority derives from the physician's superior knowledge of diseases and their management. These four virtues, just like the virtues in Gregory, represent anti-egalitarian, *moral-aristocratic virtues* – in Gregory's cases with exemplars provided by women of learning and virtue, and in Percival's case, I believe, by Christ.

In his *Medical Ethics* Percival also makes reference to and applies the virtues of "attention, steadiness, and humanity" (Percival, 1803). Baker sees these as Gregorian, but not *tenderness* in the earlier passage. This is a mistake, as we have just seen. Gregory makes constant use of 'attention', and 'humanity' can be readily exchanged with 'sympathy'.

How does this Gregorian language of virtues fit with moral realism? This is a large question, a detailed answer to which would require a lengthy reply. Suffice it to say the following, therefore: Moral realists, such as Richard Price, and his student, John Taylor, and his student, Thomas Percival, understood human relationships, such as father-to-son or physician-to-patient, to be real, constitutive metaphysical elements of individuals. Individual relations are among the category of accidents in Aristotle (1984) and persist into modern philosophy, e.g., in Leibniz (McCullough, 1996). Moreover, these real, constitutive elements can be discovered by the disciplined use of reason. Reason discloses that duties constitute human relationships – in the case of medicine, the obligation to care for the sick defines the physician. Virtues such as attention, steadiness, tenderness, condescension provide the jointly sufficient and individually necessary character traits for being the sort of person who lives by and thus is really and reliably constituted by the obligation to care for the sick.

Percival, like Gregory, goes on to apply this ethical framework to a variety of topics, including, as in Gregory's case, *consultation*. Influenced by the work of Leake, the sociologists Jeffrey Berlant and Ivan Waddington view Percival's discussion to be concerned only with *etiquette*, i.e., intraprofessional concerns, and thus devoid of the ethical content that medical ethics is supposed to have, namely, the physician's obligations to the patient. More recently, Paul Starr (1982) and David Rothman (1991) continue this misreading of the history of medical ethics. This is indeed a misreading, because Gregory sees the ethical problems of consultation precisely in terms of the issue of making sympathetic regard for the patient the basis of intraprofessional behavior. Percival bases his medical ethics on the constitutive relational reality that defines the social role of the physician – the obligation to care for the sick. In this he builds crucially on Gregory's accomplishment. Baker has done a superb job of demolishing this reading as utterly unfounded in the texts of either Percival or Gregory: (1) neither Gregory nor Percival confused ethics with etiquette, contra Leake and company; and (2) the thesis that medical ethics served a drive for monopolization cannot be proved (Baker, 1993a, 1995b). To which we can add that the "etiquette" criticism suffers from the *presentism* of those who live in a time when the relationships among various health care providers are fairly well worked out and stable, and who assume, contrary to fact, that these matters were well worked out in the late eighteenth century. The recent changes in American medicine, most particularly the large-scale introduction of the concept of managed practice, have destablized these relationships, raising once again intra- and interprofessional issues as central to the agenda of medical ethics. (More on this in the next section of this chapter.)

While it may well be the case – the historical explication of texts remains to be done – that Percival's influence became larger than Gregory's on Anglo-American medical ethics through the middle of the nineteenth century, Gregory's work had its impact. The authors of the "Boston Medical Police," for example, report that they developed their document, a guide to physician conduct in clinical practice, on the following basis:

> That having examined the different publications of Gregory, Rush and Percival upon this subject, they first selected from them such articles, as seemed the most applicable to the circumstances of the profession in

this place (Warren, Hayward, and Fleet, 1995, p. 41).

In a number of places they copy text from Gregory in tact. Consider, for example, the following:

> If a physician cannot lay his hand to his heart and say, that his mind is perfectly open to conviction, from whatever quarter it may come, he should in honour decline the consultation (Warren, Hayward, and Fleet, 1995, p. 42).

Compare this to the following passage from *Lectures*:

> If a physician cannot lay his hand to his heart, and say that his mind is perfectly open to conviction, from whatever source it shall come, he should in honour decline the consultation (Gregory, 1772c, p. 37).

In the United States, the "Boston Medical Police," according to Chester Burns, "became *the* model for codes of medical ethics adopted between 1817 and 1842 by at least thirteen societies in eleven states, New York not included" (Burns, 1995, p. 138). Burns notes that the New York code went beyond the medical police, incorporating the medical jurisprudence of Percival and Gregory's ideals about cooperation among physicians, surgeons, and apothecaries (Burns, 1995, p. 138).

In Britain, Michael Ryan writes a compendium, as Burns calls it, of medical ethics, laws relating to medicine, and forensic medicine (Ryan, 1831). Ryan notes that Gregory and Percival wrote "the only works we have on the subject" (Ryan, 1831, p. 38) of medical ethics, "but these are not deemed authority, nor are they perused by medical students" (Ryan, 1831, p. 38). In contrast, Ryan also claims that *Lectures* "has justly received the universal approbation of the profession for more than half a century" (Ryan, 1831, p. 39). Ryan misreads Gregory's *Lectures* as "an excellent abridgement of the maxims of preceding writers, and on many occasions their language is quoted ipsissimus verbis" (Ryan, 1831, p. 39), e.g., Roderigo de Castro. Ryan includes a translation of passages from Castro's (1662) *Medicus-Politics* and some verbal formulations are similar, e.g., concerning consultations. Castro, however, bases his work on the concept of prudence and its related virtues, which include "circumspection, foresight, caution, perspicuity, whence also self-control, sobriety, mildness, modesty or moderation, pleasing costume ... " (Castro, 1662, p. 121),[58] and so on. Castro's basis for his views bears no resemblance to what Gregory first employs in the English-language literature. Thus, while some phrasing may be similar, their intent and meaning differs from the sources from which Gregory might have taken these phrasings.

The watershed event of nineteenth-century Anglo-American medical ethics was the development and publication of the American Medical Association's "Code of Medical Ethics" and responses to it in both the United States and Britain. The sources of influence on the "Code" remain to be sorted out. Nonetheless, Percival's influence looms large. Gregory's influence was felt, no doubt, through Percival, as well as through acquaintance with Gregory's work by authors of the "Code" (Baker, 1995). This is evident in the first section of the first article, where the Percivilian-Gregorian list of four virtues appears.

John Bell's introduction to the "Code" begins in a way that displays the distance that had opened between Gregory's medical ethics and that of the framer's of the "Code:"

> Medical ethics, as a branch of general ethics, must rest on the basis of religion and morality. They comprise not only the duties, but also the rights of a physician ... (Bell, 1995, p. 65).

The only time that Gregory uses the term 'right' is with reference to the patient, never to the physician. Gregory would say that talk of physicians' rights echoes the concerns of the medical corporations and guides, not a medical *profession*. The AMA Code of 1847 represents lost ground in this respect. Gregory also bases his medical ethics on a secular source, sympathy, which, while consistent with natural religion, was not equivalent to it. The distinct American accent in medical ethics on religion, beginning with Rush, continues to express itself in the "Code."

By the middle of the nineteenth century in the English-speaking world, Gregory's influence began to wane; it became attenuated. Other intellectual resources came into play, in a complex mix with social, economic, scientific, and national resources, that future scholarship will have to identify and analyze. Still, Gregory's influence was considerable. As we shall now see, his importance for contemporary bioethics and medical ethics has been underestimated.

III. GREGORY'S IMPORTANCE FOR BIOETHICS

A. *Gregory's Enlightenment Project*

We have seen that Gregory's medical ethics comes out of and reflects the distinctive features of the Scottish Enlightenment; his medical ethics relies on Hume's moral sense account of sympathy, and his philosophy of medicine (based on Baconian philosophy of medicine and science of man) reflects the Scottish commitment the "experimental method of reasoning," including its metaphysics and epistemology. Alasdair MacIntyre (1981, 1988) has argued that there was something called the "Enlightenment project" and that is was bound to fail. If MacIntyre's critique applies to Gregory, then his medical ethics will be of greatly diminished significance for contemporary bioethics and medical ethics. I therefore turn to a consideration of MacIntyre's account of the Enlightenment project, and why it had to fail. I then show that Gregory escapes this critique.

MacIntyre commits himself to the view that there was a *single* Enlightenment, which displayed a "unity and coherence" of a single culture that reached across national boundaries and identities (MacIntyre, 1981, p. 36). During the Enlightenment – a largely eighteenth-century phenomenon – and starting earlier, according to MacIntye:

> ... 'morality' became the name for that particular sphere in which rules of conduct which are neither theological nor legal nor aesthetic are allowed a cultural space of their own (MacIntyre, 1981, p. 38).

These rules of conduct require explanation for their authority, i.e., why anyone should take them seriously, live by them, and judge themselves and others by them. This created the need for "the project of an independent rational justification of morality" (MacIntyre, 1981, p. 38). This Enlightenment project relies on a particular intellectual structure, involving, according to MacIntyre, "valid arguments which will move from premises concerning human nature as they [moralists who undertake this project] understand it to be to conclusions about the authority of moral rules and precepts" (MacIntyre, 1981, p. 50). This project depends on some underlying assumptions, which MacIntyre states in the following terms:

> We thus have a threefold scheme in which human-nature-as-it-happens-to-be (human nature in its untutored state) is initially discrepant and discordant with the precepts of ethics and needs to be transformed by the instruction of practical reason and experience into human-nature-as-it-could-be-if-it-realized-its-*telos* (MacIntyre, 1981, pp. 50-51).

The problem, as MacIntyre sees it, is twofold. First, Enlightenment thinkers took the view that reason could not regulate the passions, from which MacIntyre draws the implicit, unstated conclusion that the passions therefore could not be regulated. Thus, the *telos* of human nature could never be achieved. Second, Enlightenment thinkers eliminated "any notion of essential human nature and with it the abandonment of any notion of a *telos* ... " (MacIntyre, 1981, p. 52).

Both claims are false for the moral sense theorists of Scotland, including Hume and Gregory. Both hold an *essentialist metaphysics of human nature*; they just don't believe that there is evidence that unaided reason will tell us reliably what that essence is. They also have a view of the *telos* for human beings. As we saw in the previous chapter, the end or goal of medicine, as of all other human activities, is "the convenience and happiness of life" (Gregory, 1772c, p. 151).

Gregory has no more to say on this subject. However, based on the discussion of the previous two chapters, we can fill in Gregory's very abstract formulation. One can contribute to human happiness by maintaining and strengthening its requisites. Medicine contributes to the human *telos*, Gregory could say, by addressing health, one of the requisites of human happiness. In an age of staggering mortality rates from acute diseases and injuries, exercising the capacities of medicine such as preserving health can relieve man's estate by reducing these mortality rates and thus easing the harsh reality of early mortality.

Moreover, it seems that human happiness concerns the improvement of our nature: the passions functioning under proper regulation and judgment, to which goal reason makes significant contributions. Reason uninformed and uncorrected by experience – in the proper sense of that technical term of art in the new experimental method – has little contribution to make to happiness thus understood, because reason is an inert, inactive principle of mind. It turns out that we can do old-fashioned Aristotelian metaphysics just fine; the goal is noble. But the Cartesian and Scholastic methods no longer can be taken to be adequate to that goal.

Thus, *contra* MacIntyre, the Scottish moralists have all of the ingredients for the Scottish Enlightenment project in place and it was decidedly *not* bound to fail. The Scottish understanding of human nature and their moral precepts are far from being "discordant" (MacIntyre, 1981, p. 53) with each other. Human nature as it happens to be involves the weak principle of reason and the principle of sympathy, neither properly developed although both are fully capable of becoming properly developed, i.e., properly functioning. Human nature if it realized its *telos* would be properly functioning and regulated sympathy, with reason in its secondary place. The "instruction" of experience does the transformation of the first into the second. Thus, each of the three elements of the scheme – the conception of untutored human nature, the conception of the precepts of experience-based, not rational, ethics, and the conception of properly realized human nature (MacIntyre, 1981, p. 51) – makes reference to the other two and so the "status and function" of each do indeed count as "intelligible" (MacIntyre, 1981, p. 51). In short, MacIntyre's views seem trapped by the concept of a *single* Enlightenment based on reason and rational morality, a concept alien to the Scottish Enlightenment. But there was *more* than one Enlightenment, a historical fact with non-trivial philosophical consequences.

H. Tristram Engelhardt (1986, 1996) makes much of this MacIntyrean necessary failure of the Enlightenment for contemporary bioethics and its possibilities.[59] In addition to MacIntyre's critique, Engelhardt argues that the Enlightenment project of justifying morality must fail because it must appeal to a "content full" concept of the human good, just as do the religious and aesthetic views of morality that secular morality intends to displace, to use MacIntyre's language. Engelhardt argues that reason cannot produce an authoritative content-full concept of the good, including the goods or ends of human beings that medicine should seek. Gregory would agree; this is instinct's province.

Moreover, in Engelhardt's terms, the ethics of medicine that Gregory develops from his concepts of disease and sympathy is not content-*full*; it is content-*minimal*. Indeed, the "convenience and happiness of life" (Gregory, 1772c, p. 151) is nearly empty of content, it is so "abstract," so to speak. Gregory thus makes no reference to some detailed, overall account of human happiness in terms of rank-ordered goods and values – *the* good for humans. Indeed, his concepts of disease and health do not require him to do so. These concepts require reference only to improvement of the human condition, to the "convenience" of life, i.e., making it less full of early death, disease, injury, pain, and distress. Medicine possesses limited capacities to serve this goal, mainly in the form of relief from afflictions to "convenience and happiness." So Gregory does require an account of medicine's capacities and these are the first three "Capitall Inquirys" he set out as a medical student and then developed in his philosophy of medicine and medical ethics. But he does not require any account of *the* good for humans and so on this count, too, Gregory's enlightenment project succeeds in avoiding Engelhardt's critique.

In the language of contemporary bioethics, it is possible to have a virtue-based account of medical ethics in which the physician can form reliable beneficence-based clinical judgments about what is in the interest of the patient quite independent from any need for an account of what *the* good for humans is. Medicine's capacities remain directed to the "relief of man's estate," by reducing mortality and morbidity and protecting functions related to having a quality of life – however patients define it – from deterioration that accompanies most chronic illness. Beneficence obligates the physician to seek for the patient the greater balance of goods over harms as these are understood from and balanced within a rigorous clinical perspective on the patient's interests (McCullough and Chervenak, 1994; Beauchamp and Childress, 1994). The content of beneficence is based in the social role that the capacities of medicine define and, therefore, there is *no need for a content-full concept of the good or ends of medicine*, in order to give meaning to beneficence.

By contrast, any method for bioethics that concerned itself with the ends or goals of medicine, rather that its capacities and their proper limits, *would* be subject to the Engelhardtian critique, as some of my own earlier work would be (Beauchamp and McCullough, 1984). Edmund Pellegrino and David Thomasma, for example, claim that the *ends* of medicine can be identified and "are ultimately the restoration or improvement of health and, more proximately, to heal, that is, to cure illness and disease or, when this is not possible, to care for and help the patient to live with residual pain, discomfort, or disability" (Pellegrino and Thomasma, 1993, pp. 52-53). On such an account "[i]f these ends are to be achieved, the good of the patient provides the architectonic of the relationship" (Pellegrino and Thomasma, 1993, p. 53). But for this sort of project – which is *not* Gregory's – to succeed, there must be a persuasive account of what the *good of the patient* is, and Engelhardt is correct to claim that all such accounts founder on the inability of reason to provide persuasive content for this good. Gregory would say, correctly in my view, that Pellegrino and Thomasma have *confused the capacities of medicine with its end*.

The lesson from Gregory for contemporary bioethics is that both should simply abandon any attempt to provide philosophical accounts for the end or ends of medicine, not simply because all such accounts are indeed bound to fail but for the more persuasive, practical reason that bioethics does not require any such account. Whatever the ends of human life might be, the profession of medicine can provide limited capacities to influence our ability to achieve those ends – in the eighteenth century, and now – mainly by relieving us of affliction that results in reduced capacity to identify, seek, and enjoy "convenience and happiness" in our lives, the content of which we can fill in in all sorts of ways. Medicine has no capacity to tell us how to do this. Thus, with Pellegrino and Thomasma, the Hastings Center project on the ends of medicine (Stuber, 1995) involves deep and intractable conceptual incoherence. This is one area where previous historical figures understand better than we do what the problem is – setting the proper limits on medicine's capacities, not its ends – and how to go about addressing that problem. The problem for American medicine, at least, is that it lost this sense of limits on medicine's capacities, even as they were increased and improved, and so bioethics of the last three decades in the United States can be read as undertaking the task of rediscovering Gregory's wisdom, without knowing that that was the task and, in the case of the Hastings Center project, continuing to misconstrue it.

B. The Persistence of Pre-Modern Ideas in Modern Medical Ethics: Paternalism and the Physician as Fiduciary

The reader will perhaps have noticed that I have nowhere in this text characterized Gregory or his medical ethics as "modern." This term is used in the history of philosophy to signify that a philosopher has made a decisive break from medieval or scholastic philosophy, which is somehow seen as a failure in its inability to respond to "modern" developments, especially in physics in the seventeenth and eighteenth centuries. The "modern" represents the attempt to understand the world in secular terms, using reason. Thus, as we saw above, MacIntyre points out that during the Enlightenment a sense of morality developed independently of theological concerns. This development paralleled that of physics or natural philosophy. The very phrase signals the autonomy of the undertaking from a variety of sources. These include theological sources and religious authorities and so the modern breaks sharply, so it is thought, from medieval ways of thinking in which such sources and authorities loomed very large indeed.[60]

Enlightenment thinkers were well aware that reason could not claim autonomy from the passions. Indeed, the proper relationship between reason and the passions became a central topic of psychology, epistemology, and moral philosophy. MacIntyre and others who see the Enlightenment as a "unified and coherent" cultural phenomenon take the view that "the" Enlightenment solution to this problem was to have reason control and subordinate the passions. The Scottish moral sense philosophers took just the opposite view. Reason, as we saw in Chapter Two, is the weaker and therefore less reliable of the two main principles of mind. *Instinct* shows itself upon investigation to be both universal – Gregory in *Comparative View* finds it in non-human animals – and reliable, if it is trained to its proper function.

Historians of philosophy dispute when the distinctively *modern* appeared in modern philosophy. Traditionally, Descartes has been thought to be the first modern philosophers. More recently, scholars such as John Deely (1994) have located the modern earlier, in the work of Vitoria, de Soto, or Suarez. No one, it seems, would dispute that eighteenth century thinkers were modern, unless they had fallen back on older, discredited, Scholastic ways of thinking. Thus, it would seem altogether fitting to characterize Gregory as a modern thinker. After all, his philosophy of medicine draws crucially from the Baconian experimental method and natural philosophy and so exhibits one of the principal characteristics of the modern in the history of ideas. Gregory's medical ethics also draws deeply on Hume's concept of sympathy, a central discovery of Hume's application of the experimental method to "moral subjects" (Hume, 1978, p. xi). Don't these two considerations together make Gregory a modern thinker?

Baker answers in the affirmative and claims that Gregory "formulated the first modern theory of medical ethics" (Baker, 1993b, p. 863). This answer suffices, but only if these two elements of Gregory's medical ethics constituted the whole of Gregory's medical ethics; they do not. Baker also characterizes Gregory as the "inventor of medical paternalism" (Baker, 1993b, p. 863).

> For Gregory's objective was that of the classic medical humanism: that is, to use sympathy to 'manage [patients] properly'; and when humane management required, he even permitted a 'deviation from the truth' (Baker, 1993b, p. 863).

By 'paternalism' Baker means the imposition of beneficence-based clinical judgment on patients without sufficient ethical justification and therefore without moral authority (Beauchamp and Childress, 1994). This reading fails to appreciate that Gregory does, indeed, provide justification for the physician's "governance" of patients and how much information to share with them and under what circumstances, as we saw in the previous chapter.

There is, however, another sense of 'paternalism' in which Gregory does, indeed, count as a fully self-conscious paternalist, the sense in which the clan chief and laird owe duties of paternalism to their social subordinates. That is, certain social roles are defined by asymmetrical obligations *downward*, an idea with deep roots in medieval, Highland Scotland. In this respect, Gregory is no modern thinker.

Indeed, he finds two central social developments of modernism, commercial and city life, abominable. In *Comparative View* he bemoans the replacement of a society "in which Mankind appear to the greatest advantage" in which natural religion prevails with a society marked by the "state of commerce" (Gregory, 1772a, p. viii, p. x (one)).[61] In commercial society

> ... men leave the plain road of nature, superior knowledge and ingenuity, instead of combating a vitiated taste and inflamed passions, are employed to justify and indulge them ... (Gregory, 1772a, p. x (one) - x (two)).[61]

In particular, money and its accumulation come to dominate life with disastrous consequences, including particularly

> ... an universal passion for riches, which corrupts every sentiment of Taste, Nature, and Virtue. This at length reduces human nature to its most unhappy state in which it can ever be beheld (Gregory, 1772a, p. x (two)).[61]

This is because

> ... money becomes the universal idol, to which every knee bows, to which every principle of Virtue and Religion yields ... (Gregory, 1772a, p. x (two) - xi).[61]

Bacon's several idols have been replaced by one, towering, malignant, and wholly destructive idol – commerce or the pursuit of money. It should come as no surprise that Gregory in the *Lectures* reacts so strongly against medicine as a trade and proposes ways to insulate medicine from commerce, e.g., by having men of independent means practice medicine, so as to eliminate commercial aspects from medical practice.

Gregory's characterization of the solution to this problem is very interesting:

> It was this consideration of Mankind in the progressive stages of society, that led to the idea, perhaps a very romantic one, of uniting together the peculiar advantages of these several stages, and cultivating them in such a manner as to render human life more comfortable and happy (Gregory, 1772a, p. xv).

The romantic solution, in other words, draws heavily, even exclusively, from the earlier, pre-commercial and therefore pre-modern stages of society.[62] Doing so will produce the "most advanced state of society" in which individuals "may possess all the principles of genuine taste ... delicacy of sentiment and sensibility of heart ... " (Gregory, 1772a, p. xv). As we saw in the previous two chapters, this medieval idea means that taste, sentiment, and sensibility of heart – all of the virtues – are acquired as a result of the hard work of committing oneself to a way of life that was then fading out of existence.

Gregory infuses his concept of medicine as a profession and the physician as a gentleman with the romantic, medieval, conservative, moral-aristocratic idea of the moral life of service in which interest never corrupts duty. As Haber notes, a concept of duty and service cuts against the grain of modern, egalitarian, democratic societies (Haber, 1991). In summary, then, Gregory blends the distinctively modern experimental method and its results in Hume's moral psychology and philosophy with the pre-modern, medieval idea of a life of service uncorrupted by interest.

In doing so, Gregory forges the concept of the physician as a fiduciary. In law a fiduciary is understood to be "a person holding the character of a trustee, in respect to the trust and confidence involved in it and the scrupulous good faith and candor which it requires" and also to be a "person having a duty, created by his undertaking, to act primarily for another's benefit in matters connected with such undertaking" (Black, 1979, p. 563). As a fiduciary, the physician (1) must be in a position to know reliably the patient's interests, (2) should be concerned primarily with protecting and promoting those interests; and (3) should be concerned only secondarily with protecting and promoting his or her own interests (McCullough and Chervenak, 1994). A physician who practices medicine consistently according to these requirements is worthy of the trust of patients.

For Gregory, the physician knows reliably the interests of the patient. Baconian method applied to medicine results in well-founded concepts of health and disease, diagnoses, and treatment plans, thus satisfying the first of these three conditions. Properly functioning sympathy in the form of tenderness directs the physician's concern primarily to the patient and motivates the physician to relieve the patient's suffering, thus satisfying the second condition. Properly functioning sympathy in the form of steadiness and in its anathema toward the pursuit of interest satisfies the third condition. Without using the term 'fiduciary' Gregory invents the concept in response to the crisis of intellectual and moral trust in the physicians of his time.

This helps us to see that the concept of a fiduciary in Gregory's formulation of it involves, at its core, a pre-modern idea. Gregory meant for this idea to become a living reality for physicians; otherwise they could not claim to be gentlemen or professionals in any ethically meaningful sense of the terms. Mark Rodwin has pointed to some aspects in which physicians have yet to realize Gregory's ambition for them, e.g., in accepting gifts that might influence clinical judgment and practice (Rodwin, 1993). In an extension of his analysis of commercialism in medicine Rodwin more recently has characterized the concept of being a fiduciary as a metaphor in American law and has suggested that "[t]here are no right or wrong metaphors" because the "law adopts and uses metaphors, in part, to solve social and legal problems" (Rodwin, 1995, p. 256). If this analysis if correct, it helps us to see that two centuries after Gregory forged the idea, the actual social institution and practice of medicine has yet fully to realize it. Moreover, it may enjoy only tenuous existence of a metaphor, rather than the living normative reality that Gregory surely meant it to become. If Buchanan's analysis is correct, then there is "strictly speaking no medical profession" in the United States "in the legitimating or normatively adequate sense" (Buchanan, 1996, p. 130).

When bioethics began to come to prominence three decades ago, it did so in large measure by attacking medical paternalism, in the sense of imposing beneficence-based clinical judgment on patients without sufficient ethical justification and therefore moral authority, usually in violation of autonomy. The ethical principle of respect for autonomy obligates the physician to seek the greater balance of goods over harms for the patient, not as these goods and harms are understood and balanced from a rigorous clinical perspective, but from the *patient's perspective*, over which the patient should be presumed sovereign (McCullough and Chervenak, 1994; Beauchamp and Childress, 1994). Looking back now, it seems that writing on medical paternalism in the 1970s became a right of passage into the exciting new field of bioethics.

Not only was the field not new, as we then thought, it did not appreciate fully the conceptual dimensions of medical paternalism. Tom Beauchamp reports that 'paternalism' "refers loosely to acts of treating adults as a benevolent father treats his children" (Beauchamp, 1995, p. 1915). Paternalism thus provided an obviously antiegalitarian approach, to be criticized as such in a United States that had come closer to its ideals of freedom as a result of the civil rights movement of the 1950s and 1960s and other social changes. Paternalism as the benevolent behavior of one's father is far removed indeed from Gregory's sympathy-based, medieval concept. More important, the egalitarian attacks on paternalism, e.g., Robert Veatch's triple contract theory (Veatch, 1981), represent, at a deeper and – at the time – unappreciated level, an attack on the pre-modern idea of the moral-aristocracy of the gentleman professional. It is no surprise, therefore, that the phrase 'physician-patient relationship' became reversed ib bioethics to the 'patient-physician relationship.'

This linguistic turn marks a profound conceptual change. For Gregory, when the physician possessed both reliable medical knowledge and clinical skills governed by the intellectual virtue of diffidence and regard for the patient governed by the virtues of sympathy, we can now say, there existed the social role of the physician as a fiduciary, in which sick individuals could confidently place their intellectual and moral trust. They could put aside fears of exploitation and abuse. The new social role of the physician as a professional in the ethically meaningful sense of the term that Gregory provides for it creates, as *its* derivative, the social role of being a patient, someone who stands to benefit from the physician's care and can trust in this. The role of being a patient derives from the role of being a physician, because asymmetrical obligations downward constitute the latter. In forging the social role of being a patient Gregory uses the term 'patient' but he also uses 'the sick', whereas Hoffmann used only *'aegrotus'*. Such obligations provide powerful solutions indeed to the problems confronting Gregory and sick people, as we saw in the previous chapter. Egalitarian bioethics removed these solutions, making physicians and the sick, once again "equal," as well as wary, competitive, interested strangers to each other, an implication ignored by its advocates.

One singular advantage of Gregory's account, of which he was aware in his own terms, is that no patient is a stranger to the physician. It is not the case that social differences open unbridgeable gulfs between the physician and the patient, as Rothman (1991) wants to claim. Indeed, Gregory would (rightly) say that Rothman's account is ethically superficial and not worthy of further serious consideration.

Engelhardt's concept of the patient and physician as "moral strangers" makes a deeper challenge, but it too can be addressed in Gregorian terms. Engelhardt argues, in effect, that because beneficence-based clinical judgment must appeal to some content-full concept of the good for the patient, because any such attempt must fail, and because such content-full concepts of the good are nonetheless required for the moral authority of beneficence-based clinical judgment, the physician lacks intellectual and therefore moral authority to make any claims about what is in the *patient's interests*. Only communities that can provide such content, most notably faith communities, provide the needed authority. Thus, Engelhardt (1986, 1996) proposes a *utopian solution* in which medicine is "subscripted," with Roman Catholic-medicine, Orthodox Catholic-medicine, or other hyphenated forms of medicine flourishing in a fully pluralistic, tolerant society. But this simply recreates the competition of concepts of disease and health and of warranted forms of diagnosis and treatment that led to the crisis of trust in eighteenth-century British medicine.

The persuasive power of egalitarian approaches to bioethics such as Veatch's and utopian approaches such as Engelhardt's both depend for their persuasive power on the assumption that the sick – there are no more "patients" – can rely on and trust in physicians. In other words, the contemporary attack on paternalism is ultimately parasitic at its core on what it rejects, because it assumes the trustworthiness of the physician as a pre-modern, moral-aristocratic fiduciary animated by a moral life of service. But this involves a contradiction, because the force of these contemporary approaches to bioethics is to radically deconstruct the concepts of a fiduciary, medicine as a profession, moral-aristocracy, and the social role of being a patient – right out of existence.

There are now changes in medicine that also promise – or threaten – to replicate the conditions of eighteenth-century British medicine. I refer now to the changes in American medicine, brought about by the new managed practice of medicine (Chervenak and McCullough, 1995). Medicine is no longer a cottage industry of small-scale, fee-for-service, entrepreneurial, face-to-face practice with the physician owing little or no accountability to peers, institutions, and society. Instead, medicine is rapidly becoming an institutional, large-scale, pre-paid phenomenon, to which the management concepts of business can be applied more effectively than when medicine was largely practiced as a cottage industry. The need to do so has been created by payers – both private and public, especially the former in the United States – who now demand efficiency in the form of reduced growth of medical care expenditures, or even absolute reduction of costs. For payers, the main problem in American medicine is its inefficiency. Among the factors that contribute to this inefficiency are the oversupply of physicians, hospitals, and technology and the "carrying costs" that this oversupply imposes, as well as the extraordinary variability in the diagnosis and treatment of patients, a variability that seems unconnected to clinical scientific discipline (captured in part by outcomes measured in terms of morbidity, mortality, and physiologic functions). These problems of inefficiency will no longer be tolerated by payers, and the "solution" is to shift medicine to an industrial model and apply to medicine the two main business tools of efficient industrial practice: (1) creating economic conflicts of interest in how physicians are paid, which directly and indirectly influences clinical judgment and practice; and (2) regulating clinical judgment and practice directly through such mechanisms as practice guidelines, prospective utilization review, physician report cards, etc., the latter deeply in debt to the work of W. Edwards Deming (1986).

Gregory's medical ethics help us to see what is at stake for physicians and patients in this turbulent change of American medicine and in other countries, as managed practice becomes an increasingly global phenomenon. Creating economic conflicts of interest poses a direct threat to the physician as a fiduciary, because this management tool uses the physician's self-interest – 'interest' in Gregory's language – to manipulate the physician's clinical judgment and behavior. The third of the three conditions for being a fiduciary – keeping one's own interests systematically in a secondary status – becomes unstable, at best. But then the second condition, keeping the patient's interests as the physician's primary consideration, destablizes. As a consequence, the physician-patient relationship destablizes and threatens to become a commercial transaction, precisely the morally disastrous problem that Gregory meant to solve.

Practice guidelines – when they are well developed and scientifically reliable – hold out the promise of making clinical judgment and practice conform to the rigorous Baconian standards that Gregory proposes in his philosophy of medicine. Neither Gregory, nor physicians now, can complain if clinical judgment and practice are obliged to become more *scientific*, in the sense of being evidence-based, than the unscientific variability they so often exhibit, especially in the fee-for-service setting.

However, if the regulation of clinical judgment and practice has little or no support in reliable science – a problem exacerbated if practice guidelines or other forms of management are driven mainly by price competition – then Gregory would rightly object to them as destructive of the profession of medicine in its ethical sense.

The combination of economic conflicts of interest with scientifically weak management of clinical judgment and practice will result, Gregory would warn us, in the destruction of medicine in a form that rightly can claim intellectual and moral authority and thus in the destruction of the physician as a fiduciary and, with it, the physician-patient relationship. Whereas Gregory attempted to forge that relationship as a solution to his problem list, we may be in the process of *undoing that relationship*, to our peril; for, when we become "the sick," Gregory would warn us, we find ourselves without the *intellectual and moral protection* of the social role of being a patient created by a fiduciary profession of medicine. If Buchanan (1996) is correct, then managed care will simply propel American medicine, at least, down the road it has already taken to self-destruction as a profession worthy of the name.

Contemporary approaches to bioethics such as those of Veatch and Engelhardt and changes in American medicine to a managed practice model combine to transform the physician-patient relationship into the medical institution-the sick relationship. Gregory helps us to see what would then be at stake. The sick could have no confidence in the concepts of disease and health, the diagnoses, and the treatment plans around which these medical institutions would be organized. These institutions would be mainly, if not exclusively, *commercial* in character, meaning that they would exist for the pursuit of their stakeholders' – whether for profit or not there will be stakeholders – interests. There will still be an oversupply of institutions and physicians, leading to a brutal competition, the eventual results of which will include underemployed or even unemployed physicians and defunct health care institutions. Respect for autonomy, the main conceptual tool of contemporary bioethics, will be asserted in defense of the sick. However, in the face of institutional power now developing in American medicine – we are in the middle stages of the formation of an oligopolistic market – asserting the autonomy of individuals may be the equivalent – in the politics of actual power – of shouting into the whirlwind.

In short, only at our peril do we give up the pre-modern, medieval, Scottish Highland, moral-aristocratic concept of paternalism and the fiduciary concept of medicine as a profession that it helped to create. Percival, who, recall, wrote the first ethics for medical institutions, shows us one way to avoid this peril.

> The physicians and surgeons should not suffer themselves to be restrained, by parsimonious considerations, from prescribing *wine*, and *drugs* even of *high price*, when required in diseases of extraordinary malignity and danger. The efficacy of every medicine is proportionate to its purity and goodness; and on the degree of these properties, *caeteris paribus*, both the cure of the sick, and the speediness of its accomplishment must depend. But when drugs of an inferior quality are employed, it is requisite to administer them in larger doses, and to continue the use of them a longer period of time; circumstances which, probably, more than counterbalance any savings in their original price. If the case, however, were far otherwise, no oeconomy, of a fatal tendency, ought to be admitted into institutions, founded in principles of the purest beneficence, and which, in this age and country, when well conducted, can never want contributions adequate to their liberal support (Percival, 1803, pp. 13-14).

Percival limits his argument about the management of institutional resources to the case of grave, life-threatening illness. He first makes a cost-benefit argument – as near as I can tell, the first such argument in the history of medical ethics. But, being aware that such an argument might fail, he then makes the case for institutional obligations of "purest beneficence," calling to mind Verlac's language (Gregory, 1787). That is, the physician's clinical judgment and practice should be unalloyed with economic considerations. Clinical judgment, properly formed, should be a function solely of disciplined judgment of what is in the patient's interest – thus taking full advantage of and extending the implications of the physician-patient relationship that Gregory forged.

We cannot make this Percivilian argument in response to managed practice. Nor can we appeal to Gregory's "romantic" solution – that's just, indeed, what it is – to return to a pre-modern, pre-commercial social order. Both solutions are unrealistic and to adopt them would be to make oneself irrelevant, as some have done (Emanuel and Dubler, 1994; American Medical Association, 1994). We can, though, attempt to preserve the pre-modern idea of *paternalism*, the physician as fiduciary, and seek to identify, create, and sustain manageable forms of the economically responsible fiduciary practice of medicine. Gregory would, at least and correctly, warn us that this constitutes an unstable, even combustible mix. Our challenge, if we are to avoid recreating Gregory's problem list – it is a *problem* list, because market solutions ultimately do not work for fiduciary social institutions, because these institutions are intrinsically other-regarding, not self-regarding, and, as a consequence, are economically inefficient – is to find ethically justified form*s* of this combustible mix. The choice of the plural is deliberate; there is bound to be a variety of ethically justified forms of economically responsible fiduciary medicine. As we seek to identify these forms of a reforged fiduciary profession of medicine we would do well to remember another lesson from Gregory. Market solutions will appeal to freedom of choice as a fundamental moral category. Gregory would warns us against this move, as corrosive of the life of service that even a reforged profession of medicine will still require. Responsibility remains a more fundamental moral category for fiduciary professions than freedom, a fundamental ethical implication of medieval moral-aristocracy of the Scottish type. In other words, the market is a permanent moral problem for fiduciary professions and societies that want them; markets should not be regarded as solutions. This deeply conservative dimension of Gregory's medical ethics cannot be avoided two centuries later. Unless market forces are restrained and physicians make self-conscious choices to sufficiently insulate clinical judgment and behavior from conflicts of interest inherent in the market of medicine (Rodwin, 1993; Buchanan, 1996), then the transformation of medicine that Gregory initiated will have ended in spectacular failure only two centuries after, conceptually, he set it in motion.

C. Virtue-Based and Care Approaches to Bioethics

For a variety of reasons – including what seems to be a felt need to be new and different that has come to characterize bioethics in the last decade or so; it is in this respect remarkably nineteenth century in character – virtue-based and care approaches to bioethics have come to prominence in recent years. Both distinguish themselves from principle-based approaches, as set out in Tom Beauchamp and James Childress' *Principles of Biomedical Ethics* (1994). Now in its fourth edition this is perhaps the most influential text in bioethics in the United States, certainly, and of considerable importance internationally.

In the course of rewriting the four editions, Beauchamp and Childress have come to the view that *virtues* play an important role in any complete bioethics, especially in an ethics for the health care professions (Beauchamp and Childress, 1994, pp. 462 ff). They define virtues in the following terms:

> Virtues in practice are habitual character traits that dispose persons to act in accordance with worthy goals and role expectations of health care institutions (Beauchamp and Childress, 1994, p. 463).

This understanding seems to be in some debt to MacIntyre's formulation:

> *A virtue is an acquired human quality the possession and exercise of which tends to enable us to achieve those goods which are internal to practices and the lack of which effectively prevents us from achieving any such goods* (MacIntyre, 1981, p. 178).

Both definitions emphasize that we are not born with the character traits that count as virtues, for both definitions assume that we have the capacity to develop the virtues. Both see a close connection between social roles and virtues. This understanding of virtue contrasts with that of Pellegrino and Thomasma who take what they call an Aristotelian and medieval approach:

> For Aquinas, as for Aristotle, ethics was teleological. The moral quality of human acts derived from their relationship to the final end of human life. Virtues are habitual dispositions to perform actions that accord with this end (Pellegrino and Thomasma, 1993, p. 8).

This concept of virtue involves the problem of defining *the* end for all human beings, discussed above, and the difficulty of sustaining a unitary teleology of human nature in the face of biological variability – a problem that Pellegrino and Thomasma do not address. This should come as no surprise; given biological variability, any talk of *the telos* or *the end* of human existence that is more than content-minimal becomes spurious. Gregory's "convenience and happiness" of life can readily accommodate biological variability.

Gregory understood the virtues of the physician to be those habitual traits of character required for the proper functioning of diffidence and sympathy in the care of patients. Two differences between Gregory and contemporary accounts of the virtues are worth noting: First, his great advantage over us was that he could base his account of the virtues of the physician in what he took to be scientific fact, namely, the results of Hume's physiologic, moral-psychological investigations in the *Treatise*.

He could thus provide his reader with a foundation or origin of the virtues of the physician in the facts of a demonstrable, transparent, interpersonal physiologic function.

MacIntyre helps us to understand why this path of inquiry is probably no longer open to us. For Hume, MacIntyre says, sentiments involve "simple conjunctions of judgment and passion" (MacIntyre, 1988, p. 302). We can see that this is a bit too simplified. Sentiments, properly functioning and regulated involve reliable conjunctions of judgment and passion. By the early part of the twentieth century, MacIntyre claims, this robust view of sentiments and the moral judgment and action they generate had become greatly attenuated, even unrecognizable. Sentiments had become emotions and emotions were little more than "moments of consciousness" (MacIntyre, 1988, p. 302). Moments of conscious come and go, they lack the durability of habitual character traits. Moments of consciousness therefore are no longer psychological capacities or functions, as sympathy was for Hume and Gregory. Without functionality moments of consciousness lack normativity; they become neither intrinsically good nor bad and have only instrumental value.

MacIntyre attributes this change in the culture to the novelists of the early twentieth century, notably Joyce (MacIntyre, 1988, p. 302). Joyce's novels are usually treated as stream of consciousness, without any given coherence. I rather think of them and of much of the music, painting, and sculpture of our century, especially in the middle third of the century, as expressionist. The French and other Impressionists caught vividly coherent moments. The expressionists, especially the American abstract expressionists after World War II, under the perduring influence of the Surrealists, present vivid moments at great internal tension, close to disintegration, as in Jackson Pollock's drip paintings. The viewer must bring to them whatever *coherence* these works of art have and that will vary greatly from viewer to viewer. Moreover, given the abstractness of these works, e.g., of the mature Rothko or Kline or Frankenthaler, the moments of consciousness are presented detached from content and from the artist. They make reference to themselves, almost exclusively.

Add to this the impact of Freud, who taught us that, contrary to what Descartes might have written and thought, we do not have clear and distinct ideas of ourselves. We elude ourselves and even the most intense introspection, following strict Baconian, experimental method, will not escape some level of opacity. Thus, we cannot expect – as, obviously, did Hume and Gregory and all the Scottish moral sense philosophers, as well as Kant and Hegel later – that we can read ourselves out without error if only we follow the intellectual and scientific discipline of the experimental method.

Put together, these reflections, if they are reliable, strongly suggest that we cannot look to the facts of human psychology for the basis or foundations of the virtues. We remain opaque to ourselves and so we must always be open to conviction that any account we presently subscribe to carries, at best, moderate probability or truth-value. And we do not discover coherence upon examination of consciousness and its products; we create and sustain, often poorly, whatever coherence – unity is a fleeting ideal of utopian philosophy – of consciousness that we claim for ourselves. This seems the bedrock condition of life in the twentieth century in developed countries.

This means that the project for a virtue-based ethics of the Pellegrino-Thomasma stripe is multiply doomed to failure. To some extent Beauchmp, Childress, and MacIntyre survive, because they look to *social roles* as the basis of the virtues. In a sense, so did Gregory: he asked himself what sort of social role needed to be created for people who called themselves physicians, aspired to be or were gentlemen, and who also wanted to be known as professionals, i.e., worthy of the intellectual and moral trust of patients. The *virtues* of sympathy, tenderness, and steadiness, define this new social role, in conjunction with the intellectual *virtue* of openness to conviction generated by diffidence.

Gregory also understood, however, that any virtue-based account of medical ethics required moral exemplars. In this he reflects learning as old as that of the ancient Greeks. After all, Plato goes on at great length in *Republic* about the training of the select few guardians to be philosopher kings, providing a detailed – though utopian – exemplar of the virtues of the good ruler. Gregory has *real exemplars* in mind, women of learning and virtue, such as Mrs. Montagu, who exhibit habitually the proper functioning of sympathy in response to both the pleasure of others and, of far greater importance, to suffering and loss. None of the contemporary accounts of virtue-based bioethics offers us *moral exemplars*, a striking and damaging incompleteness in these accounts.

Virtue-based approaches to bioethics are often seen by their proponents as complementary to principle-based approaches. Approaches to bioethics based on concepts of care also arise from criticism of the deficiencies of principle-based approaches. These criticisms come from both psychology, notably the work of Carol Gilligan (1982), and from philosophy, notably Nel Noddings (1984) and Sara Ruddick (1989). Nancy Jecker and Warren Reich (1995) provide a concise account of these developments, while Reich (1995a, 1995b) has done pioneering work on the historical development of an ethic of care prior to what we call "bioethics."

Contemporary care approaches to bioethics emphasize feminine ethics, i.e., a distinctive – though not exclusive – concern of women with creating, sustaining, and enriching relationships. This contrasts with more male oriented, principle-based approaches, in which relations count for very little, or not at all. The advantage of the former is that it takes account of the concrete and particular in the moral life, which principle-based approaches, because of their abstractness, cannot.

Gregory, too, we saw, writes a *feminine* medical ethics, indeed the first such in the English-language literature of medical ethics. An important critique of care theories can be advanced on Gregorian grounds. Care theorists tend to treat distinctively feminine caring as a natural capacity that always functions well. Gregory helps us to see that this is a mistake. Experience teaches that no natural capacity always functions well. Sometimes the capacity functions too little and sometimes it functions in excess. That caring may be a distinctive capacity of women provides no insurance against this problem.

Caring, in other words, becomes normative only when it is a *trained* capacity, not a natural capacity. Left untutored the capacity for caring could either lead to dissipation, opting for languor and so not caring when one should, or to excess, caring so much that the passions run unchecked. Both must count as degradations or deformations of our nature and are therefore only of negative normative standing. There is an observed variability of caring, just as there is of every human capacity, Gregory would say. It would be an error, he would add, to assume that every capacity – even valued capacities such as caring – always function properly. That caring is a crucial part of feminine consciousness does not count as a reply to this Gregorian critique.

There is, then, no uncomplicated path from caring to virtues. Instead, caring should be understood as a capacity that when functioning properly generates virtues. That is, a trained capacity is the key to acquiring virtues. Care theorists fail to see the importance of *trained capacity* for acquired virtues, and virtue theorists do not attend to the capacities that, when well functioning, generate the virtues of the social role of the physician.

D. *Sympathy and Empathy*

One doesn't see much use of 'sympathy' in the recent bioethics literature and the medical humanities, but one does encounter 'empathy'. This use reflects a concern to correct the abstracting, objectivizing tendencies of scientific thinking to re-establish a connection with and concern for the individual patient. Indeed, scientific ways of thinking about patients challenge the possibility of direct, individuated concern for each patient, one at a time, in his or her full particularity.

> A century and a half after the beginnings of medicine's bacteriological revolution, physicians are still seeking to reconcile "scientific" styles of thought and practice with the interpersonal skills so central to successful healing. The stress on "scientific" objectivity – the physician as diagnostician, the patient as "object of study" – has not made this an easy task. Modern medicine's stringent epistemological standards often constrain the possibility of empathic practice (Milligan and More, 1994, p. 3).

A concept of empathy rather than a concept of sympathy was called upon to do the work of correcting the one-sidedness of scientific clinical thinking, it seems, because 'sympathy' connoted "vagueness or sentimentality," reflecting the "negative cultural valence" that became attached to 'sympathy' during the nineteenth and into the twentieth century (More, 1994, p. 21). This occurred in part because sympathy was gendered *feminine*, thus decreasing its utility in male-dominated medicine, as Ellen More explains in her very useful scholarly study of this history (More, 1994).

In the literature on empathy one immediately encounters a serious problem: the lack of a stable meaning for the term, with variation between accounts of it as a cognitive state, as an affective state, and as a combination of the two (Deigh, 1995, p. 759, n. 14). Indeed, in one volume on the subject the understandings of 'empathy' by contributors to the volume display a marked and conceptually disabling variation (Spiro, *et al.*, 1993). Some definitions are so vague as to be bereft of any conceptual or clinical utility, e.g., "Empathy is the feeling that persons or objects arouse in us as projections of our feelings and thoughts" (Spiro, 1993, p. 7). There does appear to be agreement that empathy differs from "sympathy and tenderness" and links to "awakening compassion for the sick" (Seltzer, 1993, p. ix), reflecting the disdain for feminized concepts that

More (1994) and Regina Morantz-Sanchez (1985, 1994) document. There also appears to be agreement that the cognitive dimension of empathy involves one using imagination to grasp the immediacy of the situation of another individual and his or her affective response to it. On this account, empathy requires some level of intellectual discipline in its exercise. Maureen Milligan and Ellen More propose the definition from Alexandra Kaplan, who understands empathy as "the capacity to take in and appreciate the affective life of another while maintaining a sufficient sense of self to permit cognitive structuring of the experience" (Milligan and More, 1994, p. 3; Kaplan, 1991). Thus, in a bow to scientific medicine, empathy itself takes on a scientific mantle. More reports on Michael Lief and Renée Fox's classic study on "detached concern," in which appears a definition of empathy that bows in precisely this direction:

> Empathy essentially involves an emotional understanding of the patient, "feeling into" and being on the same "affective wave length" as the patient; at the same time, it connotes awareness of enough separateness from the patient so that expert medical skills can be rationally applied to the patient's problems. The empathic physician is sufficiently detached or objective in his attitude toward the patient to exercise sound medical judgment and keep his equanimity, yet he also has enough concern for the patient to give him sensitive, understanding care (Lief and Fox, 1963; in More, 1994, p. 31).

It seems plain that this definition, which has been widely influential, aims to escape the debilitating nineteenth- and twentieth-century history of sympathy's descent into the netherworld of concepts marginalized because of their being gendered feminine. This definition also reflects one's opacity of oneself to oneself and to others, as well as the "expressionist" account of self as momentary. Finally, this definition reflects a fear of loss of control and exhaustion of one's emotional reserves, from too affective a response to patients that might occur if one actually attempted to enter into their suffering (Milligan and More, 1994, pp. 3-4).

Gregory was well aware of this last problem, as we have seen in the previous two chapters. For Gregory, one does not enter into the care of patients with untutored or untrained sympathy. Only sympathy left in its natural, untrained state creates the problem of loss of control, what Gregory calls being "unmanned." Sympathy properly trained, patterned on the exemplar of women of learning and virtue, means disciplined engagement with the patient with the more reliable human faculty, instinct or passion.

Reason, it seems, cannot do this work.

Present day accounts of empathy are thus at a considerable remove from Gregory's account of sympathy; they are also more modest. These accounts do not accept that one can enter fully into the experience of another, which Gregory (with Hume) thought that one could. These accounts take the view that one can have an *imaginative* conception of another's experience that captures enough of that experience to give one a reliable sense of what that other individual is experiencing. In short, current accounts of empathy hark back – without realizing it, apparently – to Adam Smith's (1976) account of sympathy. Recall that for Smith, when A sympathizes with B, A imagines E, the experience that B appears to be having. Mercer's account of sympathy nicely captures what 'empathy' has come to mean in the literature of bioethics and the medical humanities:

> I maintain that if it is correct to make the statement 'A sympathizes with B' then the following conditions must be fulfilled:
>
> (a) A is aware of the existence of B as a sentiment subject;
>
> (b) A knows or believes that he knows B's state of mind;
>
> (c) there is a fellow-feeling between A and B so that through his imagination A is ale to realize B's state of mind; and
>
> (d) A is altruistically concerned for B's welfare (Mercer, 1972, p. 19).

The second condition (b) reflects the opacity of oneself to oneself and others, which is offset by (c), which states the assumption that we are enough like each other so that we can reliably interpret behavior in terms of its associations with such internal states as sentiments or feelings. Gregory would alter (b) to read, "A experiences B's feelings," and (c) to read in its second part "so that through the double relation of impressions and ideas A has the same state of mind as B."

In the effort to "reclaim the validity of empathy as intersubjective knowledge without simultaneously marginalizing it" More turns to "affiliative" or "relational" psychology, work by Jean Miller (1976) that predates Gilligan's (1982) more well-known work. More, in effect, proposes to link empathy to care theories, to cash in the clinical bioethical implications of empathy. In making this move, More mirrors earlier nineteenth-century efforts to model empathy on the relationship of nurturing that occurs between mother and child (Morantz-Sanchez, 1994) that anticipate contemporary efforts along similar lines (Held, 1993). This modeling, of course, should not be restricted to women: female nurturance provides the exemplar that both women and men follow. Empathy joined to care theory becomes a powerful form of feminine ethics.

The problem with this *feminine* ethics is that it threatens the concerns and agenda of *feminist* ethics (Jecker and Reich, 1995). At the social and institutional levels of human activity this empathy-based ethic of care could lead to the oppression of women by limiting them to nurturing roles, thus reinforcing the oppression of women that feminist ethics, rightly, wants to correct. This is a problem with care theories generally, since they tend to build their relational model of ethics on what appear to be ways of thinking that women distinctively display. But women may display this way of thinking because of their oppression. To raise one of the causes of oppression to the level of the normative would constitute a set-back to women's legitimate interests in freedom and equality with men, to say the very least.

Gregory is also a *feminine medical ethicist*, but of a different stripe than that based on relational or affiliative psychology or concepts of nurturance. Gregory's moral exemplars of tenderness and steadiness, the virtues of sympathy, are women of learning and virtue who, because they are women of learning and virtue, are capable of having valuable relationships with women – as the women of the Bluestocking displayed in their own lives. Such women are also capable of having valuable and disciplined – in the form of asexual – relationships with men. These are women of worth, worthy equals in the company of learned men, as Mrs. Montagu set out to show in her salons and as the members of the Bluestocking Circle did in their publications.

Gregory's views on women of learning and virtue as moral exemplars of correctly functioning sympathy are also consistent with his own feminist views on marriage. A woman of learning and virtue had worth other than, and more valuable than, her dowry. She will, if she is patient, attract men who find her worthy of his company and she knows how to distinguish true from false suitors – essential survival skills in a world populated by men of false manners and ambitious to increase their wealth and political standing through marriage. The very same skills that so suit a woman in her pursuit of learning and in the asexual virtues requisite to Mrs. Montagu's salons suit a woman during courtship. Gregory's feminine ethics, because he builds them on the virtues requisite for adult-to-adult relationships, lead directly to his feminist ethics.

Gregory does not take the view that women by virtue of their biological capacity, e.g., for birth and nurturance, have advantage over men as exemplars. This way of doing ethics is utterly alien to his way of thinking. Rather, some women by virtue of their learned, acquired traits – the result of determined effort, as we saw in the case of Mrs. Montagu (Myers, 1990) – provide the needed exemplar. Women were advantaged, we saw earlier, by being sheltered from the world of commerce and its brutal – then and now – pursuit of interest. Women of learning and virtue such as Mrs. Montagu, moreover, brought their virtues to the world of commerce, e.g., in Mrs. Montagu's struggle to maintain the tradition of paternalism in her collieries after her husband's death.

Now, a contemporary feminist might well say that Gregory's women of learning and virtue can function as exemplars precisely because of oppression. But this critique involves a vicious presentism, because it denies the possibility that one could with powerful philosophical justification reach the view that men were oppressed. But Gregory could and did reach such a view. As someone unswervingly committed – in a way typical of Scottish moral sense theorists – to the views that instinct is a more reliable faculty when instinct is properly tutored and functioning, that properly functioning sympathy is the expression of properly functioning instinct, that the pursuit of self-interest is anathema to properly functioning sympathy, and that commerce is defined by the unbridled pursuit of self-interest and therefore is destructive of responsibility as the fundamental moral category, Gregory could reach no other view. In his context and time, Gregory properly reaches the conclusion that women are socially advantaged to tutor sympathy to its proper functioning, especially women of means – their own or their husband's or father's. Only presentist late-twentieth-century concepts of oppression of woman enjoy canonical status, but they do so at the price of being built on ignorance, Gregory (we can imagine) might say to a late twentieth-century feminist critique. Moreover, he could add, his views are feminist in that he seeks to protect his daughters and women generally from one of the chief forms of oppression of sympathy, namely, marriage for convenience and economic security – all forms of self-interest.[65] The presentism of such a later twentieth-century critique involves failure to appreciate that (unlike Gregory's economy and society) our society in the United States supports – even expects – women to enter the work force, and provides educational opportunities for them to prepare to do so.

E. National Bioethics

Gregory wrote his medical ethics in the distinctive context of the Scottish Enlightenment, partly in response to the crisis in Scottish national identity, and utilizing a method for ethics that came from the pre-eminent Scottish moral sense philosopher, David Hume. A generation later Thomas Percival writes the first ethics of medical institutions on the basis of the ethics of the pre-eminent philosopher of the English Enlightenment, Richard Price. Thus, the two giants of eighteenth-century Anglo-American medical ethics introduce a methodological pluralism into medical ethics that reflects the particular features of two national Enlightenments. There was no single, over-arching, "unified and coherent" Enlightenment or Enlightenment project in medical ethics.

At the time, Gregory would surely be the first to claim that his philosophical method, the experimental method of Bacon in his philosophy of medicine and Hume's variant of that method in his medical ethics, is *the true method*. As such, it transcends national identities and the cultural boundaries and limits that such identities create. But, then, so would Percival. The problem here is that Gregory's moral sense philosophy and Percival's moral realism are ultimately fundamentally incompatible.

Thus, an important historical legacy to contemporary bioethics of eighteenth-century medical ethics – a legacy that only grows richer and more complex in the nineteenth century – is methodological pluralism, with each method claiming to be *the* true method for medical ethics. A similar methodological pluralism exists in contemporary bioethics, with multiple methods, each making powerful claims for its canonical intellectual and moral authority.

Surely, not all such claims can be true. There seem to be two responses available: continue the search and arguments for *the* canonical method or accept the *methodological pluralism* as an irremediable fact. The history of Western philosophy gives vivid testimony to the succession – and failure – of claims for *the* canonical method in philosophy generally and in ethics in particular. I have suggested elsewhere, that if we taken an evolutionary perspective on the history of Western philosophy, this multiplicity of claims to priority can be interpreted as a mapping of the variability of reason and experience and how they play themselves out (McCullough, 1996). I am therefore inclined to the second response.

One consequence of this move is that contemporary bioethics will have a particular character, as a function of, among other factors, national cultural contexts. It is no accident, for example, that *respect for autonomy* came to prominence in the United States thirty years ago, at a time in our history when we were finally realizing the full promise of human rights in the United States and becoming, at last, a more complete, mature democratic republic. Thus, the claim by colleagues in other countries that there is a distinctively American bioethics needs to be taken seriously – although there are almost certainly multiple variants on what's distinctively *American* in American bioethics.

On such a view a central task for bioethics as an international intellectual phenomenon becomes mapping the national variants of methods and their results. Of course, these should be then be subjected to sustained critical scrutiny. This work, I hope, will set the stage for a serious discussion of the possibilities of a transnational bioethics. Gregory could be confident that his method was *the true method*; after all, it was proving itself again and again in science and in "the relief of man's estate." We, however, cannot so readily partake of Gregory's confidence. This is one area in which the passage of two centuries challenges Gregory's confidence in his method and its results; it is no longer a live option for us.

V. CONCLUDING WORD

I remarked in Chapter One that bioethics has yet to achieve maturity as a field of the humanities because it typically is written, taught, and practiced as if it had a history that began three, perhaps four, decades ago in the United States. This ahistorical stance typically understands bioethics to have been developed in response to scientific and technological advances in biomedical science, clinical research, and clinical practice that outstripped our capacity – or threatened to outstrip our capacity – to understand and manage well their ethical implications. The power of this ahistorical stance in bioethics – it amounts to an ideology[64] – can be seen most recently in the arguments given and accepted for supporting the study of the ethical, legal, and social implications of the Human Genome Project.

Gregory would, I believe, reject this claim. He would, I think, claim instead, that if medical ethics is based firmly in both a philosophy of medicine and in a reliable method for critical ethical analysis and argument, then no scientific or clinical advance could threaten to outstrip our capacity to understand and manage well its ethical implications. Providing bioethics with its history will, in my judgment, strengthen this Gregorian claim for the capacities of bioethics.

An historically grounded bioethics would also insist on an unyielding skepticism regarding a fundamental assumption of ahistorical bioethics, namely, that there is an abundance of novelty in the advances of biomedical and clinical sciences. There is, in my judgment, far less novelty in the "rapid changes" of late twentieth-century medicine and life generally than we commonly say and think that there are. This is not to deny the pace or quantity of such change; it is to claim that much of it is temporary, fleeting, and so does not run deep.

Consider, for example, Gregory's problem list, in response to which he wrote his medical ethics. There was then an oversupply of health care providers, physicians included, with resultant fierce competition for market share, prestige, and power. There was then little or no scientific discipline in much of clinical judgment and practice. Indeed, medical interventions displayed a high failure rate. The sick self-physicked, invoking lay concepts of health and disease often at a remove from and therefore in competition with medical concepts. Society had created a new institution, the Royal Infirmary, to care for the select sub-population. In this institution, it was claimed, patients were abused, including abuse as research subjects. All of this occurred in the context of a crisis in Scottish national identity. Gregory's response was to invent medicine as a profession in its ethical sense, and he forged the ethical concept of medicine as a fiduciary profession.

Now, in the United States, at least, the changes of managed practice are creating an oversupply of physicians, because a principal goal of managed practice is to reduce demand to a rational, lower level and then bring supply into line with this lower level of demand. In such a market, there is already intense competition as physicians try to avoid underemployment and unemployment, and health care institutions struggle to survive, perhaps with some or even many failing to do so. There is still insufficient scientific discipline in clinical judgment and practice, as evidenced in the striking regional variation in the diagnosis and management of common diseases. Medical interventions, especially new ones, exhibit a non-trivial failure rate, most recently with the new genetic technologies that, we thought, would bring an instant revolution, with no downside, no limits to their capacities to control and direct human biology. Patients self-physick and lay concepts of health and disease compete with medical concepts, especially in the United States where we have created the first universal culture in the history of the planet. Americans are still learning how to do this genially, if at all, reflecting a deep crisis of national identity. We have created a new institution of medicine, the vast corporate enterprise that is rapidly replacing the old cottage industry of fee-for-service enterprise.

In short, changes that we are now making in society and medicine are recreating the conditions of eighteenth-century British medicine that Gregory judged anathema and set out to change. His legacy to us – *medicine as a fiduciary profession in its intellectual and moral senses* – means that we are *not* outstripped in our capacity to understand and manage well the changes that we are making. To be sure, we cannot follow Gregory and Percival and argue for a fiduciary medicine unalloyed and untainted by economic concerns. As Haavi Morreim (1991) has persuasively argued, this option has forever been closed to us.

Our task in attending to American medicine and medicine in countries with economically finite resources for health care, as historically informed students and practitioners of bioethics, is to reforge medicine as a fiduciary profession practiced in an economically responsible way. Economic variables have always shaped clinical judgment; the change in the United States has been from an economics of relative abundance (roughly during the three decades following World War II) to an economics of relative scarcity created as a deliberate policy of private and public payers – as an economically rational response to slower, historically more normal economic growth. A similar change in the direction of relative scarcity had already occurred, earlier than in the United States, in other developed countries. From a Gregorian perspective we can see that the concept of an economically disciplined and responsible fiduciary profession of medicine involves an ethically unstable, even combustible mix, but not necessarily an unmanageable one. We can indeed understand this complex, ethically unstable situation for what it is and set about managing it well by building on the conceptual tools that Gregory forged, and unwittingly left for us to reforge. Ahistorical bioethics would have us be flummoxed and forced to start from scratch, or nearly so, because all change is assumed to be novel and therefore unprecedented. This poverty of intellectual and practical response becomes unnecessary given a mature bioethics, aware of and shaped in critical dialogue with and response to its history.

NOTES

CHAPTER ONE

1. They are now available for the first time in a contemporary edition (McCullough, 1998).
2. The qualifying phrases "English-speaking" or "English-language" are not used in what follows; such phrases should be understood to apply.
3. Gregory's account of sympathy contrasts in important ways with Osler's concept of equanimity. In his essay, "Aequanimitas," Sir William Osler (1995) explains that by this term he means both a "bodily" and a "mental" virtue. In its sense as a bodily virtue 'aequanimitas' means "coolness and presence of mind under all circumstances, calmness amid storm, clearness of judgment in moments of great peril, immobility, impassiveness, or, to use an old and expressive word, *phlegm*" (Osler, 1995, p. 23). In one's physical expression and behavior, the physician is to be imperturbable: "In a true and perfect form, imperturbability is indissolubly associated with wide experience and an intimate knowledge of the varied aspects of disease" (Osler, 1995, p. 24). The mental virtue that parallels this is the ability "to bear with composure the misfortunes of our neighbors" (Osler, 1995, p. 28). In all of this, Osler captures what Gregory means by steadiness, one of the virtues of sympathy. (See Chapter Three). Osler goes on to add to *aequanimitas* the capacity for a measure of "insensibility" and even "callousness" (Osler, 1995, p. 25). In this he departs – even regresses – from Gregory, who rejects insensibility explicitly and also emphasizes the other virtue of sympathy, tenderness, with which insensibility – or dissipation in Gregory's time – and callousness – what Gregory calls hardness of heart – are wholly incompatible. If this analysis is correct it helps us to see that Osler genders *aequanimitas* male, while Gregory genders sympathy female. (See Chapters Two and Three.) Osler's library included Gregory's works on medical ethics, as well as student notes of Gregory's lectures. The extent to which Osler on *aequanimitas* might have been influenced by Gregory and the reasons why he differs from Gregory are topics beyond the scope of this book, but worthy of serious investigation by Osler scholars. The parallels seem striking, suggesting that, in formulating his idea of aequanimitas, Osler may not have acknowledged the influence of Gregory.
4. Osler is often given credit by his followers for taking medical students "to the bedside," suggesting that Osler thereby introduced a major innovation in the history of medical education. If this is the claim of those who take this view, the credit is misplaced and should be accorded much earlier, at least to the mid-eighteenth century at the medical school of the University of Edinburgh and the Royal Infirmary of Edinburgh and, before them, to Boerhaave in Leiden. Gregory in his medical ethics lectures, no doubt following the example of his own clinical teacher, Rutherford, emphasized the indispensability of clinical teaching for the proper education of a physician. Gregory follows the example of Rutherford: taking medical students to the teaching ward at the Royal Infirmary of Edinburgh.

CHAPTER TWO

5. I have not been able to identify or locate manuscript materials that Gregory might have left from this period of his academic life.

6. Did, perhaps, Lochhead's connection between self-physicking and *Pilgrim's Progress* suggest the choice of the title for Dorothy Porter and Roy Porter's (1989), *Patient's Progress*?

7. I defer discussion of how Gregory addresses questions of jurisdiction and authority in the care of dying patients to Chapter Three.

8. Consider, as well, other works. Gainsborough's "The Hon. Frances Duncombe" was made c. 1777:

> The subject's blue satin dress is looped with white ribbons over a white satin underskirt. Her standing lace collar also is white, as are the lace beneath her slashed sleeves and at her elbows, her plumed hat [in her hand], and her satin shoe. The dress, skirt, and shoe are trimmed with pearls, and in the center of her corsage she wears a square red stone (The Frick Collection, 1968, p. 48).

Gainsborough made "The Mall in St. James Park" around 1783 (The Frick Collection, 1968, p. 58). Groups of women pass by in a kind of fashion parade. The women wear layered finery and hats and are all looking at each other. A woman on the far right, seen walking away from the viewer on the arm of a man, eyes the three women to her – and the viewer's – left, appraisingly. These three look to away to their right, almost as if passing in review. Sir Joshua Reynolds made "Elizabeth, Lady Taylor" around 1780.

> Lady Taylor wears a cream-white silk dress, a gold-embroidered sash, and a filmy white fichu striped with gold ... Her hat is decorated with ostrich plumes, broad blue satin ribbon trimmed with black-and-gold cord, and loops of narrower blue ribbon (The Frick Collection, 1968, p. 95).

She is obviously a woman of means; of what accomplishment the viewer cannot tell.

9. Read, at bottom of the page: "*Du Bon ton* is a cant phrase in the modern French language for the fashionable air of conversation and manners."

10. I will examine this correspondence below, in section VIII on the death of Gregory's beloved wife, and in section IX on his years in Edinburgh, where his life ended.

11. Reid takes this from Terrence (Reid, 1991, p. 315 n. 5).

12. I will present Gregory's contributions to the Aberdeen Philosophical Society in an order that makes conceptual sense, rather than chronologically.

13. I cannot find the letter from Mrs. Montagu to Gregory to which he makes reference. Presumably Mrs. Montagu wrote first, upon learning of the death of Gregory's wife. She exemplifies the same "Goodness" as did Mrs. Gregory, the very goodness of heart , i.e., character, that well-developed sympathy should produce as one of its main moral effects.

14. Lisbeth Haakonsen (1997) gives more prominence to Reid's influence on Gregory than it deserves, perhaps misled by the linguistic dress in which Gregory presents his ideas on sympathy. The latter, to repeat, come from Hume. Haakonssen also makes much of a "mutual contract" and "implied" contract as "the central point in Enlightenment medical ethics" (as quoted in Baker, 1996, p. 93). This smacks of Locke and consent-based political philosophy. This is alien to Humean political philosophy and, to repeat, alien to the Humean concept of sympathy, which binds us to one another naturally, i.e., whether or not we consent. Moreover, Gregory nowhere in his two medical ethics books uses 'contract', 'agreement', 'consent'

or their cognates to describe the basis of the physician's duties to his patient. Gregory writes his medical ethics for a *physician-patient* relationship, not a *patient-physician* relationship, which would indeed be contractual and may well be what Reid had in mind. I have relied on a review of Haakonssen's forthcoming book by Robert Baker (1996). Professor Haakonseen declined to share the manuscript of her book with me and it had not appeared by the time that the present book went to press.

15. Note the agricultural metaphors in this passage.

16. "Omnio humani corporis conditio, quae actiones vitales, naturales, vel & animales, laedit, Morbus vocatur" (Boerhaave, 1728, p. 1).

17. "... item inde scitur, Sanitatem esse facultatem corporis aptam omnibus actionibus perfecte exercendis" (Boerhaave, 1747, p. 362).

18. Tytler reflects on Gregory's character in the passage with which this chapter began.

19. Compare this text to a set of student notes from the year before:

> I have now resigned the case of the pat[ients] in the Clin.[ical] Ward, to my ingenious & worthy colleague Dr Cullen, who is to contunue y^e [the] Lecture upon the pat. cases. I have never seen the Ward better supplied w^t [with] matter for useful observ^n [observation] & I have marked w^t [with] pleasure the partic^r [particular] attention given by most of you to these cases. – I have formerly obs^d [observed] y^t [that] Medicine is not to be considered merely as a Speculative Science, that is to be acquired by reading & attending the Lectures of the Professors, but y^t it is a practical art, y^t is to be taught by accurate observ^n & extensive experience. No man can become a tolerable Physician or Surgeon w^{t}out [without] this experience, w^c [which] he must acquire by closs attention to the practice of others, or fm [from] those unhappy persons who are destined to be ye first victims of his ignorance, of his rashness or timidity. Some Gentlemen seem to think y^t a libral educat^n [education] supersedes the necesity of attending to practice, but this is a mistake; for no acuteness of understanding or Solidity of Judgem^t [judgment] can expect [should read 'exempt'?] them fm this necessity.
>
> On the Contrary, a liberal educ^n & ye Studying Physic upon a Systematic plan, show the disadvantages under w^c the Science labours fm the neglect of practice & lead us to distinguish between real & pretended facts, such an educ^n enlightens & directs the conduct of obser^n w^c w^{t}out the assistance of Science is commonly defective, erroneous, blind, & inconclusive. So y^t instead of being at variance w^t one anot^r [another] and independ^t [independent] they are intimately & inseparably connected. What I mean by Theory, in the gen^l [general] principals of the Science founded upon accurate induction fm undoubted facts, and not crude conjectures substituted in the place of facts, and Speculations w^c can have no more influence upon the practice of Physic, than on ship-building. I hope none of you will look on the simplicity of my practice as less useful, than if I had varied my plan of cure and prescriptions. The only method of improving the practice of Physic, is to give a fair trial of each remedy; otherwise our inferences from ye effects are inconclusive. And if we persist when they do not succeed it argues stupidity; but if we desert them when they do succeed, we want only support w^t the lives of our fellow creatures (NLM MS B6, 1771-1772, pp. 347-349).

CHAPTER THREE

20. "Est enim res objecta arti nostri, Hominis vita, sanitas, morbus, mors; horum causae, quibus oriuntur; eorundem media, quibus diriguntur" (Boerhaave, 1747, p. 8)
21. "Proinde est Medicina Scientia eorum, quorum applicatorum effectu, vita sana (1) conservatur, aegra (2) vera in priorem restituitur salubritatem" (Boerhaave, 1747, p. 8).
22. "Cujus itaque Necessitas, Utilitas, Nobilitas, sponte patent" (Boeerhaave, 1747, p. 8).
23. "De Prudentia circa personam ipsius medicine" (Hoffmann, 1749, p. 3).
24. "Regula I. *Medicus sit Christianus.* Itaque tamquam Christianus humanitatem exercibit, ad quem optimam occasiorum exhibebit medicina, quando consideramus, quid sit homo, quam caducus, quam fragilis, unde orum habeat, nascitur enim inter stercus et urinam. Quid igitur vita hominis? nil nisi umbra
    Regula II. *Medicus sit moderatus, nec multum de rebus religionem et fidem spectantibus disputet. ...*
    Regula III. *Medicus non sit Atheus*" (Hoffmann, 1749, p. 3).
25. "*Deus est causa, per quam cuncta conservantur*" (Hoffmann, 1749, p. 3).
26. "*Medicus sit Philosophus.* Per Philosophiam aquirimus sapientiam" (Hoffmann, 1749, p. 4).
27. "Deus est colendus" (Hoffmann, 1749, p. 4).
28. "Societas est servenda" (Hoffmann, 1749, p. 4).
29. "Ordo naturae est servandus" (Hoffmann, 1749, p. 4).
30. "Juxta hand regulam ipse homo temperanter vivit, non facile turbatur motibus praeternaturalibus, sed vitam longam metamque a Deo propositam acquirit" (Hoffmann, 1749, p. 4).
31. "*De virtutibus peculiaribus ad famam conservandam summe necessariis*" (Hoffmann, 1749, p. 7).
32. "*Medicus sit himilis, non superbus ... Medicus sit tactiturnus*" (Hoffmann, 1749, p. 7).
33. "Sit itaque misericors, modestus, humanus, fugiat ceu pestam vitam dissolutam, verba obscoena, ebriatatem, omnemque ludum illicitum, alias omnem aegrotorum amittit fiduciam" (Hoffmann, 1749, p. 8).
34. "*Obstetrix ante omnia sit pia, casta, sobria, non temaria, taciturna, perita*" (Hoffmann, 1749, p. 13).
35. "*Incurabiles morbus praestat non attingere*" (Hoffmann, 1749, p. 25).
36. "*De prudentia Medici circa foeminas aegrotantes*" (Hoffmann, 1749, p. 25).
37. "*Medicus debet esse castus.* Castus sit in verbis & factis, quando aegrotantes visitare debet foeminas; non enim deest Medico libidinem exercendi occasio, praesertim apud incastas foeminas ... " (Hoffmann, 1749, p. 25).
38. "*Ante omnia Medicus se defendere debet a contagio morborum malignorum*" (Hoffmann, 1749, p. 29).
39. " ... abstineat ergo Medicus circa tempus mortis a visitatione" (Hoffmann, 1749, p. 29).
40. "*Medicus semper caustissime agat in morbis malignis quoad medicamenta heroica*" (Hoffmann, 1749, p. 29).
41. "Regula V. *Medicus prognosin cautam debet formare in morbis malignis & acutis.* Optime dixit HIPPOCRATES. In certae sunt omnes praedictiones circa spem salutis in morbis acutis. Circumspecte itaque loqui debet Medicus, ne se prostituat, nec statim absolute decernat, quod necesse sit mori, quod non evasurus sit, sed semper restricive & cum conditione, licet periculum adsit dicat ... sed tamen mortis meditationem non esse negligendam ... hora ad moriendum paratus sit" (Hoffmann, 1749, p.29).
42. He hopes, he says earlier, that his lectures will "at least in some measure contribute to make

you good Physicians & not Quacks."

43. We will see later in this chapter that Gregory differs even more sharply from Hoffmann on this topic.

44. "Nullus collega alterum, vel ignoranitae, vel malae praxis, vel alicujus sceleris, aut ignominiosi criminis nomine accusabit; vel publice contumeliis afficiet" (*Statua Moralia*, in Clark, 1964, p. 30).

45. Robert Veatch's (1981) reading of Bacon as therapeutically aggressive does not apply to Gregory – and probably not to Bacon, either.

46. Baker has recently advanced additional, decisive arguments against the ethics/etiquette distinction (Baker, 1993a, 1995). I return to the topic of medical etiquette at greater length in the next chapter.

CHAPTER FOUR

47. "... les deux premiers discours exposent quels sont ces devoirs & des différentes nuances imperceptibles pour quiconque n'a pas été doué de la plus grande sensibilité, in connues a ces ésprits grossiers & mercénaires qui ne voient qu'un état lucratif dans leur profession ..." (Gregory, 1787, p. xxiii).

48. "Il ne s'occupe point de blesser l'amour propre, mais il a l'art d'attacher le plairir a la practique des devoirs" (Gregory, 1787, p. xxiv).

49. " ... & instruire le public des devoirs du médicin, n'est-ce pas le seul moyen qui reste pour rendre justice au vrai mérite, & pour fermer la voie à des réputations trop souvent usurpées par des docteurs courtisans; démasquer enfin de tels hommes, & leur pagnés de modestie & de franchise ... " (Gregory, 1787, pp. xxv-xxvi).

50. " ... des opinions & des discussions de la profession sur la matiere de l'art... " (Gregory, 1787, p. xxix).

51. "... ils y respireront ces sentimens d'humanité, de noblesse & de generosité ... " (Gregory, 1787, p. xxix).

52. " ... connaitre quels sont les devoirs du médicin ... " (Gregory, 1787, p. xxx).

53. " ... l'objet de la bienfaisance publique ... " (Gregory, 1787, p. xxxiv).

54. " ... vers cette classe digne de compassion ... " (Gregory, 1787, p. xxxiv).

55. "... e che la qualita virtuose, da lui giudicate necessarie in un medico ... apprendere con maggior sicurezza la medicina ... " (Gregory, 1789, pp. viii-viii).

56. " ... abbandando per un momento quall' amor proprio ... " (Gregory, 1789, p. xx).

57. "Eine Schrift, die man mit Recht allen jungen Aerzten empfehlen kann" (Gregory, 1778, p. 4v).

58. " ... circumspectio, prividentia, cautio, perspicacia; deinde etiam continentia, sobrietas, mansuetudo, modestia sive moderatio, decens ornatus ... " (Castro, 1662, p. 121).

59. Engelhardt (1996, p. 41) also argues that there will be many sympathies, not just one. This argument could readily be advanced against contemporary definitions of sympathy, such as James Wilson's: "... *sympathy*, by which I mean the human capacity for being affected by the feelings and experiences of others" (Wilson, 1993, p. 30). This definition is so vague as to be open to multiple instantiations and so is vulnerable to Engelhardt's critique. So, too, for current literature on empathy, with its lack of any stable definition (Deigh, 1995). Not so for Gregory. 'Sympathy' has a precise definition, namely, Hume's double relation of impressions and ideas. Moreover, Gregory can distinguish deformed expressions of sympathy from

properly functioning sympathy. Thus, Engelhardt's argument on this score does not quite reach Gregory.

60. Natural philosophy also claims autonomy from political authority, inasmuch as knowledge gained naturally precedes and provides the foundation for political authority, just as Hobbes and, later, Locke taught.

61. There are two page x in this text. The references therefore states which p. x to consult.

62. To what extent the work of Veatch (1981) and Engelhardt (1986, 1996), as well as others who emphasize respect for autonomy and a contractual relationship between the sick and physicians, can be read as "romantic" is a topic worth exploring, but is beyond the scope of this book.

63. Mary Wollstonecraft attacks Gregory's views on women (Hardt, 1982). Wollstonecraft gets Gregory wrong, seriously wrong (as do, in my judgment, many recent feminists who have commented on Gregory). For example, Wollstonecraft notes that Gregory warns his daughters that they will not often encounter men of "genuine sentiments," because so many men of false sentiments or false manners preyed on women. Wollstonecraft writes, in reply, "Hapless woman" (Hardt, 1982, p. 210) and continues in a sarcastic vein (echoing the views of the Bluestocking Circle, it is worth pointing out). Gregory did *not* think women hapless. Any fair reading of *all* that he says about women (Wollstonecraft is selective in her reading of Gregory) shows that he holds just the opposite view. Women – and his medical students, we saw – need instruction, however, to deal with the social pathology of the man of false sentiment or manners. These predators can be hard to distinguish from men of genuine sympathy (a problem care theorists need to take up; the pathology has not disappeared in an era of "customer relations" training in health care institutions). Gregory arms his daughters against such men, a move that makes no sense if he thought them hapless.

64. Arthur Caplan has recently given expression to this ahistorical ideology in bioethics: "I have not always believed that research in the history of bioethics and medical ethics is crucial for undertaking the study of contemporary problems. But some of my own work on the Holocaust and bioethics and the Tuskegee study has persuaded me that bioethics needs to attend much more seriously to its history." "Bob [Baker] doubts that eighteenth and nineteenth century medical ethics was either nothing but 'footnotes' to Hippocratic thought, or a 'fig' leave' [sic] for exploitation; it was exactly the same endeavor in which bioethicists are currently engaged – the endeavor to discover ethical solutions to the moral problems vexing the practice of medicine. Hmmm; we shall see" (Caplan, 1996, p.1, p. 6).

BIBLIOGRAPHY

I. BIBLIOGRAPHICAL NOTE

This bibliography is divided into three sections. The first contains the published work of John Gregory. I include the first editions of his works, editions actually cited in the text, translations of the medical ethics lectures, and subsequent editions of the ethics lectures. The second section lists the manuscript sources cited in the text. There is an introductory note to this section, explaining the abbreviations used for manuscript citations in the text. The third section lists all other published sources, including primary and secondary sources.

II. PUBLISHED WORK BY OR OF JOHN GREGORY

Gregory, J.: 1765, A Comparative View of the State and Faculties of Man with those of the Animal World, J. Dodsley, London.

Gregory, J.: 1770, *Observations on the Duties and Offices of a Physician, and on the Method of Prosecuting Enquiries in Philosophy*, W. Strahan and T. Cadell, London. Reprinted in L.B. McCullough (ed.), *John Gregory's Writings on Medical Ethics and Philosophy of Medicine*, Kluwer Academic Publishers, Dordrecht, The Netherlands, 1997, pp. 000-000.

Gregory, J.: 1772a, *A Comparative View of the State and Faculties of Man with Those of the Animal World*, 5th ed., J. Dodsley, London.

Gregory, J.: 1772b, *Elements of the Practice of Physic. For the Use of Students*, Printed by Balfour and Smellie, for J. Balfour, Edinburgh.

Gregory, J.: 1772c, *Lectures on the Duties and Qualifications of a Physician*, W. Strahan and T. Cadell, London. Reprinted in L.B. McCullough (ed.), *John Gregory's Writings on Medical Ethics and Philosophy of Medicine*, Kluwer Academic Publishers, Dordrecht, The Netherlands, 1997, pp. 000-000.

Gregory, J.: 1774, *A Father's Legacy to His Daughters*, W. Strahan & T. Cadell, London, J. Balfour, Edinburgh.

Gregory, J.: 1778, anonymous (trans.), *Vorlesungen über die Pflichten und Eigenschaften eines Arztes. Aus dem Englischen nach der neuen und verbesserten Ausgabe*, Caspar Fritsch, Leipzig.

Gregory, J.: 1779, *A Father's Legacy to his Daughters, by the late Dr. Gregory*, A New Edition, T. Strahan, W. Cadell, and R. Creech, Edinburgh. (The visitor to the Library of Congress can enjoy the special pleasure of reading from the copy owned by Thomas Jefferson.)

Gregory, J.: 1787, B. Verlac (trans.), *Discours sur les dévoirs, les qualités et les connaissances du médicin, avec un cour d'études*, Crapart & Briands, Paris.

Gregory, J.: 1788, *The Works of the Late John Gregory, M.D.*, A. Strahan and T. Cadell, London, W. Creech, Edinburgh.

Gregory, J.: 1789, F.F. Padovano (trans.), *Lexioni Sopra i Doveri e la Qualita di un medico*, Gaetano Cambiagi, Florence, Italy.

Gregory, J.: 1805, *Lectures on the Duties and Qualifications of a Physician. Revised and Corrected by James Gregory, M.D.*, W. Creech, Edinburgh, and T. Cadell and W. Davies, London.

Gregory, J.: 1817, *Lectures on the Duties and Qualifications of a Physician*, M. Carey and Son, Philadelphia, Pennsylvania.

Gregory, J.: 1820, *On the Duties and Qualifications of a Physician, New Edition*, J. Anderson, London.

III. MANUSCRIPT AND UNPUBLISHED SOURCES

Texts include contractions, with transcriptions at their first appearance. AUL 37, 1762 is an exception, because it is so heavily contracted. It is transcribed with all contractions eliminated. Each manuscript source is identified by a capitalized abbreviation, which indicates the library possessing the manuscript. The code is as follows:

    AUL = Aberdeen University Library

    CPP = College of Physicians, Philadelphia, Pennsylvania

    EUL = Edinburgh University Library

    HL = The Huntington Library, San Marino, California

    NLS = National Library of Scotland, Edinburgh

    MBO = McGill University, Montreal, Biblioteca Osleriana

    RCPE = Royal College of Physicians of Edinburgh

    RCPSG = Royal College of Physicians and Surgeons of Glasgow

    RCSE = Royal College of Physicians of Edinburgh

    WIHM = Wellcome Institute for the History of Medicine, London

Manuscripts are listed alphabetically by abbreviation.

Aberdeen University Library

Note: The information regarding titles and dates of material for the Aberdeen Philosophical Society is based on the information in Ulman (1990).

AUL 37: 1758, 'A discourse on the different branches of philosophy particularly the philosophy of the mind', November 15, 1768, by David Skene for the Aberdeen Philosophical Society.

AUL 37: 1762, 'Dr Gregory – Whether the art of medicine, as it has been usually practised, has contributed to the advantage of mankind', July 12, 1761, Question 59 for the Aberdeen Philosophical Society. Reprinted in L.B. McCullough (ed.), *John Gregory's Writings on Medical Ethics and Philosophy of Medicine*, Dordrecht, The Netherlands, Kluwer Academic Publishers, 1994, pp. 000-000.

AUL 129: 1769 & 1770, 'The clinical lectures of Dr Gregory in the Royal Infirmary at Edinburgh Annis 1769 & 1770', 'Taken by G. French'.

AUL G404a: 1766, 'Address of the students of medicine to the Right Hon. Lord Provost, Magistrates, and Town-Council of the City of Edinburgh'. Printed broadside.

AUL 2131/6/1/12: 1760, 'Mr Reid – Whether moral character consists in affections wherein the will is not concerned; or in fixed habitual and constant purposes?', April 15, 1760, Question 44 for the Aberdeen Philosophical Society.

AUL 2206/22: 1738, 'Notebook' of 1738-1739 by John Gregory, also titled 'John Gregory: His book'.

AUL 2206/45: 1743, 'Medical notes', including 'A proposal for a medicall society' by John Gregory.

AUL 3107/1/1: 1758, 'The difficulty of a just philosophy of mind; General prejudices against D-d Humes system of the mind; & some observations on the perceptions we have by sight', October 11, 1758, by Thomas Reid, a Discourse for the Aberdeen Philosophical Society.

AUL 3107/1/3: 1758a, 'A discourse on the different branches of philosophy particularly the philosophy of the mind', November 15, 1768, by David Skene for the Aberdeen Philosophical Society.

AUL 3107/1/3: 1758b, 'The state of man compared with that of the lower creation', October 11, 1758, by John Gregory, a Discourse for the Aberdeen Philosophical Society.

AUL 3107/1/4: 1759, 'An inquiry into those faculties which distinguish man from the rest of the animal creation', August 28, 1759, by John Gregory, a Discourse for the Aberdeen Philosophical Society.

AUL 3107/1/6: 1760, 'A discourse on the study of mankind' November 10, 1760, by David Skene for the Aberdeen Philosophical Society. (Continuing the subject of the discourse in AUL 37, 1758.)

AUL 3107/1/10: 1765: 'Moral evidence', November 26, 1765 and February 11, 1766, by John Stewart, a Discourse for the Aberdeen Philosophical Society.

AUL 3107/2/1: 1758: 'Mr Campbell – What is the cause of that pleasure we have from representations or objects which excite pity of other painful feelings?', February 8, 1758, Question 4 for the Aberdeen Philosophical Society.

AUL 3107/2/4: 1761, 'Mr. Gordon – Whether slavery be in all cases inconsistent with good government', November 24, 1761, Question 45 for the Aberdeen Philosophical Society.

AUL 3107/2/6: 1764, 'Mr Thomas Gordon – How far the profession of a soldier of fortune is definsible in foro conscientiae', January 24, 1764 Question 64 for the Aberdeen Philosophical Society.

BIBLIOGRAPHY

College of Physicians, Philadelphia

CPP 10a44: 1771, 'Lectures on clinical medicine 1771-1772'. This includes material from 1773, as well.

Edinburgh University Library

EUL 2106 D: 1773, 'Lectures on the institutions of medicine, Dr Gregory Sr, 1773'.

EUL D.C.6.125: 1772, 'Lectures on the practice of medicine by John Gregory Professor of Physick in the University of Edinburgh 1772" by John Bacon. The following note appears at the beginning: "N.B. These lectures were written at Edinburgh in the years 1772 and 1773. The Manuscripts from which I copied them, were lent me by my ingenious and worthy Fried Doctor Remmet of Exeter."

EUL Dc.7.116: 1769-1770. 'The practice of physic deliver'd in the College of Edinburgh in 1769-70 by Doctor Gregory, Volume the second, wrote by Alexander Dick'.

EUL E.B. 6104 ED 1: 1744: 'Practical remarks on the sympathy of the parts of the body by the late Dr. James Crawford Professor of Medicine in the University of Edinburgh', article XLV in *Medical Essays and Observations, Revised and Published by a Society in Edinburgh*, Vol. V, Part II.

EUL La.II. 647/98: 1766, Letter of J. Coutts to Provost J. Stewart of Edinburgh about the appointment of Dr. J.G. as his Majesty's first physician. London, May, 13, 1766.

EUL mic. m. 28: 1766, Benjamin Rush's 'Journal Commencing Aug 31 1766' (microfilm).

The Huntington Library

HL MO 949: 1760, Letter to Edward from Dorothy Forbes, Baroness Forbes, December 20, 1760.

HL MO 1063: 1756, Letter to Elizabeth Montagu from Elisabeth Gregory, June 28, 1756. HL mistakenly notes the original of this letter to be Kings College, Cambridge. The Gregorys were then in Aberdeen, at King's College.

HL MO 1064: n.d., Letter to Mrs. Montagu from Dr. Gregory, no date, incomplete, last portion missing. Given contents of latter, probably 1761, after death of Gregory's wife on September 29, 1761 (Lawrence 1871, Vol I., p. 157, 167).

HL MO 1065: 1766, Letter to Mrs. Montagu from John Gregory in Edinburgh, October 2, 1766.

HL MO 1066: 1766, Letter to Mrs. Montagu from John Gregory in Edinburgh, October 11, 1766.

HL MO 1067: 1766, Letter to Mrs. Montagu from John Gregory in Edinburgh, October 25, 1766.

HL MO 1068: 1766, Letter to Mrs. Montagu from John Gregory in Edinburgh, November 18, 1766.

HL MO 1071: 1767, Letter to Mrs. Montagu from John Gregory in Edinburgh, January 3, 1767.

HL MO 1072: 1767, Letter to Mrs. Montagu from John Gregory in Edinburgh, February 12, 1767.

HL MO 1075: 1768, Letter to Mrs. Montagu from John Gregory in Edinburgh, August 14, 1768.

HL MO 1077: 1769, Letter to Mrs. Montagu from John Gregory in Edinburgh, August 1, 1769.

HL MO 1078: 1770, Letter to Mrs. Montagu from John Gregory in Edinburgh, June 3, 1770.

HL MO 1083: n.d., Letter to Mrs. Montagu from John Gregory. This is after 1766. HL places it c. 1770?.

HL MO 1085: 1771, Letter to Mrs. Montagu from John Gregory in Edinburgh, May 3, 1771.

Osler Library, McGill University, Montreal, Quebec, Canada

MBO 1768: 1771 'Clinical Lectures by Dr. John Gregory Edinr An. Dom. 1771'. Ms. 7568 in *Biblioteca Osleriana*.

National Library of Medicine, Bethesda, Maryland

NLM MS B6: 1771, 'Clinical Lectures by Dr John Gregory, begun Nov 28 1771'.

NLM MS B7: 1768, 'The practical course delivered by Dr. Gregory at Edinburgh 1768/9'.

National Library of Scotland, Edinburgh

NLS 3648: 1770, Letter to a woman not names from John Gregory in Edinburgh, January 20, 1770. The closing of this letter makes it clear that it is addressed to Mrs. Montagu: "Miss Gordon & your young friends join in their affectionate compliments to you & Mr. Montagu ... "

Royal College of Physicians of Edinburgh

RCPE Gregory, John 1:1766, 'Lectures on medicine', by John Gregory. This librarian's note reads: "The maybe the copy used by Gregory to lecture, 1766-73. The additions and interpolations strongly suggest original work."

Royal College of Physicians and Surgeons of Glasgow

RCPSG 1/9/3: 1750, 'Clinical lectures delivered in the Royal Infirmary Edinburgh 1750-1756 by Professor John Rutherford'.

RCPSG 1/9/4: n.d., 'Clinical lectures by Doctor Robert White', reporter unknown.

RCPSG 1/9/5: 1767, 'The Practice of Physic, As Delivered in the University of Edinburgh by Doctor Gregory, Years 1767 & 1768', by 'J. Foster, student'. One of several extant student-note versions of Gregory's medical ethics lectures. Reprinted in L.B. McCullough (ed.), *John Gregory's Writings on Medical Ethics and the Philosophy of Medicine*, Kluwer Academic Publishers, Dordrecht, The Netherlands, 1997, pp. 000-000.

RCPSG 1/9/7: 1768, 'Clinical lectures of Doct^r Wm Cullen ... delivered in the Royal Infirmary of Edinburgh 1768 & 1769', reporter unknown.

RCPSG 1/9/9: 1768, 'Lectures by Dr. Young Professor of Midwifery in the College of Edinburgh. Nov. 22, 1768', Two Vols., by J.M. Foster.

Royal College of Surgeons of Edinburgh

RCSE C 12: 1771-1772: 'Cases of patients in the clinical wards. Royal Infirmary Edin(burgh 1771-72', 'Dr Gregory', Two Vols.

RCSE C 36: 1771, 'Clinical lectures by Dr Gregory 1771 & Dr Cullen 1772'.

RCSE D 27: 1769, 'Lectures on medicine' of John Gregory. One of several extant student-note versions of Gregory's medical ethics lectures. Reprinted in L.B. McCullough (ed.), *John Gregory's Writings on Medical Ethics and Philosophy of Medicine*, Kluwer Academic Publishers, Dordrecht, The Netherlands, 1997, pp. 000-000.

RCSE D 27: 1770, 'Lectures on the pathology by Dr Gregory Edinburgh 1770'.

Wellcome Institute for the History of Medicine

WIHM 2617: 1771-1772: 'Clinical lectures' by John Gregory, notetaker unknown.

WIHM 2618: 1767-1768, 'Lectures on the Practice of Physic', by John Gregory. One of several extant student-note versions of Gregory's medical ethics lectures. Reprinted in L.B. McCullough (ed.), *John Gregory's Writings on Medical Ethics and Philosophy of Medicine*, Kluwer Academic Publishers, Dordrecht, The Netherlands, 1997, pp. 000-000.

IV. OTHER PUBLISHED SOURCES

American Medical Association, Council on Ethical and Judicial Affairs: 1995, 'Ethical issues in managed care', *Journal of the American Medical Association*, 273, 330-335.

Aristotle: 1984, 'Categories', in J. Barnes (ed.), *The Complete Works of Aristotle*, Princeton University Press, Princeton, New Jersey, Vol. I, pp. 3-24.

Bacon, F.: 1875a, 'Catalogue of Particular Histories and Titles', in J. Spedding, R.L. Ellis, and D.D. Heath (eds.), *The Works of Francis Bacon*, Vol. IV, Longmans, Cumpers, and Co., pp. 265-270.

Bacon, F.: 1875b, 'The New Organon', in J. Spedding, R.L. Ellis, and D.D. Heath (eds.), *The Works of Francis Bacon*, Vol. IV, Longmans, Cumpers, and Co., pp. 39-248.

Bacon, F.: 1875c, 'Of the dignity and advancement of learning', in J. Spedding, R.L. Ellis, and D.D. Heath (eds.), *The Works of Francis Bacon*, Vol. IV, Longmans, Cumpers, and Co., pp. 273-498.

Bacon, F.: 1875d, 'Preface, the Great Instauration', in J. Spedding, R.L. Ellis, and D.D. Heath (eds.), *The Works of Francis Bacon*, Vol. IV, Longmans, Cumpers, and Co., pp. 13-21.

Bacon, F.: 1875e, 'Preparative toward a natural and experimental history', in J. Spedding, R.L. Ellis, and D.D. Heath (eds.), *The Works of Francis Bacon*, Vol. IV, Longmans, Cumpers, and Co., pp. 249-263.

Bacon, F.: 1977, *The Historie of Life and Death with Observations Naturall and Experimentall for the Prolongation of Life*, Arno Press, New York, facsimile of 1638 edition published by I. Okes, London.

Baker, R.: 1993a, 'Deciphering Percival's Code', in R. Baker, D. Porter, and R. Porter (eds.), *The Codification of Medical Morality: Historical and Philosophical Studies of the Formalization of Western Medical Morality in the Eighteenth and Nineteenth Centuries. Volume One: Medical Ethics and Etiquette in the Eighteenth Century*, Kluwer Academic Publishers, Dordrecht, The Netherlands, pp. 179-211.

Baker, R.: 1993b, 'The history of medical ethics', in W.F. Bynum and R. Porter (eds.), *Companion Encyclopedia of the History of Medicine*, Routledge, London and New York, Vol. II, pp. 852-887.

Baker, R. (ed.): 1995a, *The Codification of Morality: Historical and Philosophical Studies of the Formalization of Western Medical Morality in the Eighteenth and Nineteenth Centuries. Volume Two: Anglo-American Medical Ethics and Medical Jurisprudence in the Nineteenth Century*, Kluwer Academic Publishers, Dordrecht, The Netherlands.

Baker, R.: 1995b, 'Introduction' in R. Baker (ed.), *The Codification of Morality: Historical and Philosophical Studies of the Formalization of Western Medical Morality in the Eighteenth and Nineteenth Centuries. Volume Two: Anglo-American Medical Ethics and Medical Jurisprudence in the Nineteenth Century*, Kluwer Academic Publishers, Dordrecht, The Netherlands, pp. 1-22.

Baker, R.: 1996, 'Recent work in the history of medical ethics and its relevance to bioethics', *APA Newsletters*, 96, 90-96.

Baker, R, Porter, D., and Porter, R. (eds.): 1993, *The Codification of Medical Morality: Historical and Philosophical Studies of the Formalization of Western Medical Morality in the Eighteenth and Nineteenth Centuries. Volume One: Medical Ethics and Etiquette in the Eighteenth Century*, Kluwer Academic Publishers, Dordrecht, The Netherlands.

Bard, S.: 1769, *A Discourse upon the Duties of a Physician with some Sentiments on the Usefulness and Necessity of a Public Hospital: Delivered before the President and Governors of King's College at the Commencement Held the 16th of May, 1769, as Advice to Those Gentlemen Who Received the First Medical Degrees Conferred by that University*, A. & J. Robertson, New York.

Barfoot, M.: 1990, 'Hume and the culture of science', in M.A. Stewart (ed.), *Studies in the Philosophy of the Scottish Enlightenment*, Clarendon Press, Oxford, pp. 151-190.

Beattie, J.: 1876, J.R. Irvine (intro.), *Elements of a Moral Sense (1790)*. A Facsimile Reproduction, Scholars' Facsimiles & Reprints, Delmar, New York.

Beauchamp, T.L.: 1993, 'Common sense and virtue in the Scottish moralists', in R. Baker, D. Porter, and R. Porter (eds.), *The Codification of Medical Morality: Historical and Philosophical Studies of the Formalization of Western Medical Morality in the Eighteenth and Nineteenth Centuries. Volume One: Medical Ethics and Etiquette in the Eighteenth Century*, Kluwer Academic Publishers, Dordrecht, The Netherlands, pp. 99-121.

Beauchamp, T.L.: 1995, 'Paternalism', in W.T. Reich (ed.), *Encyclopedia of Bioethics*, 2nd ed.,

Macmillan, New York, pp. 1914-1920.

Beauchamp, T.L. and Childress, J.F.: 1994, *Principles of Biomedical Ethics*, 4th ed., Oxford University Press, New York.

Beauchamp, T.L., and McCullough, L.B.: 1984, *Medical Ethics: The Moral Responsibilities of Physicians*, Prentice-Hall, Inc., Englewood Cliffs, New Jersey.

Bell, J.: 1995, 'Introduction to the Code of Medical Ethics', in R. Baker (ed.), *The Codification of Morality: Historical and Philosophical Studies of the Formalization of Western Medical Morality in the Eighteenth and Nineteenth Centuries. Volume Two: Anglo-American Medical Ethics and Medical Jurisprudence in the Nineteenth Century*, Kluwer Academic Publishers, Dordrecht, The Netherlands, pp. 65-72.

Bell, J., Hays, I, Emerson, G., Morris, W.W., Dunn, T.C, Clark, A., and Arnold, R.D.: 1995, 'Note to the convention', in R. Baker (ed.), *The Codification of Morality: Historical and Philosophical Studies of the Formalization of Western Medical Morality in the Eighteenth and Nineteenth Centuries. Volume Two: Anglo-American Medical Ethics and Medical Jurisprudence in the Nineteenth Century*, Kluwer Academic Publishers, Dordrecht, The Netherlands, pp. 73-74.

Berlant, J.: 1975, *Profession and Monopoly: A Study of Medicine in the United States and Great Britain*, University of California Press, Berkeley, California.

Biro, J.: 1993, 'Hume's new science of mind', in D.F. Norton (ed.), *The Cambridge Companion to Hume*, Cambridge University Press, Cambridge, England, pp. 33-63.

Black, H.C.: 1979, *Black's Law Dictionary*, West Publishing Co., Minneapolis, Minnesota.

Blunt, R. (ed.): 1923, *Mrs. Montagu, "Queen of the Blues:' Her Letters and Friendships from 1762 to 1800*, Constable and Company Limited, London.

Boerhaave, H.: 1728, *Aphorismi de Cognescendis et Curandis morbis ... ab Hermanno Boer-haave*, apud Samuelem Luchtmans et Theodorum Haak, Batavorum, Lugduni (The Netherlands).

Boerhavve, H.: 1744, *Consultationes Medicae; sive Sylloge Epistolarum cum Responsis Hermani Boerhaave in Britannia primum editae nunc aliquot exemplis auctores*, apud A. Vandenhoeck, Gottingae (Germany).

Boerhaave, H.: 1745, *... Medical Correspondence; containing the various symptoms of chronical distempers; the professor's opinion, method of cure, and remedies*, J. Nourse, London.

Boerhaave, H.: 1747, *Prolegomena* in H. Boerhaave, *Institutiones Medicae: in usus annuae exercitationes domesticos digestae ...*, apud Guillelmum Cavelier, Pariis.

Boerhaave, H.: 1751a, *Dr. Boerhaave's Academic Lectures on the Theory of Physic, Being a Genuine Translation of his Institutes and Explanatory Comment, Collated and Adjusted to Each Other, as they were Dictated to his Students at the University of Leyden*, 2nd ed., W. Innys, London.

Boerhaave, H.: 1751b, *Institutiones Medicas*, in H. Boerhaave *Opera Omnia Medica Complectentia*, Expensis Stephani Abbate, Neopoli, Vol. I.

Brown, T.M.: 1993, 'Mental diseases', in W.F. Bynum and R. Porter (eds.), *Companion Encyclopedia of the History of Medicine*, Routledge, London and New York, Vol. I, pp. 438-463.

Buchan, W.: 1769, *Domestic Medicine*, Balfour, Auld & Smellie, Edinburgh.

Buchanan, A.E.: 1996, 'Is there a medical profession in the house?', in R.G. Spece, D.S. Shimm, and E.E. Buchanan (eds.), *Conflicts of Interest in Clinical Practice and Research*, Oxford University Press, New York, pp. 105-136.

Burns, C.: 1995, 'Reciprocity in the Development of Anglo-American Medical Ethics, 1765-1865', in R. Baker (ed.), *The Codification of Morality: Historical and Philosophical Studies of the Formalization of Western Medical Morality in the Eighteenth and Nineteenth Centuries. Volume Two: Anglo-American Medical Ethics and Medical Jurisprudence in the Nineteenth Century*, Kluwer Academic Publishers, Dordrecht, The Netherlands, pp. 135-143.

Burton, J.H.: 1846, *Life and Correspondence of David Hume*, William Tail, Edinburgh.

Butterfield, L.H. (ed.): 1851, *Letters of Benjamin Rush*, (2 Vols.) Princeton University Press, for the American Philosophical Society, Princeton, New Jersey.

Bynum, W.F.: 1993, 'Nosology', in W.F. Bynum and R. Porter (eds.), *Companion Encyclopedia of the History of Medicine*, Routledge, London and New York, Vol. I, pp. 335-356.

Callahan, D.: 1995, 'Bioethics', in W.T. Reich (ed.), *Encyclopedia of Bioethics*, 2nd ed., Macmillan, New York, pp. 247-259.

Caplan, A.L.: 1996, 'Rediscovering our past: Work on the history of medical ethics at the center', *Center for Bioethics Newsletter*. University of Pennsylvania, 1, pp. 1 & 6.

Carter, J.J. and Pittock, J.H. (eds.): 1987a, *Aberdeen and the Enlightenment: Proceedings of a Conference Held at the University of Aberdeen*, Aberdeen University Press, Aberdeen Scotland.

Carter, J.J. and Pittock, J.H.: 1987b, 'Introduction', in J.J. Carter and H.J. Pittock (eds.), *Aberdeen and the Enlightenment: Proceedings of a Conference Held at the University of Aberdeen*, Aberdeen University Press, Aberdeen Scotland, pp. 1-6.

Castro, R.: 1662, *Medicus-Politicus, sive de officiis medico-politicus tractatus, quattuor distinctis Libris ...*, ex Bibliopolio Zachariae Hertelli, Hamburgi (Germany).

Chapone, H.: 1806, *Letters Addressed on the Improvement of the Mind, Addressed to a Lady*, Daniel Johnson, Portland, Maine.

Chervenak, F.A. and McCullough, L.B.: 1995, 'The threat to autonomy of the new managed practice of medicine', *Journal of Clinical Ethics*, 6, 320-323.

Chitnis, A.C.: 1987, 'The eighteenth-century Scottish intellectual inquiry: Context and continuities versus civic virtue', in J.J. Carter and H.J. Pittock (eds.), *Aberdeen and the Enlightenment: Proceedings of a Conference Held at the University of Aberdeen*, Aberdeen University Press, Aberdeen Scotland, pp. 77-92.

Clark, Sir G.: 1964, *A History of the Royal College of Physicians of London*, Clarendon Press, Oxford, Volume I.

Climenson, E.: 1906, *Elizabeth Montagu: The Queen of the Bluestockings. Her Correspondence from 1720-1761* (2 Vols.), John Murray, London.

Comrie, J.D.: 1927, *History of Scottish Medicine to 1860*, Bailliere, Tindall & Cox, for the Wellcome Historical Medical Museum, London.

Craig, W.S.: 1976, *History of the Royal College of Physicians of Edinburgh*, Blackwell Scientific Publications, Oxford.

Daiches, D.: 1986, *The Scottish Enlightenment: An Introduction*, The Saltire Society, Edinburgh, Scotland.

Davie, G.:1991, *The Scottish Enlightenment and Other Essays*, Polygon, Edinburgh, Scotland.

Deely, J: 1994, *Early Modern Philosophy and Postmodern Thought*, University of Toronto Press, Toronto, Ontario, Canada.

Deigh, J.: 1995, 'Empathy and universalizability', *Ethics*, 105, 743-763.

Deming, W.E. : 1986, *Out of Crisis*, Massachusetts Institute of Technology, Center for Advanced Engineering Study, Cambridge, Massachusetts.

Descartes, R: 1979, *Meditations on First Philosophy*, Hackett Publishing Company, Inc., Indianapolis, Indiana.

Digby, K.: 1669, *Of the Sympathetick Powder: A Discourse in a Solemn Assembly at Montpelier*, John Williams, London.

Dunn, J.: 1964, 'Authorship of Gregory's critique of Hume', *Journal of the History of Ideas*, 25, 125-129.

Eccles, A.: 1982, *Obstetrics and Gynaecology in Tudor and Stuart England*, Kent State University Press, Kent, Ohio.

Emanuel. E. and Dubler, N.: 1995, 'Preserving the physician-patient relationship in the era of managed care', *Journal of the American Medical Association*, 273, 323-329.

Engelhardt, H.T., Jr.: 1986, *The Foundations of Bioethics*, Oxford University Press, New York.

Engelhardt, H.T., Jr.: 1996, *The Foundations of Bioethics*, 2nd ed., Oxford University Press, New York.

Engelhardt, HT., Jr. and McCullough, L.B.: 1980, 'Confidentiality in the consultation-liaison process: Ethical dimensions and conflicts', in C.C. Kimball (ed.), *Psychiatric Clinics Symposium on Liaison Psychiatry*, W.B. Saunders Co., Philadelphia, Pennsylvania, pp. 407-417.

Faden, R.R., and Beauchamp, T.L.: 1986, *A History and Theory of Informed Consent*, Oxford University Press, New York.

Ferguson, F.: 1990, *Scotland: 1689 to the Present*, Mercat Press, Edinburgh. *The Edinburgh History of Scotland*, Vol. 4.

Fissell, M.E.: 1993, 'Innocent and honorable bribes: Medical manners in eighteenth-century Britain', in R. Baker, D. Porter, and R. Porter (eds.), *The Codification of Medical Morality: Historical and Philosophical Studies of the Formalization of Western Medical Morality in the Eighteenth and Nineteenth Centuries. Volume One: Medical Ethics and Etiquette in the Eighteenth Century*, Kluwer Academic Publishers, Dordrecht, The Netherlands, pp. 19-45.

Fletcher, J.F.: 1954, *Morals and Medicine: The Moral Problems of the Patient's Right to Know the Truth, Contraception, Artificial Insemination, Euthanasia*, Princeton University Press, Princeton, New Jersey.

Forbes, Sir W.: 1824, *An Account of the Life and Writing of James Beattie, LL.D. Late Professor of Moral Philosophy and Logic in the Marischal College and University of Aberdeen. Including many of his Original Letters*, E. Roper, London.

French, R.: 1993a, 'The anatomical tradition', in W.F. Bynum and R. Porter (eds.), *Companion Encyclopedia of the History of Medicine*, Routledge, London and New York, Vol. I, pp. 81-101.

French, R.: 1993b, 'Ethics in the eighteenth century: Hoffmann in Halle', in A. Wear, J. Geyer-Kordesch, and R. French (eds.), *Doctors and Ethics: The Earlier Historical Setting of Professional Ethics*, Rodopi, Amsterdam and Atlanta, pp. 153-180.

French, R.: 1993c, 'The medical ethics of Gabriele Zerbi', in A. Wear, J. Geyer-Kordesch, and R. French (eds.), *Doctors and Ethics: The Earlier Historical Setting of Professional Ethics*, Rodopi, Amsterdam and Atlanta, pp. 72-97.

The Frick Collection, 1968: *The Frick Collection: An Illustrated Catalogue*, The Frick Collection, New York.

Gainsborough, T. c.1757, 'Sarah, Lady Innes', The Frick Collection, New York. Reproduced in E. Munhall, 1971, *The Frick Collection: Handbook of Paintings*, The Frick Collection, New York, p. 54.

Gay, P.: 1966: *The Enlightenment: An Interpretation. The Rose of Modern Paganism*, W.W. Norton & Company, New York.

Gay, P.: 1969, *The Enlightenment: An Interpretation. The Science of Freedom*, W.W. Norton & Company, New York.

Gelfand, T.: 1993, 'The history of the medical profession', in W.F. Bynum and R. Porter (eds.), *Companion Encyclopedia of the History of Medicine*, Routledge, London and New York, Vol. II, pp. 1119-1150.

Geyer-Kordesch, J. 1993a, 'Natural law and medical ethics in the eighteenth century', in R. Baker, D. Porter, and R. Porter (eds.), *The Codification of Medical Morality: Historical and Philosophical Studies of the Formalization of Western Medical Morality in the Eighteenth and Nineteenth Centuries. Volume One: Medical Ethics and Etiquette in the Eighteenth Century*, Kluwer Academic Publishers, Dordrecht, The Netherlands, pp. 123-139.

Geyer-Kordesch, J.: 1993b, 'Women and Medicine', in W.F. Bynum and R. Porter (eds.), *Companion Encyclopedia of the History of Medicine*, Routledge, London and New York, Vol. II, pp. 888-914.

Gevitz, N.: 1993, 'Unorthodox medical theories', in W.F. Bynum and R. Porter (eds.), *Companion Encyclopedia of the History of Medicine*, Routledge, London and New York, Vol. I, pp. 603-633.

Gilligan, C.: 1982, *In a Different Voice: Psychological Theory and Women's Development*, Harvard University Press, Cambridge, Massachusetts.

Gisborne, T.: 1794, *An Enquiry into the Duties of Men in the Higher and Middle Classes of Society in Great Britain Resulting from their Respective Stations, Professions and Employment*, B. & J. White, London.

Gisborne, T.: 1795, *Principles of Moral Philosophy Investigated and Briefly Applied to the Constitution of Civil Society*, B. & J. White, London.

Granshaw, L.: 1993, 'The hospital', in W.F. Bynum and R. Porter (eds.), *Companion Encyclopedia of the History of Medicine*, Routledge, London and New York, Vol. II, pp. 1180-1203.

Grant, A.: 1884, *The Story of the University of Edinburgh During its First Three Hundred Years*, Longmans, Green, and Co., London.

Gregory, J(ames).: 1800, *Memorial to the Managers of the Royal Infirmary*, Murray & Cochrane, Edinburgh.

Haakonssen, K.: 1990, 'Introduction' in T. Reid, *Practical Ethics: Being Lectures and Papers on Natural Religion, Self-Government, Natural Jurisprudence, and the Law of Nations*, Princeton University Press, Princeton, New Jersey, pp. 1-99.

Haakonssen, L.: 1997, *Medicine and Morals in the Enlightenment: John Gregory, Thomas*

Percival and Benjamin Rush, Editions Rodopi, Amsterdam, The Netherlands.

Haber, S.: 1991, *Authority and Honor in the American Professions, 1750-1900*, University of Chicago Press, Chicago.

Hardt, U.H.: 1982, *A Critical Edition of Mary Wollstonecraft's A Vindication of the Rights of Woman: With Strictures on Political and Moral Subjects*, The Whitston Publishing Company, Troy, New York.

Hauerwas, S.: (in press) 'Engelhardt theologically considered', in B. Minogue and G. Palmer-Fernandez (eds.), *Reading Engelhardt*, Kluwer Academic Publishers, Dordrecht, The Netherlands, pp. 31-44.

Hays, I.: 1995, 'Code of Ethics', in R. Baker (ed.), *The Codification of Morality: Historical and Philosophical Studies of the Formalization of Western Medical Morality in the Eighteenth and Nineteenth Centuries. Volume Two: Anglo-American Medical Ethics and Medical Jurisprudence in the Nineteenth Century*, Kluwer Academic Publishers, Dordrecht, The Netherlands, pp. 75-87. Reprinted from *Minutes of the Proceedings of the National Medical Convention, held in the City of New York, 1846*.

Held, V.: 1993, *Feminist Morality: Transforming Culture, Society, and Politics*, University of Chicago Press, Chicago, Illinois.

Hesse, M. 1967, 'Action at a distance and field theory', in P. Edwards (ed.), *Encyclopedia of Philosophy*, Macmillan, New York.

Hobbes, T.: 1968, C.B. MacPherson (ed.)., *Leviathan*, Penguin Books, Hamondsworth, England. Based on the 1651 "Head" edition.

Hoffmann, F.: 1971, L.S. King (trans.), *Fundamenta Medicinae*, MacDonald, London.

Hoffmann, F.: 1749, *Medicus Politicus, sive Regulae Prudentiae secundum quas Medicus Juvenis Studia sua et Vitae Rationem Dirigere Debet*, in F. Hoffmanni, *Operum Omnium Physico-Medicorum Supplementum in Duas Partes Distributum*, apud Fratres de Tournes, Genevae.

Horn, D.B.: 1967, *A Short History of the University of Edinburgh*, Edinburgh University Press, Edinburgh.

Hume, D.: 1978, P.H. Nidditch (ed.), *A Treatise of Human Nature*, 2nd ed., The Clarendon Press, Oxford, England. Based on the 1739-1740 edition.

Hume, D.: 1987a, 'Of national characters', in E.F. Miller (ed.), *David Hume: Essays Moral, Political, and Literary*, Liberty Classics, Indianapolis, Indiana, pp. 197-215.

Hume, D.: 1987b, 'Of the origin of government', in E.F. Miller (ed.), *David Hume: Essays Moral, Political, and Literary*, Liberty Classics, Indianapolis, Indiana, pp. 37-41.

Hutcheson, F.: 1730, *An Essay on the Nature and Conduct of the Passions and Affections with Illustrations on the Moral Sense by the author of the Inquiry into the Original of Our Ideas of Beauty and Virtue*, J & J. Knapton, London.

Jecker, N.S. and Reich, W.T.: 1995, 'Care: Contemporary ethics of care', in W. T. Reich (ed.)., *Encyclopedia of Bioethics*, 2nd ed., pp. 336-344.

Johnston, D.B.: 1987, 'All honourable men? The award of irregular degrees in King's College and Marischal College in the eighteenth century', in J.J. Carter and H.J. Pittock (eds.), *Aberdeen and the Enlightenment: Proceedings of a Conference Held at the University of Aberdeen*, Aberdeen University Press, Aberdeen Scotland, pp. 136-145.

Jonsen, A.R.: 1990, *The New Medicine and the Old Ethics*, Harvard University Press, Cambridge, Massachusetts.

Jonsen, A.R. and Jameton, A.: 1995, 'Medical ethics, history of: The United States in the Twentieth Century', in W.T. Reich (ed.), *Encyclopedia of Bioethics*, 2nd ed., Macmillan, New York, pp. 1616-1632.

Kames, Lord. (Home, H): *Essays on the Principles of Morality and Religion. In Two Parts*, R. Fleming, for A. Kincaid and A. Donaldson, Edinburgh.

Kaplan, A.G.: 1991, 'Male or female psychotherapists for women: New formulations', in J.V. Dordan, *et al.* (eds.), *Women's Growth in Connection: Writings from the Stone Center*, Guilford Press, New York.

King, L.S.: 1970, *The Road to Medical Enlightenment, 1650-1695*, MacDonald, London, American Elsevier, Inc., New York.

King, L.S., 1978, *The Philosophy of Medicine: The Early Eighteenth Century*, Harvard University Press, Cambridge, Massachusetts.

Laurence, A.: 1992, 'Review of *The Bluestocking Circle*', *Women's History Review*, 1, 162-164.

Lawrence, C.J.: 1975, 'William Buchan: Medicine laid open', *Medical History*, 19, 20-35.

Lawrence, P.D.: 1971, 'The Gregory family: A biographical and bibliographical study. To which is annexed a bibliography of the scientific and medical books in the Gregory Library' (Two Vols.) (Ph.D. Dissertation), Aberdeen University, Aberdeen, Scotland.

Leake, C.: 1927, *Percival's Medical Ethics*, Williams and Wilkins, Baltimore, Maryland.

Lerner, G.: 1986, *The Creation of Patriarchy*, Oxford University Press, New York.

Lief, H.I. and Fox, R.C.: 1963, 'Training for 'detached concern' in medical students', in H.I. Lief (ed.), *The Psychological Basis of Medical Practice*, Harper and Row, New York, pp. 12-35.

Lochhead, M.: 1948, *The Scottish Household in the Eighteenth Century: A Century of Scottish Domestic and Social Life*, The Moray Press, Edinburgh, Scotland.

Loudon, I.S.L.: 1993, 'Childbirth', in W.F. Bynum and R. Porter (eds.), *Companion Encyclopedia of the History of Medicine*, Routledge, London and New York, Vol. II, pp. 1050-1071.

Lyttleton, G.: 1760, *Dialogues of the Dead*, W. Sandby, London. (The visitor to the Library of Congress can enjoy the special pleasure of reading from the copy owned by Thomas Jefferson.)

MacIntyre, A.: 1981, *After Virtue*, University of Notre Dame Press, Norte Dame, Indiana.

MacIntyre, A.: 1984, 'The relationship of philosophy to its past', in R. Rorty, J.B. Schneewind, and Q. Skinner (eds.), *Philosophy in History: Essays on the Historiography of Philosophy*, Cambridge University Press, Cambridge, England, 1984, pp. 31-48.

MacIntyre, A. 1988, *Which Justice? Whose Rationality?*, University of Notre Dame Press, Notre Dame, Indiana.

MacIntyre, A.: 1990, *Three Rival Versions of Moral Inquiry: Encyclopedia, Genealogy, and Tradition*, University of Notre Dame Press, Notre Dame, Indiana.

MacKenzie, H.: 1953, *Third Statistical Account of Scotland: The City of Aberdeen*, Oliver and Boyd, Edinburgh.

Mackie, J.D.: 1987, *A History of Scotland*, 2nd. ed., Penguin Books, Hammondsworth, England.

McCullough, L.B.: 1978a, 'Medical ethics, history of: Britain and the United States in the eighteenth century', in W.T. Reich (ed.), *Encyclopedia of Bioethics*, Macmillan, New York, pp. 957-963.

McCullough, L.B.: 1978b, 'Historical perspectives on the ethical dimensions of the patient-physician relationship: The medical ethics of Dr. John Gregory', *Ethics in Science and Medicine*, 5, 47-53.

McCullough, L.B.: 1981: 'Justice and health care: Historical perspectives and precedents', in E.E. Shelp (ed.)., *Justice and Health Care*, D. Reidel Publishing Company, Dordrecht, The Netherlands, pp. 51-71.

McCullough, L.B.: 1983: 'The legacy of modern Anglo-American medical ethics: Correcting some misperceptions', in E.E. Shelp (ed.)., *The Clinical Encounter*, D. Reidel Publishing Company, Dordrecht, The Netherlands, pp. 47-63.

McCullough, L.B.: 1984, 'Virtues, etiquette, and Anglo-American medical ethics in the eighteenth and nineteenth centuries', in E.E. Shelp (ed.)., *Virtue and Medicine*, D. Reidel Publishing Company, Dordrecht, The Netherlands, pp. 81-92.

McCullough, L.B.: 1995, 'Laying clinical ethics open', *Journal of Medicine and Philosophy*, 18, 1-8.

McCullough, L.B.: 1996, *Leibniz on Individuals and Individuation: The Persistence of Pre-modern Ideas in Modern Philosophy*, Kluwer Academic Publishers, Dordrecht, The Netherlands.

McCullough, L.B. (ed.): 1998, *John Gregory's Writings on Medical Ethics and Philosophy of Medicine*, Kluwer Academic Publishers, Dordrecht, The Netherlands.

McCullough, L.B., and Chervenak, F.A.: 1994, *Ethics in Obstetrics and Gynecology*, Oxford University Press, New York.

Mercer, P.: 1872, *Sympathy and Ethics: A Study of the Relationship between Sympathy and Morality with Special Reference to Hume's Treatise*, The Clarendon Press, Oxford, England.

Millar, J.: 1990, J.V. Price (intro.), *The Origin and Distinction of Ranks*, Thoemmes, Bristol, England.

Miller, J.: 1976, *Toward a New Psychology of Women*, 2nd ed., Beacon Press, Boston, Massachusetts.

Milligan, M.A. and More, E.S.: 1994, 'Introduction' in E.S. More and M.A. Milligan (eds.), *The Empathic Practitioner: Empathy, Gender, and Medicine*, Rutgers University Press, New Brunswick, New Jersey, pp. 1-15.

Montagu, E.: 1970, *An Essay on the Writings and Genius of Shakespear, compared with the Greek and Dramatic Poets; with some Remarks upon the Misrepresentation of Mons. de Voltaire*, A.M. Kelley, New York. Facsimile reproduction of the 1796 edition.

Montagu, E.: 1809, *The Letters of Elizabeth Montagu with Some of the Letters of her Correspondents. Published by Matthew Montagu, Esp. M.P.* (4 Vols.), T. Cadell & W. Davies, London.

Morantz-Sanchez, R.: 1985, *Sympathy and Science: Women Physicians in American Medicine*, Oxford University Press, New York.

Morantz-Sanchez, R.: 1994, 'The gendering of empathic expertise: How women physicians became more empathic than men', in E.S. More and M.A. Milligan (eds.), *The Empathic*

Practitioner: Empathy, Gender, and Medicine, Rutgers University Press, New Brunswick, New Jersey, pp. 40-58.

More, E.S.: 1994, '"Empathy" enters the profession of medicine' in E.S. More and M.A. Milligan (eds.), *The Empathic Practitioner: Empathy, Gender, and Medicine*, Rutgers University Press, New Brunswick, New Jersey, pp. 19-39.

Morreim, E.H.: 1991, *Balancing Act: The New Medical Ethics of Medicine's New Economics*, Kluwer Academic Publishers, Dordrecht, The Netherlands.

Mossner, E.C.: 1980, *The Life of David Hume*, 2nd ed., The Clarendon Press, Oxford, England.

Murdoch, A. and Sher, R.B.: 'Literary and learned culture', in T.M. Devine and R. Mitchison (eds.), *People and Society in Scotland: I: 1760-1830*, John Donald Publisher, LTD, Edinburgh, Scotland, pp. 127-142.

Myers, S.H.: 1990, *The Bluestocking Circle: Women, Friendship, and the Life of the Mind in Eighteenth-Century England*, Clarendon Press, Oxford, England.

Noddings, N.: 1984, *Caring: A Feminine Approach to Ethics and Moral Education*, University of California Press, Berkeley, California.

Nutton, V: 1993, 'Beyond the Hippocratic Oath', in A. Wear, J. Geyer-Kordesch, and R. French (eds.), *Doctors and Ethics: The Earlier Historical Setting of Professional Ethics*, Rodopi, Amsterdam and Atlanta, pp. 10-37.

Osler, W.: 1995, 'Aequanimitas', in J.P. McGovern and C.G. Roland (eds.), *The Collected Essays of Sir William Osler*, Volume I, *The Philosophical Essays*, The Classics of Medicine Library, Birmingham, AL, pp. 21-31. Originally presented as the Valedictory Address at the University of Pennsylvania (medical school), May 1, 1889.

Pellegrino, E.D.: 1991, 'Trust and distrust in professional ethics', in E.D. Pellegrino, R.M. Veatch, and J.P. Langan (eds.,), *Ethics, Trust, and The Professions*, Georgetown University Press, Washington, DC, pp. 69-89.

Pellegrino, E.D., and Thomasma, D.C.: 1981, *A Philosophical Basis of Medical Practice: Toward a Philosophy and Ethics of the Healing Professions*, Oxford University Press, New York.

Pellegrino, E.D., and Thomasma, D.C.: 1988, *For the Patient's Good: The Restoration of Beneficence in Health Care*, Oxford University Press, New York.

Pellegrino, E.D., and Thomasma, D.C.: 1993, *The Virtues in Medical Practice*, Oxford University Press, New York.

Percival, T.: 1803, *Medical Ethics, or a Code of Institutes and Precepts, Adapted to the Professional Conduct of Physicians and Surgeons*, Printed by J. Russell, for J. Johnson, St. Paul's Church Yard, & R. Bickerstaff, Strand, London.

Phillipson, N.: 1981, 'The Scottish Enlightenment', in R. Porter and M. Teich (eds.), *The Enlightenment in National Context*, Cambridge University Press, Cambridge, England, p. 19-40.

Porter, D. and Porter, R.: 1989: *Patient's Progress: Doctors and Doctoring in Eighteenth-Century England*, Stanford University Press, Stanford, California.

Porter, R.: 1981, 'The Enlightenment in England', in R. Porter and M. Teich (eds.), *The Enlightenment in National Context*, Cambridge University Press, Cambridge, England, pp.1-18.

Porter, R.: 1987, 'A touch of danger: The man-midwife as sexual predator', in G.S. Rousseau and R. Porter (eds.), *Sexual Underworlds of the Enlightenment*, Manchester University Press, Manchester, England, pp. 206-232.

Porter, R.: 1990, *The Enlightenment*, Macmillan, London.

Porter, R.: 1992, 'The patient in England, c. 1660 - c. 1800', in A. Wear (ed.,), *Medicine and Society: Historical Essays*. Cambridge University Press, Cambridge, England, pp. 91-118.

Porter, R.: 1993, 'Thomas Gisborne: Physicians, Christians and gentlemen', in A. Wear, J. Geyer-Kordesch, and R. French (eds.), *Doctors and Ethics: The Earlier Historical Setting of Professional Ethics*, Rodopi, Amsterdam and Atlanta, pp. 252-273.

Porter, R. and Teich, M. (eds.): 1981, *The Enlightenment in National Context*, Cambridge University Press, Cambridge, England.

President's Commission for the Study of Ethical Problems in Medicine and Biomedical and Behavioral Research: 1981, *Defining Death: Medical, Legal, and Ethical Issues in the Determination of Death*, U.S. Government Printing Office, Washington, D.C.

Price, R.: 1948, *A Review of the Principal Questions in Morals*, Oxford University Press, Oxford. Reprint, with a critical introduction by D.D. Raphael, of the 3rd (1787) edition.

Ramsay, J.: 1888: *Scotland and Scotsmen in the Eighteenth Century* (2 Vols.), William Blackwood & Sons, Edinburgh.

Ramsey, P.: 1970, *The Patient as Person: Explorations in Medical Ethics*, Yale University Press, New Haven, Connecticut.

Rather, L.J.: 1965, *Mind and Body in Eighteenth Century Medicine: A Study Based on Jerome Gaub's De Regimine Mentis*, University of California Press, Berkeley, California.

Reich, W.T. (ed.): *Encyclopedia of Bioethics*, 1st ed., Macmillan, New York.

Reich, W.T.: 1995a, 'Care: Historical dimensions of an ethic of care', in W.T. Reich (ed.)., *Encyclopedia of Bioethics*, 2nd ed., Macmillan, New York, pp. 331- 336.

Reich, W.T.: 1995b, 'Care: History of the notion of care,' in W.T. Reich (ed.)., *Encyclopedia of Bioethics*, 2nd ed., Macmillan, New York, pp. 319-331.

Reich, W.T: 1995, *Encyclopedia of Bioethics*, 2nd ed., Macmillan, New York.

Reich, W.T.: 1995d, 'Introduction', in W.T. Reich (ed.)., *Encyclopedia of Bioethics*, 2nd ed., Macmillan, New York, pp. xix-xxxii.

Reid, T.: 1863 (Sir W. Hamilton, ed.), *The Works of Thomas Reid*, MacLachlan and Stewart, Edinburgh, Scotland.

Reid, T.: 1990 (K. Haakonssen, ed.), *Practical Ethics: Being Lectures and Papers on Natural Religion, Self-Government, Natural Jurisprudence, and the Law of Nations*, Princeton University Press, Princeton, New Jersey. Based on manuscript sources.

Risse, G.B.: 1986, *Hospital Life in Enlightenment Scotland: Care and teaching at the Royal Infirmary of Edinburgh*, Cambridge University Press, Cambridge, England.

Rodwin, M.: 1993, *Medicine, Money, and Morals: Physicians' Conflicts of Interest*, Oxford University Press, New York.

Rodwin, M.A,: 1995, 'Strains in the fiduciary metaphor: Divided physician loyalties and obligations in a changing health care system', *American Journal of Law and Medicine*, 21, 242-257.

Rogers, K.M.: 1982, *Feminism in Eighteenth-Century England*, University of Illinois Press, Urbana, Illinois.

Rothman, D.J.: 1991, *Strangers at the Bedside: A History of How Law and Bioethics Transformed Medical Decision Making*, Basic Books, New York.

Rothschuh, K.E.: 1973, *History of Physiology*, G.B. Risse (trans. and ed.), Robert E. Krieger Publishing Company, Huntington, New York.

Royal College of Physicians of Edinburgh: 1882, *Historical Sketch and Laws of the Royal College of Physicians of Edinburgh, from its institution to August 1882*, Royal College of Physicians, Edinburgh.

Ruddick, S.: 1989, *Maternal Thinking: Toward a Politics of Peace*, Beacon Press, Boston, Massachusetts.

Rush, B.: 1805, *Observations on the Duties of a Physician, and the Methods of Improving Medicine. Accommodated to the Present State of Society and Manners in the United States*, in B. Rush, *Medical Inquiries and Observations*, 2nd ed., J. Conrad & Co., Philadelphia, Vol. I, pp. 385-408.

Ryan, M.: 1831, *A Manual of Jurisprudence, complied from the best medical and legal works: comprising an account of: The Ethics of the Medical Profession, II. The Charter and Statutes Relating to the Faculty; and III. All Medico-legal Questions, with the latest discussions. Being an Analysis of a Course of Lectures on Forensic medicine Annually Delivered in London and intended as a compendium for the use of barristers, soliciters, magistrates, coroners, and medical practitioners*, Renshaw and Rush, London.

Sefton, H.: 1987, 'David Hume and Principal George Campbell', in J.J. Carter and H.J. Pittock (eds.), *Aberdeen and the Enlightenment: Proceedings of a Conference Held at the University of Aberdeen*, Aberdeen University Press, Aberdeen Scotland, pp. 123-128.

Selzer, R.: 1993, 'Introduction', in H. Spiro, M.G.M. Curnen, E. Peschel, and D. St. James (eds.), *Empathy and the Practice of Medicine*, Yale University Press, New Haven, Connecticut, pp. ix-x.

Shakespeare, W.: 1952, 'Macbeth', in S. Barnet (ed. and intro.), *William Shakespeare, Four Great Tragedies: Hamlet, Othello, King Lear, Macbeth*, each play paginated separately.

Smellie, W.: 1800, *Literary and Characteristical Lives of John Gregory, M.D, Henry Home, Lord Kames, David Hume, Esq., and Adam Smith, L.L.D, to which are added a Dissertation on the Public Spirit and Three Essays by the Late William Smellie*, Alex Smellie, *et al.*, Edinburgh, G.G. Robinson, *et al.*, London.

Smith, A.: 1976, D.D. Raphael and A.L. MacFie (eds.), *The Theory of the Moral Sentiments*, Clarendon Press, Oxford, England.

Sperry, W.: 1950, *The Ethical Basis of Medical Practice*, P.B. Hoeber, New York.

Spiro, H.: 1993, 'What is empathy and can it be taught?', in H. Spiro, M.G.M. Curnen, E. Peschel, and D. St. James (eds.), *Empathy and the Practice of Medicine*, Yale University Press, New Haven, Connecticut, pp. 7-14.

Spiro, H., Curnen, M.G.M., Peschel, E., and St. James, D. (eds.): 1993, *Empathy and the Practice of Medicine*, Yale University Press, New Haven, Connecticut.

Starr, P.: 1982, *The Social Transformation of American Medicine*, Basic Books, New York.

Steuart, R.: 1725, *The Physiological Library. Begun by Mr. Steuart, and Some of the Students of Natural Philosophy in the University of Edinburgh, April 2, 1724. And Augmented by Some Gentlemen; and the Students of Natural Philosophy, December, 1724.*

Stewart, A.G.: 1901, *The Academic Gregories*, Oliphant Anderson & Ferrier, Edinburgh.

Stuber, J.: 1995, 'Research notes', *Hastings Center Report*, 25, 50.

Tansey, E.M.: 1993, "The physiological tradition', in W.F. Bynum and R. Porter (eds.), *Companion Encyclopedia of the History of Medicine*, Routledge, London and New York, Vol. I, pp. 120-152.

Taylor, J.: 1760, *A Sketch of Moral Philosophy; or an Essay to Demonstrate the Principles of Virtue and Religion upon a New, Natural, and Easy Plan*, J. Waugh & W. Fenner, London.

Temkin, O. and Temkin, C.L. (eds.): 1967, *Ancient Medicine: Selected Papers of Ludwig Edelstein*, Johns Hopkins University Press, Baltimore, Maryland.

Tomaselli, S.: 1991, 'Human nature', in J.W. Yolton, R. Porter, P. Rogers, and B.M Stafford (eds.), *The Blackwell Companion to the Enlightenment*, Basil Blackwell Ltd., Oxford, England, pp. 229-233.

Tong, R.: 1993, *Feminine and Feminist Ethics*, Wadsworth Publishing Company, Belmont, California.

Truman, J.: 1995, 'The compleat physician: John Gregory MD (1724-1773)', *Journal of Medical Biography*, 3, 63-70).

Tytler, A.F. (Lord Woodhouselee): 1788, 'An account of the life and writings of Dr John Gregory', in J. Gregory, *The Works of the Late John Gregory, M.D.*, A. Strahan and T. Cadell, London, W. Creech, Edinburgh, Vol. I., pp. 1-85.

Ulman, H.L.: 1990, *The Minutes of the Aberdeen Philosophical Society*, Aberdeen University Press, Aberdeen, Scotland.

Underwood, E.A.: 1977: *Boerhaave's Men at Leyden and After*, Edinburgh University Press, Edinburgh, Scotland.

Vann, R.T.: 1995, 'Theory and practice in historical study', in M.B. Norton and P. Gerardi (eds.), *The American Historical Association's Guide to Historical Literature*, Oxford University Press, pp. 1-32.

Veatch, R.M.: 1981, *A Theory of Medical Ethics*, Basic Books, Inc., Publishers, New York.

Waddington, I.: 1984, *The Medical Profession in the Industrial Revolution*, Gill & Macmilan, Dublin.

Warren, J., Hayward, L., and Fleet, J.: 1995, 'Boston Medical Police, Boston Medical Association (1808)', in R. Baker (ed.), *The Codification of Morality: Historical and Philosophical Studies of the Formalization of Western Medical Morality in the Eighteenth and Nineteenth Centuries. Volume Two: Anglo-American Medical Ethics and Medical Jurisprudence in the Nineteenth Century*, Kluwer Academic Publishers, Dordrecht, The Netherlands, pp. 41-46.

Wear, A.: 1987, 'Interfaces: Perceptions of health and illness in early modern England', R. Porter and A. Wear (eds.), *Problems and Methods in the History of Medicine*, Croom Helm, London, pp. 230-255.

Wear, A.: 1992, 'Making sense of health and the environment in early modern England', in A. Wear (ed.,), *Medicine and Society: Historical Essays*. Cambridge University Press, Cambridge, England, pp. 119-147.

Wear, A.: 1993, 'Medical ethics in early modern England', in A. Wear, J. Geyer-Kordesch, and R. French (eds.), *Doctors and Ethics: The Earlier Historical Setting of Professional Ethics*, Rodopi, Amsterdam and Atlanta, pp. 98-130.

Wear A., Geyer-Kordesch, J., and French, R. (eds.): 1993, *Doctors and Ethics: The Earlier Historical Setting of Professional Ethics*, Rodopi, Amsterdam-Atlanta.

Whytt, R.: 1765, *Observations on the Nature, Causes and Cure of those Disorders which have been commonly called nervous, Hypochondriac or Hysteric* .... 2nd ed., corrected, Balfour, Edinburgh.

Wilson, C.: 1995, *The Invisible World: Early Modern Philosophy and the Invention of the Microscope*, Princeton University Press, Princeton, New Jersey.

Wilson, J.Q.: 1993, *The Moral Sense*, The Free Press, New York.

Wilson, L.G.: 1993, 'Fevers', in W.F. Bynum and R. Porter (eds.), *Companion Encyclopedia of the History of Medicine*, Routledge, London and New York, Vol. I, pp. 382-411.

Withrington, D.J.: 1987, 'What was distinctive about the Scottish Enlightenment', in J.J. Carter and H.J. Pittock (eds.), *Aberdeen and the Enlightenment: Proceedings of a Conference Held at the University of Aberdeen*, Aberdeen University Press, Aberdeen, Scotland, pp. 9-19.

Wright, J.P.: 1991, 'Man, science of', in J.W. Yolton, R. Porter, P. Rogers, and B.M Stafford (eds.), *The Blackwell Companion to the Enlightenment*, Basil Blackwell Ltd., Oxford, England, pp. 309-310.

INDEX

Aberdeen, 20, 30, 53-57, 81, 85-86, 11, 97-122
    Poor's Hospital of, 54
    Royal Infirmary of, 54
Aberdeen Grammar School, 31
Aberdeen Philosophical Society, xiv, 16, 17, 82, 97, 123, 226, 262
    on Adam Smith, 109-110
    Discourses of, 100
    Discourses of, addressed by Gregory, 103-104
    on drinking, 98, 114
    founding of, 99
    and Hume, 100ff, 142
    on just wars, 104
    on medicine, 118-23
    Questions of, 100
    Questions of, addressed by Gregory, 103
    rules of, 100
    on skepticism, 101-102, 115-18
    on slavery, 104
    on sympathy, 104-114
    as "Wise Club," 98
abnormal, 188-, 200-201
Act of Union of 1707, 18ff, 54
active force, 190-91
Adams, Abigail, 268
Adams, John, 268
aegrotus, 179, 288
aequanimitas, 307 n. 3
affections, 89-90
age of manners, end of, 205-206
Albinus, Bernard Siegfried, 33
Alexander, Robert, 97-98
Alston, Charles, 32-33
American Medical Association
    Code of Ethics of 1847, 12, 278
amour propre, 272-73
amputation, 92
anatomy, 43
Anderson, David, 29
Anderson, Janet. *See* Gregorie, Janet
animalicula, 87
Anne, Queen, 18

apothecaries. *See* pharmacy
Aquinas, St. Thomas, 294
Argyle, Duke of, 123
Arian heresy, 102
Aristotle, 42, 87, 210, 275, 294
Athanasius, St., 101-102
attention, 15, 241, 275
authority, virtue of, 12

Bacon, Francis, 34ff, 43, 52-53, 84, 86, 115, 120, 156, 210, 233-34, 259, 262, 285
    on cure of diseases, 47
    on experiments, 35
    on geriatrics, 47
    on "Hunt of Pan," 35
    on incurable diseases, 47
    on investigation of nature, 36
    on nature of experience, 35
    on offices of medicine, 46-47
    on "outward euthanasia," 47, 234
    on preparation of the soul for death, 47, 234
    on prolongation of life, 46-47
    on regular use of medications, 47
    See also, science, Baconian; science, eighteenth-century
Baconian method, 4, 5
Baker, Robert, iv, 275-76, 284-85, 312 n. 64
Bard, Samuel, 273
Beattie, James, 82, 99, 114, 168, 182, 268-69
    controversy of, with Hume, 141-45, 269
Beauchamp, Tom, 288, 293-94, 296
Bell, John, 278
Berlant, Jeffrey, 276
Berlioz, Hector, 68
bioethics, 5
    and capacities of medicine, 282
    care-theory based, 7, 8, 12, 297
    and ends of medicine, 282
    feminine, 6-7, 16, 297, 300-302
    Gregorian critique of, 12
    history of, 14
    innocent of the history of medical ethics, 5, 14, 304-306

Philosophy and Medicine

1. H. Tristram Engelhardt, Jr. and S.F. Spicker (eds.): *Evaluation and Explanation in the Biomedical Sciences.* 1975 ISBN 90-277-0553-4
2. S.F. Spicker and H. Tristram Engelhardt, Jr. (eds.): *Philosophical Dimensions of the Neuro-Medical Sciences.* 1976 ISBN 90-277-0672-7
3. S.F. Spicker and H. Tristram Engelhardt, Jr. (eds.): *Philosophical Medical Ethics.* Its Nature and Significance. 1977 ISBN 90-277-0772-3
4. H. Tristram Engelhardt, Jr. and S.F. Spicker (eds.): *Mental Health.* Philosophical Perspectives. 1978 ISBN 90-277-0828-2
5. B.A. Brody and H. Tristram Engelhardt, Jr. (eds.): *Mental Illness.* Law and Public Policy. 1980 ISBN 90-277-1057-0
6. H. Tristram Engelhardt, Jr., S.F. Spicker and B. Towers (eds.): *Clinical Judgment.* A Critical Appraisal. 1979 ISBN 90-277-0952-1
7. S.F. Spicker (ed.): *Organism, Medicine, and Metaphysics.* Essays in Honor of Hans Jonas on His 75th Birthday. 1978 ISBN 90-277-0823-1
8. E.E. Shelp (ed.): *Justice and Health Care.* 1981
ISBN 90-277-1207-7; Pb 90-277-1251-4
9. S.F. Spicker, J.M. Healey, Jr. and H. Tristram Engelhardt, Jr. (eds.): *The Law-Medicine Relation.* A Philosophical Exploration. 1981 ISBN 90-277-1217-4
10. W.B. Bondeson, H. Tristram Engelhardt, Jr., S.F. Spicker and J.M. White, Jr. (eds.): *New Knowledge in the Biomedical Sciences.* Some Moral Implications of Its Acquisition, Possession, and Use. 1982 ISBN 90-277-1319-7
11. E.E. Shelp (ed.): *Beneficence and Health Care.* 1982 ISBN 90-277-1377-4
12. G.J. Agich (ed.): *Responsibility in Health Care.* 1982 ISBN 90-277-1417-7
13. W.B. Bondeson, H. Tristram Engelhardt, Jr., S.F. Spicker and D.H. Winship: *Abortion and the Status of the Fetus.* 2nd printing, 1984 ISBN 90-277-1493-2
14. E.E. Shelp (ed.): *The Clinical Encounter.* The Moral Fabric of the Patient-Physician Relationship. 1983 ISBN 90-277-1593-9
15. L. Kopelman and J.C. Moskop (eds.): *Ethics and Mental Retardation.* 1984
ISBN 90-277-1630-7
16. L. Nordenfelt and B.I.B. Lindahl (eds.): *Health, Disease, and Causal Explanations in Medicine.* 1984 ISBN 90-277-1660-9
17. E.E. Shelp (ed.): *Virtue and Medicine.* Explorations in the Character of Medicine. 1985 ISBN 90-277-1808-3
18. P. Carrick: *Medical Ethics in Antiquity.* Philosophical Perspectives on Abortion and Euthanasia. 1985 ISBN 90-277-1825-3; Pb 90-277-1915-2
19. J.C. Moskop and L. Kopelman (eds.): *Ethics and Critical Care Medicine.* 1985
ISBN 90-277-1820-2
20. E.E. Shelp (ed.): *Theology and Bioethics.* Exploring the Foundations and Frontiers. 1985 ISBN 90-277-1857-1
21. G.J. Agich and C.E. Begley (eds.): *The Price of Health.* 1986
ISBN 90-277-2285-4
22. E.E. Shelp (ed.): *Sexuality and Medicine.* Vol. I: Conceptual Roots. 1987
ISBN 90-277-2290-0; Pb 90-277-2386-9

Philosophy and Medicine

41. K.W. Wildes, S.J., F. Abel, S.J. and J.C. Harvey (eds.): *Birth, Suffering, and Death*. Catholic Perspectives at the Edges of Life. 1992 [CSiB-1]
ISBN 0-7923-1547-2; Pb 0-7923-2545-1
42. S.K. Toombs: *The Meaning of Illness*. A Phenomenological Account of the Different Perspectives of Physician and Patient. 1992
ISBN 0-7923-1570-7; Pb 0-7923-2443-9
43. D. Leder (ed.): *The Body in Medical Thought and Practice*. 1992
ISBN 0-7923-1657-6
44. C. Delkeskamp-Hayes and M.A.G. Cutter (eds.): *Science, Technology, and the Art of Medicine*. European-American Dialogues. 1993   ISBN 0-7923-1869-2
45. R. Baker, D. Porter and R. Porter (eds.): *The Codification of Medical Morality*. Historical and Philosophical Studies of the Formalization of Western Medical Morality in the 18th and 19th Centuries, Volume One: Medical Ethics and Etiquette in the 18th Century. 1993   ISBN 0-7923-1921-4
46. K. Bayertz (ed.): *The Concept of Moral Consensus*. The Case of Technological Interventions in Human Reproduction. 1994   ISBN 0-7923-2615-6
47. L. Nordenfelt (ed.): *Concepts and Measurement of Quality of Life in Health Care*. 1994 [ESiP-1]   ISBN 0-7923-2824-8
48. R. Baker and M.A. Strosberg (eds.) with the assistance of J. Bynum: *Legislating Medical Ethics*. A Study of the New York State Do-Not-Resuscitate Law. 1995   ISBN 0-7923-2995-3
49. R. Baker (ed.): *The Codification of Medical Morality*. Historical and Philosophical Studies of the Formalization of Western Morality in the 18th and 19th Centuries, Volume Two: Anglo-American Medical Ethics and Medical Jurisprudence in the 19th Century. 1995   ISBN 0-7923-3528-7; Pb 0-7923-3529-5
50. R.A. Carson and C.R. Burns (eds.): *Philosophy of Medicine and Bioethics*. A Twenty-Year Retrospective and Critical Appraisal. 1997
ISBN 0-7923-3545-7
51. K.W. Wildes, S.J. (ed.): *Critical Choices and Critical Care*. Catholic Perspectives on Allocating Resources in Intensive Care Medicine. 1995 [CSiB-2]
ISBN 0-7923-3382-9
52. K. Bayertz (ed.): *Sanctity of Life and Human Dignity*. 1996
ISBN 0-7923-3739-5
53. Kevin Wm. Wildes, S.J. (ed.): *Infertility: A Crossroad of Faith, Medicine, and Technology*. 1996   ISBN 0-7923-4061-2
54. Kazumasa Hoshino (ed.): *Japanese and Western Bioethics*. Studies in Moral Diversity. 1996   ISBN 0-7923-4112-0
55. E. Agius and S. Busuttil (eds.): *Germ-Line Intervention and our Responsibilities to Future Generations*. 1998   ISBN 0-7923-4828-1
56. L.B. McCullough: *John Gregory and the Invention of Professional Medical Ethics and the Professional Medical Ethics and the Profession of Medicine*. 1998   ISBN 0-7923-4917-2
57. L.B. McCullough: *John Gregory's Writing on Medical Ethics and Philosophy of Medicine*. 1998 [CiME-1]   ISBN 0-7923-5000-6

Philosophy and Medicine

58. H.A.M.J. ten Have and H.-M. Sass (eds.): *Consensus Formation in Healthcare Ethics*. 1998 [ESiP-2]   ISBN 0-7923-4944-X

KLUWER ACADEMIC PUBLISHERS – DORDRECHT / BOSTON / LONDON